T0202815

Communications
in Computer and Information Science 1748

Rationale

The CCIS series is devoted to the publication of proceedings of computer science conferences. Its aim is to efficiently disseminate original research results in informatics in printed and electronic form. While the focus is on publication of peer-reviewed full papers presenting mature work, inclusion of reviewed short papers reporting on work in progress is welcome, too. Besides globally relevant meetings with internationally representative program committees guaranteeing a strict peer-reviewing and paper selection process, conferences run by societies or of high regional or national relevance are also considered for publication.

Topics

The topical scope of CCIS spans the entire spectrum of informatics ranging from foundational topics in the theory of computing to information and communications science and technology and a broad variety of interdisciplinary application fields.

Information for Volume Editors and Authors

Publication in CCIS is free of charge. No royalties are paid, however, we offer registered conference participants temporary free access to the online version of the conference proceedings on SpringerLink (http://link.springer.com) by means of an http referrer from the conference website and/or a number of complimentary printed copies, as specified in the official acceptance email of the event.

CCIS proceedings can be published in time for distribution at conferences or as post-proceedings, and delivered in the form of printed books and/or electronically as USBs and/or e-content licenses for accessing proceedings at SpringerLink. Furthermore, CCIS proceedings are included in the CCIS electronic book series hosted in the SpringerLink digital library at http://link.springer.com/bookseries/7899. Conferences publishing in CCIS are allowed to use Online Conference Service (OCS) for managing the whole proceedings lifecycle (from submission and reviewing to preparing for publication) free of charge.

Publication process

The language of publication is exclusively English. Authors publishing in CCIS have to sign the Springer CCIS copyright transfer form, however, they are free to use their material published in CCIS for substantially changed, more elaborate subsequent publications elsewhere. For the preparation of the camera-ready papers/files, authors have to strictly adhere to the Springer CCIS Authors' Instructions and are strongly encouraged to use the CCIS LaTeX style files or templates.

Abstracting/Indexing

CCIS is abstracted/indexed in DBLP, Google Scholar, EI-Compendex, Mathematical Reviews, SCImago, Scopus. CCIS volumes are also submitted for the inclusion in ISI Proceedings.

How to start

To start the evaluation of your proposal for inclusion in the CCIS series, please send an e-mail to ccis@springer.com.

Vladimir M. Vishnevskiy ·
Konstantin E. Samouylov · Dmitry V. Kozyrev
Editors

Distributed Computer and Communication Networks

25th International Conference, DCCN 2022
Moscow, Russia, September 26–29, 2022
Revised Selected Papers

 Springer

Editors
Vladimir M. Vishnevskiy 🆔
V. A. Trapeznikov Institute of Control
Sciences of RAS
Moscow, Russia

Konstantin E. Samouylov 🆔
RUDN University
Moscow, Russia

Dmitry V. Kozyrev 🆔
V. A. Trapeznikov Institute of Control
Sciences of RAS
Moscow, Russia

RUDN University
Moscow, Russia

ISSN 1865-0929 ISSN 1865-0937 (electronic)
Communications in Computer and Information Science
ISBN 978-3-031-30647-1 ISBN 978-3-031-30648-8 (eBook)
https://doi.org/10.1007/978-3-031-30648-8

This Springer imprint is published by the registered company Springer Nature Switzerland AG
The registered company address is: Gewerbestrasse 11, 6330 Cham, Switzerland

Preface

This volume contains a collection of revised selected full-text papers presented at the 25th International Conference on Distributed Computer and Communication Networks (DCCN 2022), held in Moscow, Russia, during September 26–29, DCCN 2022 was jointly organized by the Russian Academy of Sciences (RAS), the V.A. Trapeznikov Institute of Control Sciences of RAS (ICS RAS), the Peoples'Friendship University of Russia (RUDN University), the National Research Tomsk State University, and the Institute of Information and Communication Technologies of Bulgarian Academy of Sciences (IICT BAS). Information support was provided by the Russian Academy of Sciences. The conference was organized with the support of the IEEE Russia Section, Communications Society Chapter (COM19), and the RUDN University Strategic Academic Leadership Program.

The conference is a continuation of the traditional international conferences of the DCCN series, which have taken place in Sofia, Bulgaria (1995, 2005, 2006, 2008, 2009, 2014); Tel Aviv, Israel (1996, 1997, 1999, 2001); and Moscow, Russia (1998, 2000, 2003, 2007, 2010, 2011, 2013, 2015, 2016, 2017, 2018, 2019, 2020, 2021) in the last 25 years. The main idea of the conference is to provide a platform and forum for researchers and developers from academia and industry from various countries working in the area of theory and applications of distributed computer and communication networks, mathematical modeling, and methods of control and optimization of distributed systems, by offering them a unique opportunity to share their views, as well as discuss prospective developments, and pursue collaboration in this area. The content of this volume is related to the following subjects:

1. Communication networks, algorithms, and protocols
2. Wireless and mobile networks
3. Computer and telecommunication networks control and management
4. Performance analysis, QoS/QoE evaluation, and network efficiency
5. Analytical modeling and simulation of communication systems
6. Evolution of wireless networks toward 5G
7. Centimeter- and millimeter-wave radio technologies
8. Internet of Things and fog computing
9. Cloud computing, distributed and parallel systems
10. Machine learning, big data, and artificial intelligence
11. Probabilistic and statistical models in information systems
12. Queuing theory and reliability theory applications
13. Security in infocommunication systems

The DCCN 2022 conference gathered 130 submissions from authors from 18 different countries.From these, 96 high-quality papers in English were accepted and presented during the conference, each paper secured at least three double-blind reviews. The current volume contains 29 extended papers which were recommended by session chairs and selected by the Program Committee for the Springer post-proceedings. Thus, the acceptance rate is 30.2%.

All the papers selected for the post-proceedings volume are given in the form presented by the authors. These papers are of interest to everyone working in the field of computer and communication networks.

We thank all the authors for their interest in DCCN, the members of the Program Committee for their contributions, and the reviewers for their peer-reviewing efforts.

September 2022

Vladimir M. Vishnevskiy
Konstantin E. Samouylov
Dmitry V. Kozyrev

Organization

Program Committee Chairs

V. M. Vishnevskiy (Chair)   ICS RAS, Russia
K. E. Samouylov (Co-chair)   RUDN University, Russia

Publication and Publicity Chair

D. V. Kozyrev   ICS RAS and RUDN University, Russia

International Program Committee

S. M. Abramov   Program Systems Institute of RAS, Russia
A. M. Andronov   Transport and Telecommunication Institute,
                           Latvia
T. Atanasova   IICT BAS, Bulgaria
S. E. Bankov   Kotelnikov Institute of Radio Engineering and
                           Electronics of RAS, Russia
A. S. Bugaev   Moscow Institute of Physics and Technology,
                           Russia
S. R. Chakravarthy   Kettering University, USA
D. Deng   National Changhua University of Education,
                           Taiwan
S. Dharmaraja   Indian Institute of Technology, Delhi, India
A. N. Dudin   Belarusian State University, Belarus
A. V. Dvorkovich   Moscow Institute of Physics and Technology,
                           Russia
D. V. Efrosinin   Johannes Kepler University Linz, Austria
Yu. V. Gaidamaka   RUDN University, Russia
Yu. V. Gulyaev   Kotelnikov Institute of Radio-engineering and
                           Electronics of RAS, Russia
V. C. Joshua   CMS College Kottayam, India
H. Karatza   Aristotle University of Thessaloniki, Greece
N. Kolev   University of São Paulo, Brazil
G. Kotsis   Johannes Kepler University Linz, Austria

| T. Atanasova | IIICT BAS, Bulgaria |
| I. A. Kochetkova | RUDN University, Russia |

Organizers and Partners

Organizers

Russian Academy of Sciences (RAS), Russia
V.A. Trapeznikov Institute of Control Sciences of RAS, Russia
RUDN University, Russia
National Research Tomsk State University, Russia
Institute of Information and Communication Technologies of Bulgarian Academy of Sciences, Bulgaria
Research and Development Company "Information and Networking Technologies", Russia

Contents

Distributed Systems Applications

A Survey of the Implementations of Model Inversion Attacks

Junzhe Song and Dmitry Namiot(✉)

Lomonosov Moscow State University, GSP-1, Leninskie Gory, Moscow 119991, Russian Federation
songjz@smbu.edu.cn, dnamiot@gmail.com

Abstract. Attacks on machine learning systems are usually called special manipulations with data at different stages of the machine learning pipeline, which are designed to either prevent the normal operation of the attacked machine learning systems, or vice versa - to ensure their specific functioning, which is necessary for the attacker. There are attacks that allow you to extract non-public data from machine learning models. Model inversion attacks, first described in 2015, aim to expose the data used to train the model. Such attacks involve polling the model in a special pattern and represent a major threat to machine learning as a service (MLaaS) projects. In this article, we provide an overview of off-the-shelf software tools for carrying out model inversion attacks and possible protection against such attacks.

Keywords: Model Inversion attacks · Implementations of MI attacks · Defenses against MI attacks

1 Introduction

This article is an expanded version of a paper presented at the 2022 DCCN conference [51].

Adversarial Machine Learning becoming now a threat to machine learning security, and is also an area that the software industry needs to focus on. Leaders of the software industry, such as IBM [1], Google [2], and Microsoft [3], declare that besides for securing their traditional software systems, they will also actively secure ML systems.

The topic of adversarial attacks first hit industry reports a few years ago [4] (Gartner 2019). The report stated that users of machine learning (artificial intelligence) applications should be aware of the risks of data corruption, the risks of model theft, and the emergence of hostile data samples. The main reason for this focus on the cybersecurity of machine learning (artificial intelligence) systems is the attempts to use machine learning systems in so-called critical applications (avionics, automatic driving, etc.). The possibility of adversarial attacks does not guarantee the results of machine learning systems. It is the lack of guarantees for results that is the main obstacle to the implementation of machine learning in critical applications.

V. M. Vishnevskiy et al. (Eds.): DCCN 2022, CCIS 1748, pp. 3–16, 2023.
https://doi.org/10.1007/978-3-031-30648-8_1

Model inversion attack (or MI attack) is one of the attacks on machine learning systems that try to extract private (non-public) data from working models. In model inversion attacks, the attacker tries to recover the private dataset used to train the neural network. A successful model inversion attack creates (generates) samples that describe the data in the training dataset. Such data is most often not public, therefore, a model inversion attack tries to steal the data used during network training. The first time such attacks were described in 2015. Over the past time, several strategies have been proposed for MI attacks, such as: the use of confidential information [5], an attack without the knowledge of insensitive attributes [6], the use of explainable machine learning [7], etc. At the same time, many of the proposed strategies are only proof of concept. There are not so many industrial implementations today.

At the same time, the results of surveys of various organizations [8] show that, despite the desire to protect machine learning systems, most industry practitioners are not ready to understand the security problems of machine learning systems. The vast majority of organizations do not have the right tools to secure their machine learning systems. Trusted machine learning platforms are a very hot topic these days. The purpose of this work is to collect existing implementations of MI attacks, as well as known protection methods. This kind of information should both help software industry practitioners and determine further directions for the development of software tools for the detection of MI attacks and counteraction to them.

Model inversion attacks are included in the MITRE threat atlas for machine learning systems [36]. At the same time, it was noted that for the practical implementation of such attacks, access to the Model Inference API is required.

2   A Brief Description of MI-Attacks

Suppose that we have a trained classifier. Usually, we will start it with some training data(including privacy-sensitive information) in the form of $\{X_n, y_n\}, n \in [0, n]$, where X_n is the data feature and y_n is the corresponding class label. We pick an algorithm(usually, a neural-network nowadays), and utilize these training data to train the algorithm(model) $f : y_n = f(X_n)$. After training, the model f is also able to predict the correct label of unknown feature X_{n+1}.

The *Model Inversion(MI) attack* was first proposed by *Fredrikson et al.* [5]. The attack uses a trained classifier to extract representations of training data.

Fredrikson et al. suppose that the feature $X_n = \{x_n^1, x_n^2, ..., x_n^m\}$, and the model f provides not only the corresponding label but also the confidence value of the label. Let the x_n^m be the privacy-sensitive value that we are concerned about. The methodology is to use the confidence value as guidance, at first we set $\tilde{x}_n^m$ as a random value in the range of x_n^m, then we will randomly change $\tilde{x}_n^m$. If the confidence value goes higher, then means our $\tilde{x}_n^m$ is approaching the true value x_n^m. And finally, we get the privacy-sensitive value $\tilde{x}_n^m = x_n^m$.

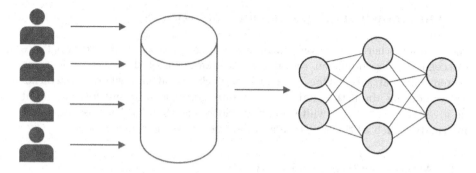

Fig. 1. A typical ML workflow [30]

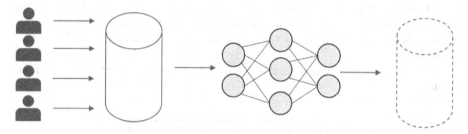

Fig. 2. The idea of a model inversion attack [30]

Figure 3 from the original work [5] gives an example of the result of a model inversion attack

Fig. 3. An original face image (right) and a restored one through model inversion attack (left) [5]

3   On Known Implementations of Attacks

As with all other attacks on machine learning systems, there are two types of scenarios, depending on the attacker's knowledge of the machine learning system being attacked. One of them is a black box attack, when the attacker knows nothing about the model and can only get a clean output for each input he gives. The other is a white box scenario where the attacker has additional information such as the structure and parameters of the target model.

3.1   Attacks by *Fredrikson et al.*

This is an unofficial implementation. In this project [9], the author implements the MI attack introduced in the original paper [5]. The author implements this attack with important dependencies **Tensorflow, Pylearn2, Matplotlib,** and **Scipy.** The target database for the implemented attack is **AT&T Database of Faces** [10].

In this program, the MI attack is realized by function **invert()** in the **class Model.**

Algorithm 1. Inversion attack for facial recognition models

1: **function** MI-FACE($label, \alpha, \beta, \gamma, \lambda$)

2:     $c(x) \stackrel{def}{=} 1 - \tilde{f}_{label}(x) + AuxTerm(x)$

3:     $x_0 \leftarrow 0$

4:     **for** $i \leftarrow 1...\alpha$ **do**

5:         $x_i \leftarrow Process(x_{i-1} - \lambda \times \nabla c(x_{i-1}))$

6:         **if** $c_{x_i} \geq max(c(x_{i-1}), ..., c(x_{i-\beta}))$ **then**

7:             break

8:         **end if**

9:         **if** $c(x_i) \leq \gamma$ **then**

10:             break

11:         **end if**

12:     **end for**

13: **return** $[argmin_{x_i}(c(x_i)), min_{x_i}(c(x_i))]$

14: **end function**

Also, the gradient step and the final output of the attack loop are pre-processed with ZCA whitening and Global Contrast Normalization with **Pylearn2**, this helps to preserve the facial features present in the input dataset. Judging by the result provided by the author, the effect of the attack strongly depends on the number of iterations. For example, for an image sample called A, the best image after an inversion attack is a blurred image A_{best}, from which we can hardly tell who he is.

The official code can be found at https://www.cs.cmu.edu/ mfredrik/.

3.2 AIJack

The **AIJack** is a **Python** library created by Github user *Koukyosyumei* [11] aimed to reveal the vulnerabilities of machine learning models. This package implements algorithms for AI security such as Model Inversion, Poisoning Attack, Evasion Attack, Differential Privacy, and Homomorphic Encryption.

The package contains Collaborative Learning (FedAVG, SplitNN), Attacks (MI-Face [5], Gradient Inversion [12–16], GAN Attack at client-side [17], Label Leakage Attack [18], Evasion Attack [19], Poisoning Attack [20]), and Defense (Moment Accountant(Differential Privacy) [21], DPSGD (Differential Privacy) [21], MID (Defense against model inversion attack) [22], Soteria (Defense against model inversion attack in federated learning) [23]).

The list of supported options is quite long. However, the possible development and maintenance of this package raise doubts. There is currently no development company behind this package.

The usage of this library is also simple, with several code lines you can get an attacker of a machine learning model. For example, a *MI-FACE* attacker (model inversion attack):

```
1 # Fredrikson [5]
2 from aijack.attack import MI_FACE
3
4 mi = MI_FACE(target_torch_net, input_shape)
5 reconstructed_data, _ = mi.attack(target_label, lam, num_itr)
```

or a *Gradient Inversion* attacker (server-side model inversion attack against federated learning):

```
1  from aijack.attack import GradientInversion_Attack
2
3  # DLG Attack [12]
4  attacker = GradientInversion_Attack(net, input_shape, distancename="12")
5
6  # GS Attack [13]
7  attacker = GradientInversion_Attack(net, input_shape,
8                          distancename="cossim", tv_reg_coef=0.01)
9
10 # CPL [15]
11 attacker = GradientInversion_Attack(net, input_shape,
12          distancename="12", optimize_label=False, lm_reg_coef=0.01)
13
14 # GradInversion (Yin, Hongxu, et al. "See through gradients: Image
15 # batch recovery via gradinversion." Proceedings of the IEEE/CVF
16 # Conference on Computer Vision and Pattern Recognition. 2021.)
17 attacker = GradientInversion_Attack(
18                      net,
19                      input_shape,
20                      distancename="12",
21                      optimize_label=False,
22                      bn_reg_layers=[net.body[1], net.body[4], net.body[7]],
23                      group_num = 5,
24                      tv_reg_coef=0.00,
25                      12_reg_coef=0.0001,
26                      bn_reg_coef=0.001,
27                      gc_reg_coef=0.001
28                              )
29
30 received_gradients = torch.autograd.grad(loss, net.parameters())
31 received_gradients = [cg.detach() for cg in received_gradients]
32 attacker.attack(received_gradients)
```

3.3   Privacy Raven

PrivacyRaven is a privacy testing library for deep learning systems [24]. Users can use it to determine the susceptibility of a model to different privacy attacks. This library lets evaluate various privacy-preserving machine learning techniques, develop new privacy metrics and attacks as well as repurpose attacks for data provenance and other use cases.

The PrivacyRaven library supports many known attack methods, such as label-only black-box model extraction, membership inference, and model inversion attacks. The developer of the library has described rather large plans for its development. In particular, it is planned to include additional attack methods, differential privacy checking, automatic hyperparameter optimization, and other functions in this library. However, the Github shows that only 3 people are involved in this development.

Deep learning systems, significantly neural networks, have proliferated in an exceedingly big selection of applications, together with privacy-sensitive use cases like identity verification and medical diagnoses. However, these models are liable to privacy attacks that focus on each intellectual property of the model and the confidentiality of the training data. All the latest research shows that the number of attacks on machine learning systems is only growing. It should also be taken into account that machine learning systems accumulate more data than traditional databases. Accordingly, machine learning systems become interesting targets for attacks. And until now, engineers and researchers haven't had the privacy analysis tools that match this trend.

Many alternative deep learning security methods are difficult to use. This complexity hinders their adoption. PrivacyRaven is intended for a general audience, which is why the authors designed it as:

- usable. Multiple levels of abstraction permit users to either alter a lot of the interior mechanics or directly manage them, betting on their use case and familiarity with the domain;
- flexible. A standard (default) style makes the attack configurations customizable and practical. It additionally permits new privacy metrics and attacks to be incorporated straightforwardly;
- efficient. The library reduces the boilerplate, affording fast prototyping and quick experimentation.

Each attack may be launched in just 10–15 lines of code.

```
1  # via extract_mnist_gpu.py
2  from privacyraven.models.victim import train_four_layer_mnist_victim
3  from privacyraven.models.four_layer import FourLayerClassifier
4  from privacyraven.utils.data import get_emnist_data
5  from privacyraven.extraction.core import ModelExtractionAttack
6  from privacyraven.utils.query import get_target
7
8
9  # Create a query function for a target PyTorch Lightning model
10 my_model = train_four_layer_mnist_victim()
11
12 def get_target_model(input_data):
13     # PrivacyRaven provides built-in query functions
14     return get_target(my_model, input_data, (1, 28, 28, 1))
15
16 # Obtain seed (or public) data to be used in extraction
17 data_mnist_train, data_mnist_test = get_emnist_data()
18
19 # Run a model extraction attack
20 my_attack = ModelExtractionAttack(
21     get_target_model, # model function
22     300, # query limit
23     (1, 28, 28, 1), # victim input shape
24     20, # number of targets
25     (3, 1, 28, 28),  # substitute input shape
26     "copycat", # synthesizer name
27     FourLayerClassifier, # substitute model architecture
```

```
28      784,  # substitute input size
29      data_mnist_train, # train data
30      data_mnist_test, #  test data
31  )
```

3.4 Adversarial Robustness Toolbox

Adversarial Robustness Toolbox (ART) is a Python library for Machine Learning cybersecurity [25]. ART provides a set of tools (Python packages) both for organizing adversarial attacks (testing machine learning models) and for organizing the protection of machine learning models (attack detection). At the same time, APT provides support for all stages of the machine learning pipeline: data poisoning attacks, evasion attacks, and data extraction attacks.

The basic packages are: **Attacks** (Evasion Attacks, Poisoning Attacks, Extraction Attacks, and Inference Attack), **Defenses** (Preprocessor, Postprocessor, Trainer, Transformer, Detector), **Estimators** (Classification, Object Detection, Object Tracking, Speech Recognition, Certification, Encoding, Generation), and **Metrics** (Robustness Metrics, Certification, Verification).

ART supports all types of data like tables, images/video, audio, all modern machine learning frameworks like TensorFlow, Keras, PyTorch, etc., and machine learning tasks like object detection, classification, speech recognition, etc.

ART toolbox supports model inversion attack (by Fredrikson et al.). The following code snippet [30] illustrates this:

```
1   import tensorflow as tf
2   from art.attacks.inference import model_inversion
3   from art.estimators.classification import KerasClassifier
4
5   # define a model.
6
7   def create_model():
8       # Define a Keras model
9       keras_model = tf.keras.Sequential([
10      tf.keras.layers.Conv2D(16, 8,
11                                  strides=2,
12                                  padding='same',
13                                  activation='relu',
14                                  input_shape=(28, 28, 1)),
15      tf.keras.layers.MaxPool2D(2, 1),
16      tf.keras.layers.Conv2D(32, 4,
17                                  strides=2,
18                                  padding='valid',
19                                  activation='relu'),
20      tf.keras.layers.MaxPool2D(2, 1),
21      tf.keras.layers.Flatten(),
22      tf.keras.layers.Dense(32, activation='relu'),
23      tf.keras.layers.Dense(10, activation='softmax')
24      ])
25      return keras_model
26
27  # compile the model
28
29  my_model = create_model()
30  my_model.compile(optimizer=tf.keras.optimizers.SGD(learning_rate=0.1),
31      loss=tf.keras.losses.SparseCategoricalCrossentropy(from_logits=True),
32      metrics=['accuracy'])
33
34  # put the model into the wrapper:
35
36  my_classifier = KerasClassifier(model=my_model, clip_values=(0, 1),
37                                  use_logits=False)
38  my_classifier.fit(train_data, train_labels,nb_epochs=20,
39                                  batch_size=254)
40  # attack the model
41
42  my_attack = model_inversion.MIFace(my_classifier)
```

4   On Comparison of Implementations

These implementations also support other types of attacks (in the case of the APT toolbox, including attacks that are not related to the theft of intellectual property). The capabilities of the considered packages are summarized in Table 1. AIJask is the leader in the number of supported attacks and defenses.

Table 1. Comparison of implementations

Implement's name	AIJack	Privacy Raven	ART
Attacks Types			
Evasion Attacks	✓		✓
Inversion Attacks	✓	✓	✓
Poisoning Attacks	✓		✓
Extraction Attacks		✓	✓
Membership Inference Attacks	✓	✓	✓
Label Leakage	✓		
Total of supported	5	3	5
Defences Types			
DPSGD(Differential Privacy)	✓		
MID(Defense against model inversion attack)	✓		
Soteria	✓		
CKKS [27]	✓		
Defensive Distillation [28]			✓
Neural Cleanse [29]			✓
Total of supported	4	0	2
Detector			
−For Evasion			
Basic detector based on inputs			✓
Detector trained on the activations of a specific layer			✓
Detector based on Fast Generalized Subset Scan [47]			✓
−For Poisoning			
Detection based on activations analysis [48]			✓
Detection based on data provenance [49]			✓
Detection based on spectral signatures [50]			✓
Total of supported	0	0	6

5   On Known MI Attack Implementations

In this section, we list some implementations of MI attacks with their target datasets.

More information, see Table 2.

where datasets are:

A - CIFAR-10

B - MNIST

C - CelebA

D - AFHQ

E - ChestX-ray8

F - Flickr-Faces-HQ

G - Fashion-MNIST

Table 2. On implementations of MI attacks

Method name	Target dataset						
	A	B	C	D	E	F	G
Knowledge-Enriched Distributional MI Attacks [37,38]	✓	✓	✓		✓	✓	
Variational Model Inversion Attacks [39,40]			✓				
Generative Model-Inversion Attacks Against DNN [41,42]	✓	✓					
UnSplit [43,44]	✓						✓
Plug and Play Attacks [45,46]			✓	✓		✓	

6 Analysis of Comparison Results and Possible Future Work

Analysis of the above comparison results of software frameworks for adversarial attacks (adversarial testing) shows that the obvious leader is the ART toolbox. Initially, this product was developed by IBM. After that, the product received support from the state project DARPA. ART toolbox now is a part of GARD (Guaranteeing AI Robustness to Deception) project [34]. The DARPA GARD program aims to build the theoretical foundations of a machine learning system to identify system vulnerabilities, describe features that will increase system reliability, and encourage the creation of effective defenses. The authors of the project confirm that currently, the protections of machine learning systems are very specific and effective only against specific attacks. In an attack-defense competition for machine learning systems, attacks win. Always (at least now) a new attack appears first and only then developers look for methods to protect against it. GARD, in turn, seeks to develop defenses that can protect against a wide range of attacks.

Model inversion attacks are a big threat, primarily for projects like machine learning as a service [35]. In addition to the fact that the disclosure of private information about training data can be very sensitive for individual projects (for example, recognition systems), any information about training data can be used to build a so-called shadow model. A shadow model is a copy of the original machine learning model that can be used to practice attacks on the target system.

Naturally, we must take into account the practical feasibility of such attacks. As can be seen from the discussion above, model inversion attacks involve multiple request-response sessions for the system being attacked. In a practical situation, especially for machine learning systems in critical areas (avionics, automatic driving, etc.), there may simply be no software interfaces for questioning the model. Model inversion attacks will be theoretically possible, but practically they could not be implemented without access to model API.

The **ART** package is one of the few that supports (contains ready-made implementations) attacks aimed at extracting data from machine learning systems [30]. As in many other cases, attacks are ahead of the defense. So far, **ART** includes only two detectors for evasion attacks (model use stage) and data poisoning attacks (learning stage). There is no detector for model inversion attacks, and the development of such a detector is a good step in the development of this package. The idea of its development will be as follows. We will start with the well-known attack by Fredrickson et al. [5] and will try to determine in the sequence of queries to the model a pattern that is characteristic of this attack.

Acknowledgement. The work was carried out as part of the development of the program of the Faculty of Computational Mathematics and Cybernetics of the Lomonosov Moscow State University "Artificial intelligence in cybersecurity" [31]. Examples of other publications within this project can be seen in the works [32] and [33].

The authors are grateful to the staff of the Department of Information Security of the Faculty of Computational Mathematics and Cybernetics of Lomonosov Moscow State University for the discussion of our work, criticism and valuable additions.

This research has been supported by the Interdisciplinary Scientific and Educational School of Moscow University "Brain, Cognitive Systems, Artificial Intelligence".

References

1. Nicolae, M.I., et al.: Adversarial Robustness Toolbox v1. 0.0. arXiv preprint arXiv:1807.01069 (2018)
2. Sambasivan, N., Holbrook, J.: Toward responsible AI for the next billion users. Interactions **26**(1), 68–71 (2018)
3. Dang, Y., Lin, Q., Huang, P.: AIOps: real- world challenges and research innovations. In: 2019 IEEE/ACM 41st International Conference on Software Engineering: Companion Proceedings (ICSE-Companion). IEEE (2019)
4. Gartner, S.A., Inc.: Anticipate data manipulation security risks to AI pipelines. https://www.gartner.com/doc/3899783
5. Fredrikson, M., Jha, S., Ristenpart, T.: Model inversion attacks that exploit confidence information and basic countermeasures. In: Proceedings of the 22nd ACM SIGSAC Conference on Computer and Communications Security, pp. 1322–1333 (2015)
6. Hidano, S., Murakami, T., Katsumata, S., Kiyomoto, S., Hanaoka, G.: Model inversion attacks for prediction systems: without knowledge of non-sensitive attributes. In: 2017 15th Annual Conference on Privacy, Security and Trust (PST), pp. 115–11509. IEEE (2017)
7. Zhao, X., Zhang, W., Xiao, X., Lim, B.Y.: Exploiting Explanations for Model Inversion Attacks (2021). arXiv preprint. arXiv:2104.12669
8. Siva Kumar, R.S., et al.: Adversarial machine learning-industry perspectives. In: 2020 IEEE Security and Privacy Workshops (SPW), pp. 69–75 (2020).https://doi.org/10.1109/SPW50608.2020.00028
9. Fredrikson, et al.: Implementation of MI attack with method. https://github.com/yashkant/Model-Inversion-Attack
10. AT&T Database of Faces. https://www.kaggle.com/datasets/kasikrit/att-database-of-faces?select=s1
11. AIJack: reveal the vulnerabilities of machine learning models. https://github.com/Koukyosyumei/AIJack
12. Zhu, L., Liu, Z., Han, S.: Deep leakage from gradients. In: Advances in Neural Information Processing Systems, vol. 32 (2019)

13. Geiping, J., et al.: Inverting gradients-how easy is it to break privacy in federated learning?. Advances in Neural Information Processing Systems, vol. 33, pp. 16937–16947 (2020)
14. Zhao, B., Mopuri, K.R., Bilen, H.: idlg: improved deep leakage from gradients. arXiv preprint arXiv:2001.02610 (2020)
15. Wei, W., et al.: A framework for evaluating gradient leakage attacks in federated learning. arXiv preprint arXiv:2004.10397 (2020)
16. Yin, H., et al.: See through gradients: Image batch recovery via gradinversion. In: Proceedings of the IEEE/CVF Conference on Computer Vision and Pattern Recognition (2021)
17. Hitaj, B., Ateniese, G., Perez-Cruz, F.: Deep models under the GAN: information leakage from collaborative deep learning. In: Proceedings of the 2017 ACM SIGSAC Conference on Computer and Communications Security (2017)
18. Li, O., et al.: Label leakage and protection in two-party split learning. arXiv preprint arXiv:2102.08504 (2021)
19. Biggio, B., et al.: Evasion attacks against machine learning at test time. In: Blockeel, H., Kersting, K., Nijssen, S., Železný, F. (eds.) ECML PKDD 2013. LNCS (LNAI), vol. 8190, pp. 387–402. Springer, Heidelberg (2013). https://doi.org/10.1007/978-3-642-40994-3_25
20. Biggio, B., Nelson, B., Laskov, P.: Poisoning attacks against support vector machines. arXiv preprint arXiv:1206.6389 (2012)
21. Abadi, M., et al.: Deep learning with differential privacy. In: Proceedings of the 2016 ACM SIGSAC Conference on Computer and Communications Security (2016)
22. Wang, T., Zhang, Y., Jia, R.: Improving robustness to model inversion attacks via mutual information regularization. arXiv preprint arXiv:2009.05241 (2020)
23. Sun, J., et al.: Soteria: provable defense against privacy leakage in federated learning from representation perspective. In: Proceedings of the IEEE/CVF Conference on Computer Vision and Pattern Recognition (2021)
24. Privacy Raven: Privacy testing for deep learning. https://github.com/trailofbits/PrivacyRaven
25. Adversarial Robustness Toolbox (ART) - Python library for machine learning security - evasion, poisoning, extraction, inference - red and blue teams. https://github.com/Trusted-AI/adversarial-robustness-toolbox
26. Advertorch: A toolbox for adversarial robustness research. https://github.com/borealisai/advertorch
27. Cheon, J.H., Kim, A., Kim, M., Song, Y.: Homomorphic encryption for arithmetic of approximate numbers. In: Takagi, T., Peyrin, T. (eds.) ASIACRYPT 2017. LNCS, vol. 10624, pp. 409–437. Springer, Cham (2017). https://doi.org/10.1007/978-3-319-70694-8_15
28. Papernot, N., McDaniel, P., Wu, X., Jha, S., Swami, A.: Distillation as a defense to adversarial perturbations against deep neural networks. In: 2016 IEEE Symposium on Security and Privacy (SP), pp. 582–597. IEEE (2016)
29. Wang, B., et al.: Neural cleanse: identifying and mitigating backdoor attacks in neural networks. In: 2019 IEEE Symposium on Security and Privacy (SP), pp. 707–723. IEEE (2019)
30. Attacks against Machine Learning Privacy (Part 1): Model inversion attacks with the IBM-ART framework. https://franziska-boenisch.de/posts/2020/12/model-inversion/
31. Sec of AI systems. https://aisec.cs.msu.ru/ (in Russian) Accessed 11 2022
32. Namiot, D., Ilyushin, E., Chizhov, I.: Artificial intelligence and cybersecurity. Int. J. Open Inf. Technol. **10**(9), 135–147 (2022). (in Russian)

33. Sneps-Sneppe, M., Sukhomlin, V., Namiot, D.: On cyber-security of information systems. In: Vishnevskiy, V.M., Kozyrev, D.V. (eds.) DCCN 2018. CCIS, vol. 919, pp. 201–211. Springer, Cham (2018). https://doi.org/10.1007/978-3-319-99447-5_17
34. GARD project. https://www.gardproject.org/ Accessed 7 2022
35. Kesarwani, M., et al.: Model extraction warning in mlaas paradigm. In: Proceedings of the 34th Annual Computer Security Applications Conference (2018)
36. Invert ML Model. https://atlas.mitre.org/techniques/AML.T0024.001/ Accessed 07 2022
37. Chen, S., Kahla, M., Jia, R., Qi, G.J.: Knowledge-enriched distributional model inversion attacks. In: Proceedings of the IEEE/CVF International Conference on Computer Vision, pp. 16178–16187 (2021)
38. Knowledge-Enriched-DMI.   https://github.com/SCccc21/Knowledge-Enriched-DMI
39. Wang, K.C., Fu, Y., Li, K., Khisti, A., Zemel, R., Makhzani, A.: Variational model inversion attacks. In: Advances in Neural Information Processing Systems, vol. 34 (2021)
40. VMI. https://github.com/wangkua1/vmi
41. Zhang, Y., Jia, R., Pei, H., Wang, W., Li, B., Song, D.: The secret revealer: generative model-inversion attacks against deep neural networks. In: Proceedings of the IEEE/CVF Conference on Computer Vision and Pattern Recognition, pp. 253–261 (2020)
42. GMI-Attack. https://github.com/AI-secure/GMI-Attack
43. Erdogan, E., Kupcu, A., Cicek, A.E.: Unsplit: Data-oblivious model inversion, model stealing, and label inference attacks against split learning (2021). arXiv preprint arXiv:2108.09033
44. Unsplit. https://github.com/ege-erdogan/unsplit
45. Struppek, L., Hintersdorf, D., Correia, A.D.A., Adler, A., Kersting, K.: Plug & play attacks: towards robust and flexible model inversion attacks (2022). arXiv preprint arXiv:2201.12179
46. Plug-and-Play-Attacks.   https://github.com/LukasStruppek/Plug-and-Play-Attacks
47. Speakman, S., Sridharan, S., Remy, S., Weldemariam, K., McFowland, E.: Subset scanning over neural network activations (2018). arXiv preprint arXiv:1810.08676
48. Chen, B., et al.: Detecting backdoor attacks on deep neural networks by activation clustering. arXiv preprint arXiv:1811.03728
49. Baracaldo, N., Chen, B., Ludwig, H., Safavi, A., Zhang, R.: Detecting poisoning attacks on machine learning in IoT environments. In: 2018 IEEE International Congress on Internet of Things (ICIOT), pp. 57–64. IEEE (2018)
50. Tran, B., Li, J., Madry, A.: Spectral signatures in backdoor attacks. In: Advances in Neural Information Processing Systems, vol. 31 (2018)
51. DCCN -2022. https://dccn.ru

Multidimensional Information System Metadata Description Using the "Data Vault" Methodology

Maxim Fomin[(✉)] [iD]

Peoples' Friendship University of Russia (RUDN University),
Miklukho-Maklaya st. 6, Moscow 117198, Russian Federation
fomin-mb@rudn.ru

Abstract. The task of determining the metadata of a multidimensional information system corresponds to the description of the parameters of cells that contain information about the facts that are included in the multidimensional data cube. Classification schemes can be used when constructing metadata. The classification scheme corresponds to certain structural component of the observed phenomenon. The cell parameters are presented in the classification scheme in a hierarchical form and are combined in metadata when connecting several classification schemes. To construct a hierarchy of elements of the classification scheme, it is necessary to identify groups of members for which there is a semantic connection with groups of members of other dimensions. Cartesian product can be applied to groups of members. As a result, clusters of members' combinations will be formed in the metadata. The complete metadata structure can be achieved by combining all clusters. In case of a large amount of aspects of analysis, a multidimensional data cube has specific properties related to sparsity. The use of classification schemes makes it possible to identify parts in the metadata that correspond to individual structural components of the observed phenomenon. If a multidimensional data cube is constructed in the process of automated data collection, the "Data vault" methodology can be used to describe the metadata. This method allows you to reflect the relationships between business objects in the metadata.

Keywords: OLAP · data warehouse · multidimensional metamodel · set of possible member combinations · cluster of member combinations

1 Introduction

The OLAP methodology has been developed to analyze large volume of well-structured data. It allows you to create multidimensional information systems that describe the data forming a multidimensional data cube. Aspects of the

This paper has been supported by the RUDN University Strategic Academic Leadership Program.

analysis of a multidimensional system correspond to the dimensions of a cube. In the OLAP methodology, their role is played by measurements related to the properties of the observed phenomenon. If many semantically different parameters are used to analyze the observed phenomenon, a multidimensional data cube gets the properties of irregular filling and sparsity [2,4,8,14,15,17,18,22]. Data analysis algorithms in this case should be based on the consideration of the semantics of the observed phenomenon [1,5,11,12,16,20,23–25]. To solve this problem, an effective information model of the analytical space should be constructed.

In the task of modeling the structure of the analytical space of an information system, the cluster modeling method can be used. The cells of a multidimensional data cube are defined by dimensions' members. To identify clusters, it is necessary to analyze the compatibility of the dimensions' members and identify their possible combinations. They correspond to possible cells and should be described in metadata. The information system contains data of the observed phenomenon related to the subject domain. This data can form several multidimensional data cubes. The task of describing the metadata of a multidimensional cube related to a separate component of the observed phenomenon, when using the cluster approach, has the following aspects: identification of semantics and classification of data of the structural components of the observed phenomenon.

2   Modeling Multidimensional Metadata

The task of data analysis is related to the subject domain. In a multidimensional information system, this relationship generates a choice of analysis aspects that correspond to the dimensions of data cube H. All dimensions of the cube form a set $D(H) = \{D^1, D^2, .., D^n\}$, there D^i is i-dimension, and $n = dim(H)$ – dimensionality of multidimensional cube [3]. In the metadata of an information system a dimension corresponds to a set of members $D^i = \{d_1^i, d_2^i, .., d_{k_i}^i\}$, there i is a number of dimension, k_i – the quantity of members. The set D^i is a subset of the elements of the base classifier that corresponds to the dimension. If large amount of aspects are used in the analysis process, there may be a semantic inconsistency between the members of different dimensions. Cells that are defined by a combination of such members cannot correspond to any fact. The result is sparsity in data cube.

A multidimensional data cube is a structured set of facts related to an observed phenomenon. Each fact corresponds to a cell, which is defined by a combination of members $c = (d_{i_1}^1, d_{i_2}^2, .., d_{i_n}^n)$ – one element for one dimension. If large amount of aspects are used in the analysis process, there may be a semantic inconsistency between the members of different dimensions. Cells that are defined by a combination of such members cannot correspond to any fact. The result is sparsity in data cube.

The absence of compatibility of members of different dimensions allows us to enter into the data model special value "Not used" for dimensions. This value is used for a dimension when it becomes semantically indeterminate when combined by a set of members from other dimensions. A multidimensional data cube has a structure that corresponds to a set of possible combinations of members. We will call this set "SPMC".

The subject of analysis in a multidimensional information system is the characteristics of the observed phenomenon corresponding to the measure values. These values are defined in possible cells of the data cube and form a set $V(H) = \{v_1, v_2, .., v_p\}$, where v_j is j-measure, p – the quantity of measures in the hypercube. In the case when any measure semantically does not correspond to the members of the cell, the value of this measure can be defined as "Not in use".

Describing SPMC requires a member compatibility analysis. An effective way to solve this problem is provided by the cluster method, which is based on the analysis of relationships between members [6]. The application of this method makes it possible to identify groups of members. The member $G_j^i = \{d_1^i, d_2^i, .., d_{m_j}^i\}$ of members in i-dimension includes m_j members ($1 \leq m_j \leq k_i$), where j is a group number. A group contains members that are equally combined with groups of members of other dimensions in SPMC.

The use of the cluster method makes it possible to identify groups of members that belong to different dimensions, but can be combined with each other. The Cartesian product, in which the operands are related groups of members from different dimensions, makes it possible to obtain a cluster $SPMC(K) = G_1 \times G_2 \times ... \times G_n$. The complete set of clusters forms the SPMC.

3 The Use of Classification Schemes in the Cluster Method

In the case of a large number of dimensions, a multidimensional data model can be successfully constructed if semantic analysis is used. It allows you to identify the links between the classification attributes corresponding to the dimensions of the data cube. Classification schemes can be used to solve the problem of analyzing data that describe the observed phenomenon [7].

System analysis makes it possible to identify individual structural components in the observed phenomenon. Each component has its own set of characteristics. When constructing a classification scheme corresponding to a component of the observed phenomenon, these characteristics are compared with aspects of the analysis of a multidimensional data cube. The classification scheme is formed by semantic analysis of members related to these aspects of analysis.

The characteristics of the structural component that relate to the observed phenomenon should be presented in a hierarchical form. At the same time, a ranking among the characteristics should be established and the main characteristic selected. Ranking determines the semantic importance of dimensions related to the structural component of the observed phenomenon, and the major dimension expresses the most important properties of this component. The remaining dimensions of the hierarchy are semantically subordinate to the main dimension and express less significant properties of the observed phenomenon related to the structural component. Dimensions that are located at the bottom of the hierarchy clarify the meaning of the main dimension and express subordinate features.

The hierarchy of characteristics of the structural component that relate to the observed phenomenon should be constructed in such a way that the members of the major dimension are described separately or in groups of elements. This will allow you to link different elements of the major dimension with different semantic aspects of the structural component.

The classification scheme of the characteristics for the observed phenomenon is an object of the multidimensional information system that describes the structural component of the observed pattern and contains the following information:

- the set of dimensions included into the classification scheme;
- the set of members of these dimensions included into the classification scheme;
- the major dimension chosen in the set of dimensions of classification scheme;
- the set of measures included into the classification scheme;
- the tree of member combinations of classification scheme which form the hierarchy of the attributes included into classification scheme.

A hierarchical method is used to classify the characteristics of the structural component that relate to the observed phenomenon. It is implemented by building a tree in the structure of which combinations of members that belong to the structural component are represented. The construction of the tree begins with the formation of the root. Groups of members of the major dimension are placed at the root of the tree. Other levels of the hierarchy are added to the root, where groups of members of other dimensions are placed. Each such group is subordinate to some group of members of the previous level. Its members clarify the semantics of the members of the group of the previous level. Each path from the root to the leaves describes a cluster of combinations of members related to dimensions belonging to the classification scheme.

4 Semantic Aspects of the Application of Classification Schemes

The structure of a multidimensional information system corresponds to the analysis tasks for which it is designed. The content of the analysis task determines the separation of the characteristics of the observed phenomenon into the subject of analysis, to which the measures correspond, and the aspects of analysis, the choice of which determines the dimensions. Relationships between dimensions values and measures can be established by applying semantic analysis. It forms the structure of a multidimensional data cube. The description of the facts is contained in the significant cells of the multidimensional cube.

If the observed phenomenon has a complex structure and a large number of aspects that are used for its analysis, the application of the described technique leads to great difficulties. Analysis of groups of values for a pair of dimensions does not allow to identify hierarchical relationships and to divide characteristics between structural components of the observed phenomenon. The task of constructing a data model of the multidimensional information system in the case of a complex structure of the observed phenomenon can be successfully solved if the classification scheme methodology is applied.

The application of semantic analysis and construction of classification scheme results in the following data:

- semantic separation of the characteristics of the observed phenomenon, their binding to the structural components of the observed phenomenon;
- ranking of characteristics, building a hierarchy of characteristics in accordance with their significance in the description of the properties of the structural component of the observed phenomenon.

If the observed phenomenon is considered as a set of structural components, classification schemes are important sources of information for building a data model of a multidimensional information system. At the same time, classification schemes play a semantic and technological role. In the first case, they establish a connection between the observed phenomenon and a multidimensional data cube, in the second - a source of data for constructing information objects.

The classification scheme includes different types of data: dimensions, measures and members. The construction of a hierarchy of characteristics describes the relationship between these data and the observed phenomenon. The analysis of the relationships that are revealed in the process of constructing hierarchies of characteristics makes it possible to determine a set of classification schemes, the use of which leads to the description of a multidimensional data cube.

5   Using the "Data Vault" Methodology to Generate Metadata

Methodology "Data vault" can be used as one of the methods of metadata development. It gives good results when it is required to modify metadata during the operation of the system. It is possible to make changes to the metadata of a multidimensional information system without modifying other subsystems. If the metadata is set in the form of a multidimensional cube, the Methodology "Data vault" allows you to establish a correspondence between the characteristics of the information system and the parameters of business processes that are data sources. A diagram of the metadata repository model is shown in Fig. 1.

When modeling a metadata structure, it is important to have correct definitions of domain elements, their attributes, and their relationships to other elements. When modeling metadata, you can use a standard approach based on the standards CIM (Common Information Model) and CWM (Common Warehouse Metamodel). CWM allows you to describe the exchange of metadata between data warehouses, CIM provides the capabilities of the universal modeling language UML to formalize the semantics of the data warehouse domain.

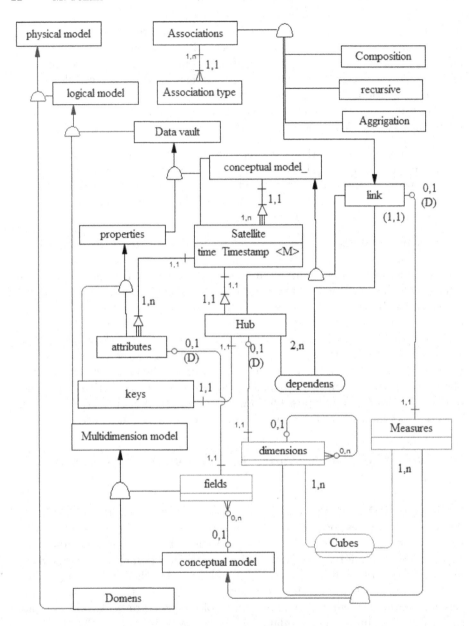

Fig. 1. Metadata repository model using the "Data vault" methodology.

Another important aspect when choosing a metadata structure is how the data warehouse architecture is organized. The classical approach uses a three-tier data representation architecture. In this case, it is necessary to allocate the data preparation area, or operational data warehouse, data warehouse and data marts. Recently, in the process of designing data warehouses at the data representation level, many developers choose the "Data vault" methodology [13].

The "Data vault" method makes it possible to build a data warehouse of an information system using a meta-model for which there is a semantic relationship with the domain. In addition, "Data vault" makes it easy to rebuild the system in case of changes to the business processes of the domain. The main objects in the model, built on the methodology of "Data vault", are: business key (in the terminology of "Data vault" – "hub"), the transaction of the business key (in the terminology of "Data vault" – "link") and the history of the business key (in the terminology of "Data vault – "sat"):

- Business key – a property of an object that uniquely identifies its place within the domain;
- Business key history – history of changes in object properties that are functionally dependent on this business key. In a multidimensional data model, the relevance of dimension attributes is maintained based on the history of the business key;
- Business key transaction – a description of an event that occurred between objects.

The business key, business key transaction, and business key history describe the conceptual level of object (entity) description of the data warehouse metadata model [9, 19, 26, 27]. At the logical and physical level of the object description, key attributes are also used, as well as information about the associative relationships of different types available in the metamodel:

- the type of Association (aggregation, composition or recursion);
- the type of arity (1:1, 1:N or M:N).

The representation of the "Data vaul" model in the form of an ER-diagram (Entity – Relationship) is given in Fig. 2.

The interpretation uses statements (rules) and facts, the relations between which are described in the form of predicates of the Prologue. The relationship between the facts can be described in the form of a diagram shown in Fig. 3.

This way of describing a data warehouse metamodel is easily extensible in terms of adding new rules that are interpreted into relationships and facts in Prolog. Covering interpreted rules in Prolog makes it possible to manage and validate the metamodel. The following section provides an example of extending and validating the current metamodel.

6   Adding a Description of a Data Cube

In the practice of developing a data warehouse, if it is used together with a multidimensional information system in which analytical thematic data marts present information in a multidimensional form, one of the key tasks in the design of a metadata repository is to organize the comparison of data marts elements with the data warehouse.

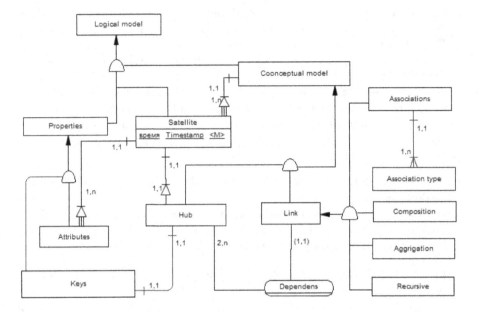

Fig. 2. The "Data vault" model.

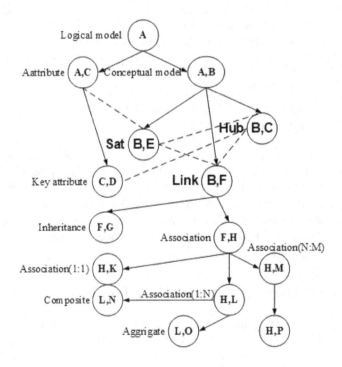

Fig. 3. The "Data vault" structure for interpretation using "Prolog" language.

The relationships between the facts presented by the "Prolog" language can be used to describe the structure of a data mart in a multidimensional view. The mapping of Prolog facts to metadata model elements should match the diagram shown in Fig. 4.

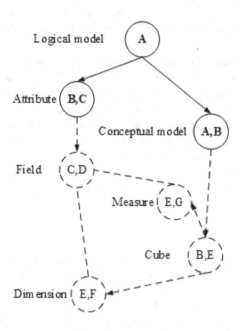

Fig. 4. The Example of adding a relationship between facts.

Thus, the syllogism between the elements of data marts and data warehouse formulated by the methodology of "Data vault" can be composed as follows:

1. As part of the analytical subsystem of the multidimensional data model, the business key of measure and business key histories can act as slow-changing measures [10, 21], because their functions are to describe the context of business key instances.
2. Business key transaction history stores detailed data that allows you to calculate measures based on dimensions' members [6, 7]. In the calculation, process uses the history of the business keys in the multidimensional data model related to the thematic data marts.

7   Conclusions

The paper considers a multidimensional approach to the design of information systems. The metamodel in such a system is semantically linked to the domain of the system using special objects: groups of members. Groups of members belonging to one dimension are semantically related to other groups of members belonging to other dimensions. Related groups form a cluster of cells.

Classification schemes can be used to solve the problem of classifying the data of the observed phenomenon. Each scheme describes a structural component of the observed phenomenon. The relationship between schemas related to different structural components determines the metadata structure. Classification schemes distribute data by hierarchy levels and can be obtained by semantic analysis using a hierarchical method.

The management of metadata of information systems in terms of business metadata was considered. We used the "Data vault" methodology and a Prolog interpreter. This approach allows to manage the data warehouse model of the system based on a metamodel, which is semantically related to the domain of the system and is easily rebuilt in case of changes in the business model of the domain. Predicates, which are described by means of Prolog, establish a correspondence between the attributes of the facts, information about which is stored in the data warehouse, and the dimensions' members in the multidimensional data marts.

References

1. Abello, A., et al.: Using semantic web technologies for exploratory OLAP: a survey. IEEE Trans. Knowl. Data Eng. **27**(2), 571–588 (2015). https://doi.org/10.1109/TKDE.2014.2330822
2. de Castro Lima, J., Hirata, C.M.: Multidimensional cyclic graph approach: representing a data cube without common sub-graphs. Inf. Sci. **181**(13), 2626–2655 (2011). https://doi.org/10.1016/j.ins.2010.05.012
3. Chen, C., Feng, J., Xiang, L.: Computation of sparse data cubes with constraints. In: Kambayashi, Y., Mohania, M., Wöß, W. (eds.) DaWaK 2003. LNCS, vol. 2737, pp. 14–23. Springer, Heidelberg (2003). https://doi.org/10.1007/978-3-540-45228-7_3
4. Chun, S.-J.: Partial prefix sum method for large data warehouses. In: Zhong, N., Raś, Z.W., Tsumoto, S., Suzuki, E. (eds.) ISMIS 2003. LNCS (LNAI), vol. 2871, pp. 473–477. Springer, Heidelberg (2003). https://doi.org/10.1007/978-3-540-39592-8_67
5. Cuzzocrea, A.: OLAP data cube compression techniques: a ten-year-long history. In: Kim, Th., Lee, Yh., Kang, BH., Slezak, D. (eds.) FGIT 2010. LNCS, vol. 6485, pp. 751–754. Springer, Heidelberg (2010). https://doi.org/10.1007/978-3-642-17569-5_74
6. Fomin, M.: Cluster method of description of information system data model based on multidimensional approach. In: Vishnevskiy, V.M., Samouylov, K.E., Kozyrev, D.V. (eds.) DCCN 2016. CCIS, vol. 678, pp. 657–668. Springer, Cham (2016). https://doi.org/10.1007/978-3-319-51917-3_56
7. Fomin, M.: The application of classification schemes while describing metadata of the multidimensional information system based on the cluster method. In: Vishnevskiy, V.M., Samouylov, K.E., Kozyrev, D.V. (eds.) DCCN 2017. CCIS, vol. 700, pp. 307–318. Springer, Cham (2017). https://doi.org/10.1007/978-3-319-66836-9_26

8. Fu, L.: Efficient evaluation of sparse data cubes. In: Li, Q., Wang, G., Feng, L. (eds.) WAIM 2004. LNCS, vol. 3129, pp. 336–345. Springer, Heidelberg (2004). https://doi.org/10.1007/978-3-540-27772-9_34

9. Gautam, V., Parimala, N.: E-metadata versioning system for data warehouse schema. Int. J. Metadata Semant. Ontol. **7**(2), 101–113 (2012). https://doi.org/10.1504/IJMSO.2012.050015

10. Giebler, C., Gröger, C., Hoos, E., Schwarz, H., Mitschang, B.: Modeling data lakes with data vault: practical experiences, assessment, and lessons learned. In: Laender, A.H.F., Pernici, B., Lim, E.-P., de Oliveira, J.P.M. (eds.) ER 2019. LNCS, vol. 11788, pp. 63–77. Springer, Cham (2019). https://doi.org/10.1007/978-3-030-33223-5_7

11. Goil, S., Choudhary, A.: Design and implementation of a scalable parallel system for multidimensional analysis and OLAP. In: Proceedings 13th International Parallel Processing Symposium and 10th Symposium on Parallel and Distributed Processing (IPPS/SPDP 1999), pp. 576–581 (1999). https://doi.org/10.1109/IPPS.1999.760535

12. Gómez, L.I., Gómez, S.A., Vaisman, A.A.: A generic data model and query language for spatiotemporal OLAP cube analysis. In: Proceedings of the 15th International Conference on Extending Database Technology (EDBT 2012), pp. 300–311. Association for Computing Machinery, New York, USA (2012). https://doi.org/10.1145/2247596.2247632

13. Inmon, W., Linstedt, D., Levins, M.: Introduction to data vault architecture. In: Inmon, W., Linstedt, D., Levins, M. (eds.) Data Architecture, 2nd edn., pp. 157–162. Academic Press (2019). https://doi.org/10.1016/B978-0-12-816916-2.00020-6

14. Jin, R., Vaidyanathan, J., Yang, G., Agrawal, G.: Communication and memory optimal parallel data cube construction. IEEE Trans. Parallel Distrib. Syst. **16**, 1105–1119 (2005)

15. Karayannidis, N., Sellis, T., Kouvaras, Y.: CUBE file: a file structure for hierarchically clustered OLAP cubes. In: Bertino, E., et al. (eds.) EDBT 2004. LNCS, vol. 2992, pp. 621–638. Springer, Heidelberg (2004). https://doi.org/10.1007/978-3-540-24741-8_36

16. Leonhardi, B., Mitschang, B., Pulido, R., Sieb, C., Wurst, M.: Augmenting OLAP exploration with dynamic advanced analytics. In: Proceedings of the 13th International Conference on Extending Database Technology (EDBT 2010), pp. 687–692. Association for Computing Machinery, New York, USA (2010). https://doi.org/10.1145/1739041.1739127

17. Luo, Z.W., Ling, T.W., Ang, C.H., Lee, S.Y., Cui, B.: Range top/bottom k queries in OLAP sparse data cubes. In: Mayr, H.C., Lazansky, J., Quirchmayr, G., Vogel, P. (eds.) DEXA 2001. LNCS, vol. 2113, pp. 678–687. Springer, Heidelberg (2001). https://doi.org/10.1007/3-540-44759-8_66

18. Messaoud, R., Boussaid, O., Loudcher, S.: A multiple correspondence analysis to organize data cubes. Front. Artif. Intell. Appl. **155**, 133–146 (2007)

19. Puonti, M., Raitalaakso, T., Aho, T., Mikkonen, T.: Automating transformations in data vault data warehouse loads. Front. Artif. Intell. Appl. **292**, 215–230 (2017). https://doi.org/10.3233/978-1-61499-720-7-215

20. Salmam, F.Z., Fakir, M., Errattahi, R.: Prediction in OLAP data cubes. J. Inf. Knowl. Manag. **15**(02), 449–458 (2016). https://doi.org/10.1142/S0219649216500222

21. Schneider, S., Frosch-Wilke, D.: Analysis patterns in dimensional data modeling. In: Kannan, R., Andres, F. (eds.) ICDEM 2010. LNCS, vol. 6411, pp. 109–116. Springer, Heidelberg (2012). https://doi.org/10.1007/978-3-642-27872-3_17
22. Thomsen, E.: OLAP Solution: Building Multidimensional Information System. Willey, New York (2002)
23. Tsai, M.-F., Chu, W.: A multidimensional aggregation object (MAO) framework for computing distributive aggregations. In: Kambayashi, Y., Mohania, M., Wöß, W. (eds.) DaWaK 2003. LNCS, vol. 2737, pp. 45–54. Springer, Heidelberg (2003). https://doi.org/10.1007/978-3-540-45228-7_6
24. Vitter, J.S., Wang, M.: Approximate computation of multidimensional aggregates of sparse data using wavelets. In: Proceedings of the 1999 ACM SIGMOD International Conference on Management of Data (SIGMOD 1999), pp. 193–204. Association for Computing Machinery, New York, USA (1999). https://doi.org/10.1145/304182.304199
25. Wang, W., Lu, H., Feng, J., Yu, J.: Condensed cube: an effective approach to reducing data cube size. In: Proceedings of the 18th International Conference on Data Engineering (ICDE 2002), pp. 155–165. IEEE Computer Society, Washington (2002)
26. Yessad, L., Labiod, A.: Comparative study of data warehouses modeling approaches: Inmon, Kimball and data vault. In: 2016 International Conference on System Reliability and Science (ICSRS), pp. 95–99 (2016). https://doi.org/10.1109/ICSRS.2016.7815845
27. Zhang, R., Pan, D.: Metadata management based on lifecycle for DW 2.0. In: 2010 8th World Congress on Intelligent Control and Automation, pp. 5154–5157 (2010). https://doi.org/10.1109/WCICA.2010.5554915

Application of Convolutional Neural Networks for Image Detection and Recognition Based on a Self-written Generator

Moualé Moutouama N'dah Bienvenue[2] and Dmitry Kozyrev[1,2(✉)]

[1] V. A. Trapeznikov Institute of Control Sciences of Russian Academy of Sciences,
65 Profsoyuznaya Street, Moscow 117997, Russia
[2] Peoples' Friendship University of Russia (RUDN University), 6 Miklukho-Maklaya
Street, Moscow 117198, Russian Federation
bmouale@mail.ru, kozyrev-dv@rudn.ru

Abstract. Object recognition is a branch of artificial vision and one of the pillars of machine vision. It consists in identifying the forms described in advance in a digital image and, in general, in a digital video stream. Although, as a rule, it is possible to perform recognition from video clips, the learning process is usually performed on images. In this paper, an algorithm for classifying and recognizing objects using convolutional neural networks is considered. The purpose of the work is to implement an algorithm for detecting and classifying various graphic objects fed from a webcam. The task is to first classify and recognize an object with high accuracy according to a given data set, and then demonstrate a way to generate images to increase the volume of the training data set by using a self-written generator. The classification and recognition algorithm used is invariant to transfer, shift and rotation. A significant novelty of this work is the creation of a self-written generator that allows using various types of augmentation (artificial increase in the volume of the training sample by modifying the training data) to form new groups of modified images each time.

Keywords: image detection · image recognition · convolutional neural networks · R-CNN (Regional Convolutional Neural Networks) model · augmentation · subdiscretization

1 Introduction

Recently, considerable attention has been paid to object recognition. The task of pattern recognition in general is the task of describing and classifying objects of a particular nature, it is the task of developing the concept of a class of

This paper has been supported by the RUDN University Strategic Academic Leadership Program.

objects [1]. The recognition of objects by a person consists in the fact that he refers them to a class of similar objects. A person divides a lot of all objects into classes and assigns a name to each class: "tables", "birds", "letter "a"", etc. A person recognizes both visual and sound signals. However, a wide range of variations of the human face due to posture, lighting and facial expression leads to a very complex distribution and impairs the effectiveness of recognition. In order to solve all these difficulties and eliminate the factor of human error, the task is to develop a face detection system that should be reliable enough to cope with changes in lighting or weather conditions. In addition, this system must process head movements and take into account the fact that the user may be farther or closer to the camera. Also, the system should be able to detect the user's head with a certain angle of rotation on each axis. The fact is that many facial recognition algorithms show good performance when the head is almost frontal. However, they cannot detect a face if it is not full-face (in profile). In addition, the system must be able to detect and recognize a face, even if the user is wearing optical or sunglasses that create an occlusion in the eye area. When developing a computer vision project, it is necessary to take into account all these factors. Therefore, there was a need for an algorithm that would well separate the objects in the picture from each other, taking into account all the difficulties. This article discusses an algorithm of convolutional neural networks that allows classifying, detecting and recognizing images while preserving spatial relationships between pixels by studying image features using small squares of input data. The efficiency of the developed algorithm and its invariance to transfer, shift and rotation are shown, which makes it possible to achieve good learning accuracy. The work is structured as follows: The second section discusses the difference between image detection and classification. The third section discusses the method of image processing and classification using convolutional neural networks. The fourth section of the section provides a practical example of image classification from the Zalando database. This part is also devoted to the problem of image detection and recognition. To test the operability of the proposed detection and recognition method, the results of experiments on recognition of unknown images are presented.

2   The Difference Between Image Detection and Classification

It is important to distinguish between image detection and classification. Knowing how to detect a circle, triangle, shape, or even a face does not necessarily belong to the field of artificial intelligence (although some algorithms are able to do this). After all, these tasks are assigned to image processing algorithms. However, the ability to distinguish one object from another (car from bicycle), to perform facial recognition is an artificial intelligence in which learning algorithms are trained to recognize (classify) objects. However, detecting shapes in an image can help a classification algorithm in the sense that one can easily extract a face from a photo or video and offer it to a classification algorithm

capable of performing facial recognition. The model that allows you to simultaneously solve the problems of object search (detection) and classification is called
the R-CNN (Regional Convolutional Neural Networks) model. When searching
for objects using the R-CNN model, the following sequence of steps is performed
(see Fig. 1) [5].

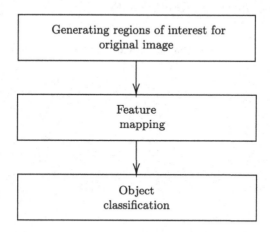

Fig. 1. Sequence of steps when using the R-CNN model

Generating areas of interest for the source image: generating areas of interest (region proposals), presumably containing the desired objects using various
algorithms (for example, Selective search or Regional-proposal function). Feature
map generation: At this step, a feature map is generated for the source image.
The formed areas of interest are scaled to a size comparable to the architecture of the CNN neural network (Convulational Neural Network or ConvNet).
The generated data is received at the input of a convolutional neural network
(CNN). Object classification: At this step, objects are classified for each area of
interest using the generated feature vector based on the Support Vector Machine
(SVM) [4].

3   Processing and Classification of Images Using Convolutional Neural Networks

3.1   Convolutional Neural Network for Image Classification

A special neural network is used to classify images: the CNN convolutional neural network. A convolutional neural network is a deep artificial neural network
with direct communication in which the neural network maintains a hierarchical structure by studying representations of internal functions and generalizing
functions in general image tasks such as object recognition and other computer
vision problems. The last CNN layers consist of neural networks whose input

data comes from the previous layers, called convolution layers. In order for a neural network to make a qualitative classification of images, it must be provided with a training base. In our case, the network training was carried out on Zalando data sets [11]. The application of a new multi-scale approach to training [11] allowed us to obtain one model for processing images of various sizes. Let us present in the form of a diagram the principles of functioning of a convolutional artificial neural network (see Fig. 2).

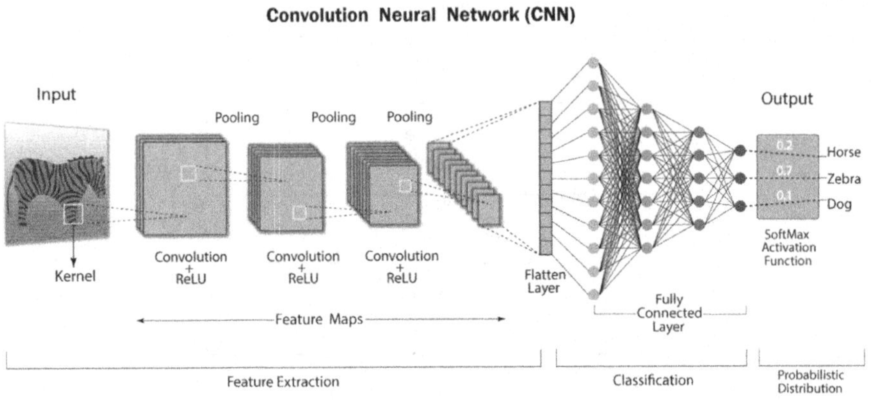

Convolution Neural Network (CNN)

Fig. 2. Architecture of convolutional artificial neural network

CNN is an architecture of a deep neural network with direct communication, consisting of several convolutional layers, each of which is followed by a pooling layer, an activation function and, possibly, a batch normalization. It also consists of fully connected layers. As the image moves across the network, it gets smaller, mainly due to the MacHpuling. The last layer outputs the prediction of class probabilities) [8].

3.2   Image Processing with Convolutional Neural Network

Image Characteristics and Image Convolution Method. It all starts with the original image. It seems that it is necessary to classify an image containing a rabbit. The goal is that when we present our machine with an image of a rabbit, the machine tells us that it is a rabbit or not. One of the solutions that intuitively comes to mind to solve this problem is to compare pixel by pixel two images, the training image and the image that needs to be classified. If these two images match, then we can say that it is a rabbit. This method has a disadvantage, since the classified image must be strictly identical to the reference image, which is completely impossible in real-world applications. The solution to this problem is to identify features in the original image (image fragments) and search for them in the image that needs to be classified. This is called the convolution method. If the image to be classified has a large number of common features with the original classified image, then there is a high probability that these images were very similar. The learning phase will consist of extracting these features, and then feeding them to the neural network so that it learns to establish a connection between the features of the original image and the marking (the name of the recognized object). At the end of the training, the neural network will be able to predict the object. To clarify this concept, consider an image with a size of 5 by 5 pixels and a feature of 2 by 2 pixels (see Fig. 3). This feature is also called a filter or core.

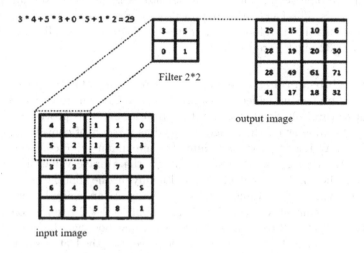

Fig. 3. Extracting a feature from an image

Convolution will consist of applying a filter to the entire image to output a new image. The following four figures explain this process: Lists are easy to create:

- We put the filter on top and to the left of the image;
- For each pixel value of the image located in the filter, it is multiplied by the value of the filter pixel;
- Let 's make the sum of these works;
- Shift by one pixel and start the product and sum again. This movement is called a Step and in our case has the value 1.

This process should be carried out until the entire image is filtered out, which will result in the creation of a new image smaller than the original image and having new values, as shown in (Fig. 3). These values are intended to identify certain features of the image. At the convolution stage, we will specify the activation function. The ReLU function is usually used, which allows you to remove negative values from the convolution image, replacing them with the value 0. Compared to sigmoid and tanh, the ReLU activation function is more reliable and accelerates convergence by six times. Unfortunately, the downside is that ReLU can be unstable during training.

The Principle of Convolution. As discussed above, the number of filters used corresponds to the features that need to be trained. Consequently, the more different filters are used at the training stage, the more effective this training stage will be; however, this may lead to retraining of the algorithm and the effect will be the opposite of what is expected. Unfortunately, there is no method for determining the optimal number of filters used, as well as their size, and it is often necessary to determine these values empirically. If we apply this convolution processing method to a single image, we will see that the number of calculations can quickly become significant if the image size is large and if we multiply the number of filters. This calculation requires a lot of resources at the processor and graphics card level. Therefore, in most cases, images with small sizes (28×28 pixels or 32×32 pixels) are used.

Pooling or Subsampling (Pooling). After receiving a filtered image, we can apply a new processing called "Pooling" to it, which will allow us to extract important features obtained as a result of convolution. This processing consists of moving the window on the image obtained as a result of convolution. In the context of "Max Pooling", we determine the maximum value contained in this window, in the context of "Average Pooling", the average value of the values contained in the window is calculated, and finally, in the context of "Stochastic Pooling", we will save the value selected randomly. As an example, let's take an image from our convolution (see Fig. 4) and apply a 2×2 pixel union window with a strict offset of 2 to it. After convolution and pooling, the original 4×4 pixel image is reduced to a 2×2 image while preserving the features of the image obtained as a result of convolution. This size reduction is especially useful for various future computations. The goal is to reduce the image while preserving important information.

4   Practical Example: Classification of an Image from the Zalando Database

4.1   Neural Network Training Based on Training and Test Data from the Zalando Database

One of the most common examples of using convolutional neural networks is the classification of handwritten letters using the MNIST (Mixed National Institute of Standards and Technology) observation collection [11]. However, in 2018 Zalando released its own collection of images called Zalando-MNIST [11]. It is this data set that is used in our practical case to train the network. Image classification is carried out using a convolutional neural network implemented on the basis of Tensorflow and Keras libraries. The application of a new multi-scale approach to training allowed us to obtain a single model for processing images of various sizes. The result of the network operation is shown in Fig. 7 and Fig. 8. Below is a diagram of the trained model with different parameters (see Fig. 5). After training is completed and the training model is formed, we get the following accuracy values on training and test data:

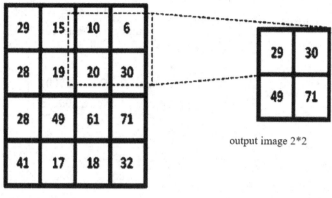

input image 4*4

output image 2*2

Fig. 4. Applying MaxPooling to a big image

Conclusion: The training phase took about 8 min in our case with a PC (laptop) equipped with an Intel Core i5 2.6 GHz processor with 8 GB of memory and an Intel NVIDIA GEFORCE 940MX CORE i5 graphics card with 64 MB of RAM. After completing the 20th epoch of training and building our training model, we obtained an accuracy of 92.26% on the test set (they may vary depending on the equipment used). The graphs in Fig. 7 and Fig. 8 show an increase in accuracy at the training stage up to 92%, as well as a decrease in error. This is an expected and logical result

4.2 Improving Classification Accuracy with the Help of a Reasoned Data Generator

It sometimes happens that there are images in the test dataset that are not in the training set. Consequently, this affects the accuracy of the model classification. A good database in terms of volume and quality is an eternal problem of an AI developer. Insufficient amount of training data can lead to both low accuracy of the neural network and retraining. Thus, to solve the problem, we will have to increase the number of images in our training set, i.e. to augment the data. Augmentation is a method of generating new images similar to samples, with changes in position, scale, brightness and some additional parameters. But the purpose of augmentation is not to generate completely new images; augmentation helps the neural network to view more variants of the same images and better identify common features when the database is too small. To do this, you do not need to search for photos on the Internet, we will use the ImageDataGenerator function of the Keras module provided for this purpose. The principle of operation of this function is as follows: Based on the image contained in the training set, the function will perform random rotation, scaling and moving actions to create new images. Based on the image contained in the training set, the function will perform random rotation, scaling and moving actions to create new images. Now

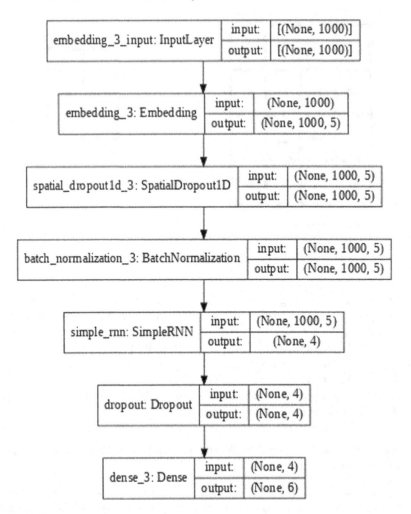

Fig. 5. Schematic of the trainable model and estimation of training

Percentage of correct answers on the training set: 92.26%
Time of processing: 8.316666666666666min

Fig. 6. Time of training and percentage of correct answers on the training set

we will generate new images and conduct new training. After completing the training and forming our training model, we get the following accuracy values on the test set.

After training our model, we saved it using a new script to make predictions on new unknown images. This approach is similar to using a model in a new application. The results have been obtained, which are shown in (Fig. 12).

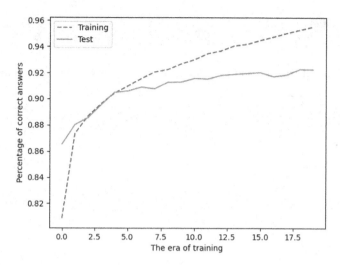

Fig. 7. Neural network accuracy on training and test data

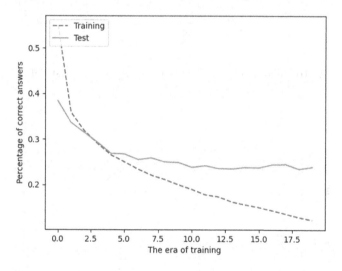

Fig. 8. Neural network errors on training and test data

Conclusion: It is noticed that the new training takes much longer (112 min, i.e. about 2 h versus 9 min with the previous training). This is due to the fact that in addition to training, new images are generated with each batch, which requires large hardware resources and, therefore, slows down the entire process. But despite this training time, we see a slight improvement in accuracy approaching 93%. At the same time, the recognition speed after training is significantly low - 0.17 min. You can also notice a decrease in the error rate between the training stage and the testing stage. Therefore, we can assume that this model is more effective than the previous one. In this part, we implemented an effective

Percentage of correct answers on the training set: 92.82%
Time of processing: 112.48333333333333min

Fig. 9. Time of training and percentage of correct answers on the training set

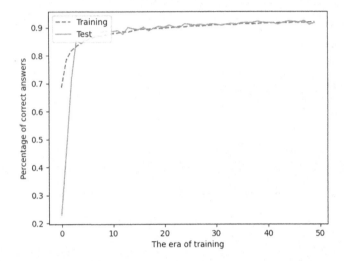

Fig. 10. Neural network accuracy on training and test data

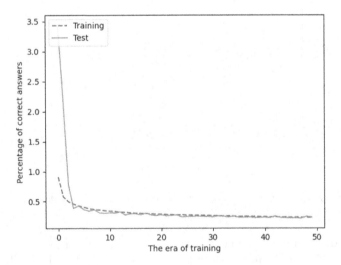

Fig. 11. Neural network errors on training and test data

image classification model with an accuracy of more than 93% using convolutional neural networks. Learning this type of algorithm requires a large number of images, considerable time and good computing resources.

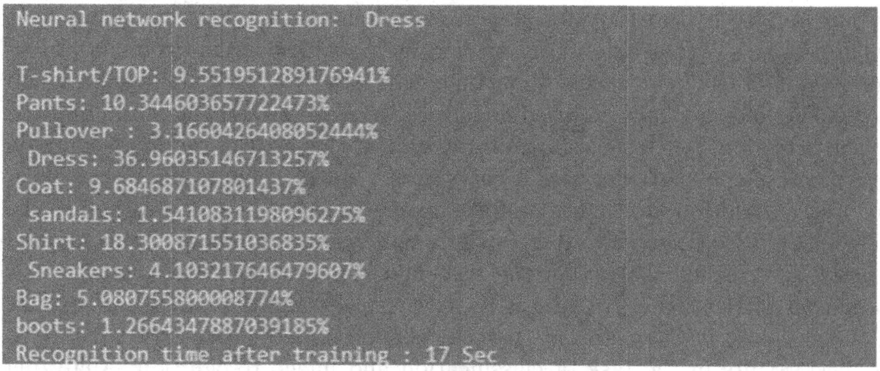

```
Neural network recognition:  Dress

T-shirt/TOP: 9.551951289176941%
Pants: 10.344603657722473%
Pullover : 3.1660426408052444%
 Dress: 36.96035146713257%
Coat: 9.684687107801437%
 sandals: 1.5410831198096275%
Shirt: 18.300871551036835%
 Sneakers: 4.103217646479607%
Bag: 5.080755800008774%
boots: 1.2664347887039185%
Recognition time after training : 17 Sec
```

Fig. 12. Model prediction results on new unknown images

The proposed method:

- on small datasets acts as a way of regularization and reduces retraining in small classes;
- allows you to make the classification resistant to some changes in the test data (for example, when submitting an image with a higher brightness relative to the training sample);
- improves the generalizing ability of the network.

This is a very powerful method for expanding the training dataset. Now you can manage to increase the volume of the dataset by adding any desired number of augmented images to it. The model was trained based on the basic method (self-written generator). This function, together with data and sample generators, made it possible to apply various types of augmentation to form new batches with modified images each time. As a result, the same model that was trained earlier without augmentation can learn deeper and better. During the entire training period, we submitted much more data, augmenting images for training at each step. The convenience of the generator is that creating augmented images on the fly takes up memory for only one batch. If you try to calculate the entire large augmented database at once, there may simply not be enough memory. To check the model at each epoch, we submitted updated versions of the test set each time. As the graphs (Fig. 10, Fig. 11 and Fig. 12) show, the model coped with them better. Finally, we evaluated the model's performance on test data that had never been presented to it.

5   Conclusions

The spread and development of computer vision technologies entails changes in other professional areas of human activity. Convolutional neural networks are used in object and face recognition systems, special medical software for image analysis, navigation of cars equipped with autonomous systems, in protection systems, and other areas. With the growth of computing power of computers, the appearance of image databases, it became possible to train deep neural networks. One of the main tasks of machine learning is the task of classifying and recognizing objects. SNS are used for optical recognition of images and objects, object detection, semantic segmentation, etc. In this article, the most common algorithms of convolutional neural networks for the problem of image classification and recognition were considered using a new method that is invariant to the transfer, shift, rotation and other random modifications of images. This method is resistant to some changes in the test data (for example, when an image is submitted with a higher brightness relative to the training sample). As a result of the training, it was revealed that the convolutional neural network showed the best results in the task of classification and image recognition. Analyzing the graphical results (Fig. 10, Fig. 11 and Fig. 12) of training and the results of the model on test data, we can conclude that the proposed model can be taught further, but we can also look for more effective neural network configurations.

References

1. Mathematical methods of pattern recognition. In: Abstracts of the 19th All-Russian Conference with International Participation (Moscow 2019), 420 p. Russian Academy of Sciences, Moscow (2019)
2. Redmon, J.: YOLO9000: better, faster, stronger. In: Redmon, J., Farhadi, A. (eds.) IEEE Conference on Computer Vision and Pattern Recognition (CVPR), pp. 6517–6525 (2017)
3. Panetto, H., Cecil, J.: Information systems for enterprise integration, interoperability and networking: theóry and applications. Enterp. Inf. Syst. **7**(1), 1–6 (2013)
4. Simonyan, K., Zisserman, A.: Very deep convolutional networks for large-scale image recognition. arXiv preprint arXiv:1409.1556 (2014)
5. Mouale, M.N.B., Kozyrev, D.V., Houankpo, H.G.K., Nibasumba, E.: Development of a neural network method in the problem of classification and image recognition. Sovremennye informacionnye tehnologii i IT-obrazovanie (Modern Inf. Technol. IT-Educ.) **17**(3), 507–518 (2021). https://doi.org/10.25559/SITITO.17.202103.507-518
6. Canny, J.: A computational approach to edge detection. IEEE Trans. Pattern Anal. Mach. Intell. **8**, 679–714 (1986)
7. Duy Thanh, N.: Face image recognition methods based on invariants to affine and brightness transformations. Diss. to the competition uch. degree cand. physical mat. Sciences, Moscow (2018)
8. Sikorsky, O.S.: Overview of convolutional neural networks for the problem of image classification. In: Sikorsky, O.S. (ed.) New Information Technologies in Automated Systems, vol. 20, pp. 37–42. Moscow (2017)
9. Erhan, D., Szegedy, C., Toshev, A., Anguelov, D.: Scalable object detection using deep neural networks. In: Proceedings of the IEEE Conference on Computer Vision and Pattern Recognition, pp. 2147–2154 (2014)
10. He, K., Zhang, X., Ren, S., Sun, J.: Spatial pyramid pooling in deep convolutional networks for visual recognition. In: ECCV (2014)
11. Krizhevsky, A., Sutskever, I., Hinton, G.E.: ImageNet classification with deep convolutional neural networks. In: NIPS (2012)

12. Hizem, W.: Capteur Intelligent pour la Reconnaissance de Visage, Télécommunications et l'Université Pierre et Marie Curie - Paris 6 (2009)
13. Guerfi Ababsa, S.: Authentification d'individus par reconnaissance de caractéristiques biométriques liées aux visages 2D/3D, Université Evry Val d'Essonne (2008)
14. Cécile Fiche, M.: Repousser les limites de l'identification faciale en contexte de vidéo surveillance, Université de Grenoble (2012)
15. Van Wambeke, M.: Reconnaissance et suivi de visages et implémentation en robotique temps-réel, Université Catholique de Louvain (2009–2010)
16. Gafarov, F.M.: G12 artificial neural networks and applications: textbook. allowance, 121 p. Gafarov, F.M., Galimyanov, A.F. (eds.) Kazan Publishing House, Kazan (2018)
17. Poniszewska-Maranda, A.: Management of access control in information system based on role concept. Scalable Comput. Pract. Experience 12(1), 35–49 (2011)
18. Arulogun, O.T., Omidiora, E.O., Olaniyi, O.M., Ipadeola, A.A.: Development of security system using facial recognition. Pac. J. Sci. Technol. 9(3), 377–384 (2008)

Estimating a Polyhedron Method Informativeness in the Problem of Checking the Automaton by the Statistical Properties of the Input and Output Sequences

S. Yu. Melnikov[1(✉)], K. E. Samouylov[1] , and A. V. Zyazin[2]

[1] Peoples' Friendship University of Russia (RUDN University), Moscow, Russia
`melnikov@linfotech.ru, ksam@sci.pfu.edu.ru`
[2] MIREA – Russian Technological University, Moscow, Russia
`ziazin@mirea.ru`

Abstract. The problem of verification of the Moore finite state machine by the statistical properties of the input and output sequences is considered. It is assumed that the initial state of the automaton is unknown. To verify the automaton, the polyhedra method is used. The results of computational experiments for four classes of non-autonomous binary shift registers are presented.

Keywords: PRNG · finite automaton · identification of automata · shift register · experiments with automata

1 Introduction. Formulation of the Problem

Pseudo-random number generation (PRNG) is one of the most important parts of all modern cryptographic protocols used for encryption and integrity. Weaknesses in PRNG have been found several times in history. A common way to test an PRNG is to run a large number of known statistical tests [14,18]. The aim of such testing is to decide whether PRNG generates sequences which may be considered as realizations of equiprobable Bernoulli sequence (for binary case). If the tests pass, then the random numbers are considered to be good enough for cryptographic use. To test PRNG, a variety of techniques are used, up to artificial neural networks [3].

When analyzing random number generators, the problem arises of identifying automata simulating PRNGs or its individual nodes [1]. If the identification of the PRNG or its individual nodes by the output and input (or intermediate) sequences available for analysis is possible, then this indicates a poor quality of this PRNG.

Similar problems of identifying individual nodes of discrete devices based on observations of output and input sequences also arise in the field of technical diagnostics [8,17].

© The Author(s), under exclusive license to Springer Nature Switzerland AG 2023
V. M. Vishnevskiy et al. (Eds.): DCCN 2022, CCIS 1748, pp. 42–51, 2023.
https://doi.org/10.1007/978-3-031-30648-8_4

Both linear and non-linear shift registers with various complication or feedback elements are widely used to design PRNGs [4,5]. This is primarily due to the fact that the de Bruijn graphs describing information transformations in such registers have a number of "good" combinatorial and structural properties and are well studied [13].

In this paper, we consider the problem of testing the hypothesis that the observed output sequence $(y_0, y_1, ..., y_{N-1})$ is obtained from the observed input sequence $(x_0, x_1, ..., x_{N-1})$, $N \geq 1$, using a given finite Moore automaton A (reference automaton), the initial state of which is unknown. One of the variants of such a problem for a special type of automata was considered in [15]. In [11], a geometric method was proposed that uses the construction of a special polyhedron R_A of the automaton A in a space of suitable dimension. Point z is considered, the coordinates of which are formed by the relative frequencies of occurrence of selected words in the observed sequences. If the Chebyshev distance between the polyhedron R_A and the point z exceeds a certain limit depending on N, then the hypothesis is rejected. Otherwise, the test can be repeated by moving to other sections of the observed sequences or by changing the sets of selected words, the frequencies of which are analyzed.

The computational complexity of this method is associated with the high complexity of constructing a polyhedron R_A for a given automaton A. However, if the polyhedron is built in advance or analytically, then in the two-dimensional case, when the frequencies of occurrence of one word in the input and one word in the output sequences are analyzed, the complexity of the described step of testing the hypothesis does not exceed the polynomial in the logarithm of the number of automaton states.

The polyhedron of the automaton is located in a real cube of suitable dimension. The effectiveness of the verification method under consideration depends, among other things, on the way the polyhedron of the reference automaton is located in this cube. There are both examples of automata whose polyhedron occupies the entire volume of the cube, and automata with polyhedra of zero volume. Let us consider several classes of automata [12] of comparable size and estimate what proportion of automata from these classes is "on average" cut off in one verification step in the proposed verification method, depending on the location of the point z.

2   Description of the Proposed Approach

Let $A_{h,f} = (Q, X = Y = \{0,1\}, h, f)$ be a binary finite Moore automaton with transition function $h : Q \times X \to Q$ and output function $f : Q \to Y$. Let Φ be the set of all functions from Q to Y. Since the automaton alphabets are binary, $|\Phi| = 2^{|Q|}$ holds. We will consider the class $K_h = \{A_{h,f}, f \in \Phi\}$ of automata with a fixed transition function h. The polygon of the automaton $A_{h,f}$, corresponding to the symbols "1" in the input and output sequences, will be denoted by $R_{A_{h,f}}$. For each point $z = (x,y)$ of the square $[0,1] \times [0,1]$, we define the function $p_{K_h}(z)$ as the probability that z belongs to the polygon of an automaton if this automaton is chosen randomly and equally probably from the class under consideration:

$$p_{K_h}(z) = \frac{\left|\{A_{h,f} | z \in R_{A_{h,f}}, A_{h,f} \in K_h\}\right|}{|K_h|}. \tag{1}$$

The value $p_{K_h}(z)$ characterizes the "informativeness" of a pair of numbers,

$$z = (x, y) = \frac{1}{N}\left(\sum_{i=0}^{N-1} x_i, \sum_{i=0}^{N-1} y_i\right), \tag{2}$$

obtained from observations of the input and output sequences. If, in particular, $p_{K_h}(z) = 1$ is true, then the proposed procedure will not provide any information useful for verification. If the value of $p_{K_h}(z)$ is close to $1/2$, then the point z belongs to the polygon of approximately every second automaton from the class under consideration, and we can expect that this observation cuts off about half of all possible options. If the value of $p_{K_h}(z)$ is small, then we can talk about the high information content of this observation.

3   Calculations for Three Classes of Automata

3.1   Shift Register

Let $f(x_1, x_2, ..., x_n)$ be a Boolean function of n binary variables, $n = 1, 2, ...$ A shift register is a Moore automaton A_f, the set of states of which is $\{0, 1\}^n$, the input and output alphabets are sets $\{0, 1\}$, the output is the value of the function f from the current state. The automaton A_f under the action of the input symbol $a_0 \in \{0, 1\}$ passes from state $(a_1, a_2, ..., a_n)$ to state $(a_0, a_1, ..., a_{n-1})$. The calculations were carried out for case $n = 4$. In the class under consideration, automata differ only in their output functions. Denote this class K_1. It consists of $2^{2^4} = 65536$ automata. For each automaton from K_1, its polygon was constructed according to the algorithms described in [7,11]. As shown in [2], the transition graph of automaton A_f has 179 cycles, the lengths of which are in the range between 1 and 16. For each cycle c, a point with coordinates $(x_c, y_{c,f})$ was calculated, where x_c and $y_{c,f}$ are the relative frequencies of the symbol "1" in the input and output markings. Next, a convex hull of 179 obtained points was built, which is the required polygon R_{A_f}. Figure 1 shows examples of polygons of randomly selected automata from the class K_1. It is easy to see that the depicted polygons do not have sides parallel to the y-axis.

There are two kinds of symmetries for shift register polygons.

Polygons of the automata with functions $f(x_1, x_2, ..., x_n)$ and $f(\bar{x}_1, \bar{x}_2, ..., \bar{x}_n)$ are symmetric with respect to the vertical axis $y = \frac{1}{2}$. Polygons of the automata with functions $\overline{f(x_1, x_2, ..., x_n)}$ and $f(x_1, x_2, ..., x_n)$ are symmetric with respect to the horizontal axis $x = \frac{1}{2}$. To prove the first symmetry, we can use the fact that the arbitrary cycle $c = (a_1, a_2, ..., a_l)$, $a_i = 0, 1$, $i = 1, 2, ..., l$, $l = 1, 2, ...$ in the transition graph of the automaton A_f has a dual cycle $\bar{c} = (\bar{a}_1, \bar{a}_2, ..., \bar{a}_l)$, formed by the inversion of all coordinates. Therefore, for the relative frequencies of the symbol "1" in the input marking, the following equality holds $x_c = 1 -$

Fig. 1. Examples of polygons of randomly selected automata from a class K_1

$x_{\bar{c}}$. From this we can conclude that polygons of the automata with functions $f(x_1, x_2, ..., x_n)$ and $f(\bar{x}_1, \bar{x}_2, ..., \bar{x}_n)$ are symmetric. The full proof is given in [10].

To prove the second symmetry, you can see that for the relative frequencies of the symbol "1" in the output marking, the equality $y_{c,f} = 1 - y_{c,\bar{f}}$ is true.

To calculate the values in formula (1), the belonging of each of the points of the form $\left(\frac{i}{66}, \frac{j}{66}\right)$, $0 \leq i, j \leq 66$ to each polygon R_{A_f} was checked.

Calculations and construction of images were carried out in the software environment Mathematica v.12 [16]. The `DeBruijnGraph` function was used to construct the automaton transition graphs. The `FindCycle` function was used to enumerate all the cycles in the automaton transition graphs. When constructing convex hulls, the `ConvexHullMesh` function was used; to obtain the polygon coordinates of the output function, the `MeshCoordinates[ConvexHullMesh [pointsOfFunction[f], AllPoints → False]]` construction was used. Cases where the convex hull was a 1-dimensional segment were handled separately. The array of received coordinates of all polygons was saved.

In the subloop iterating over the grid points of the square, the `RegionMember` function was used to check whether a point belongs to a given polygon. Surface images were plotted using the `ListPlot3D` function with default parameters. Figure 2 shows the image of the constructed surface after smoothing. The abscissa axis corresponds to the relative frequencies of the symbol "1" in the input sequence, the ordinate axis corresponds to the output sequence, the value $p_{K_1}(z)$ is plotted along the applicate axis.

3.2 Shift Register with Internal XOR

Let $f(x_1, x_2, ..., x_n)$ be a Boolean function of n binary variables, $n = 1, 2, ...$ A shift register with internal XOR is [9] an automaton $A_f^{\oplus}$, which, under the action of the input symbol $a_0 \in \{0, 1\}$, passes from state $(a_1, a_2, ..., a_n)$ to state $(a_0 \oplus a_1, a_1 \oplus a_2, ..., a_{n-1} \oplus a_n)$, where $\oplus$ is summation modulo 2.

46   S. Y. Melnikov et al.

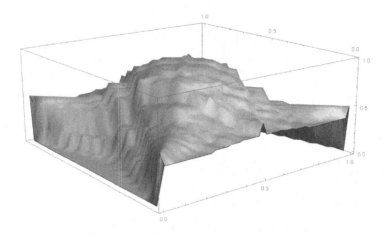

Fig. 2. Surface $p_{K_1}(z)$

The calculations were carried out for the case $n = 4$. As above, the automata in the class under consideration differ only in their output functions. Let's denote this class K_2. It consists of $2^{2^4} = 65536$ automata. The transition graphs of the automata $A_f^\oplus$ and A_f are isomorphic, therefore they have the same cycle structure. However, the labeling of the arcs of these cycles is different, and the polygons in this case turn out to be different. Figure 3 shows examples of polygons of randomly selected automata from class K_2. It can be seen that in all examples one of the sides of the polygon lies on the y-axis.

Fig. 3. Examples of polygons of randomly selected automata from a class K_2

It's easy to see that polygons of the shift registers with internal XOR with functions $f(x_1, x_2, ..., x_n)$ and $\overline{f(x_1, x_2, ..., x_n)}$ are symmetric with respect to the horizontal axis $x = \frac{1}{2}$.

Further construction of the surface was carried out in the same way as in the paragraph above. Figure 4 shows the image of the constructed surface after smoothing. The abscissa axis corresponds to the relative frequencies of the symbol "1" in the input sequence, the ordinate axis corresponds to the output sequence, the value $p_{K_2}(z)$ is plotted along the applicate axis.

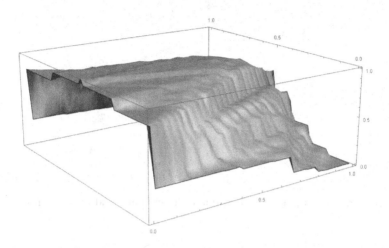

Fig. 4. Surface $p_{K_2}(z)$

3.3   Generalized Shift Register

Transition graphs of generalized shift registers are generalized de Bruijn graphs [6]. A binary generalized register of the order m, $m = 1, 2, ...$, is a Moore automaton $A_f^{(m)} = (X, Y, Q, h, f)$, where the input and output alphabets are $X = Y = \{0, 1\}$, the set of states is $Q = \{0, 1, ..., m - 1\}$, the transition function is defined by the rule $h(q, \varepsilon) = (2q + \varepsilon) \pmod{m}$, $q \in Q$, $\varepsilon = 0, 1$, the output function is some mapping $f : Q \rightarrow \{0, 1\}$. At $m = 2^t$ binary generalized shift register is a binary pass-through shift register with an accumulator of the capacity t.

The calculations were carried out for two cases, $m = 7$ and $m = 9$. We denote the corresponding classes of automata by K_3 and K_4. They consist of $2^{2*7} = 16384$ and $2^{2*9} = 262144$ automata, respectively.

The transition graphs of automata from classes K_3 and K_4 have 88 and 326 cycles, respectively. Further, the construction of polygons was carried out similarly to that indicated above. Figures 5 and 6 show examples of polygons of randomly selected automata from classes K_3 and K_4. It is noticeable that the polygons of the automata from K_4 are distinguished by a somewhat greater variety of shapes.

It's easy to see that polygons of the generalized shift registers with functions $f(x_1, x_2, ..., x_n)$ and $\overline{f(x_1, x_2, ..., x_n)}$ are symmetric with respect to the horizontal axis $x = \frac{1}{2}$.

Fig. 5. Examples of polygons of randomly selected automata from a class K_3

Fig. 6. Examples of polygons of randomly selected automata from a class K_4

For calculations by formula (1) for class K_3, the grid $\left(\frac{i}{100}, \frac{j}{100}\right)$, $0 \leq i, j \leq 100$ was used; for class K_4 grid $\left(\frac{i}{50}, \frac{j}{50}\right)$, $0 \leq i, j \leq 50$. Figure 7 and Fig. 8 show images of the constructed surfaces after smoothing. The abscissa axis corresponds to the relative frequencies of the "1" symbol in the input sequence, the ordinate axis corresponds to the output sequence, and the values $p_{K_3}(z)$ and $p_{K_4}(z)$ are plotted along the applicate axis, respectively.

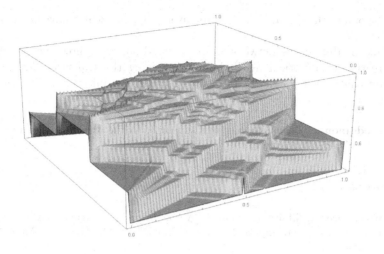

Fig. 7. Surface $p_{K_3}(z)$

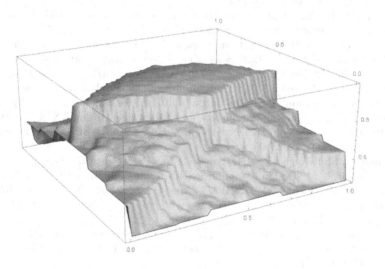

Fig. 8. Surface $p_{K_4}(z)$

4   Conclusion

1. The informativeness of the observation largely depends on which area of the square $[0,1] \times [0,1]$ the point $\frac{1}{N}\left(\sum_{i=0}^{N-1} x_i, \sum_{i=0}^{N-1} y_i\right)$ falls into. The least informative are the points near $\left(\frac{1}{2},\frac{1}{2}\right)$, in which the values of the functions $p_K(z)$ are close to the maximum.
2. The constructed surfaces differ markedly from each other. The surface $p_{K_2}(z)$ has symmetry with respect to the plane $y = \frac{1}{2}$, the other three surfaces also have symmetry with respect to the plane $x = \frac{1}{2}$. These symmetries

correspond to the symmetries of automaton polygons, which have been proved analytically.

3. Significant differences between the calculated surfaces show the possibility of creating a statistical method for recognizing the transition function of an automaton with a random output function, the initial state of which is unknown.

Acknowledgments. This paper has been supported by the RUDN University Strategic Academic Leadership Program.

References

1. Bhattacharjee, K., Maity, K., Das, S.: A search for good pseudo-random number generators: survey and empirical studies (2018). https://doi.org/10.48550/ARXIV.1811.04035
2. Bryant, P.R., Christensen, J.: The enumeration of shift register sequences. J. Comb. Theory Ser. A **35**, 154–172 (1983)
3. Fischer, T.: Testing cryptographically secure pseudo random number generators with artificial neural networks. In: 2018 17th IEEE International Conference on Trust, Security and Privacy in Computing and Communications/12th IEEE International Conference on Big Data Science and Engineering (TrustCom/BigDataSE), pp. 1214–1223 (2018). https://doi.org/10.1109/TrustCom/BigDataSE.2018.00168
4. Galustov, G.G., Voronin, V.V.: The synthesis of the correlation function of pseudorandom binary numbers at the output shift register. In: Kadar, I. (ed.) Signal Processing, Sensor/Information Fusion, and Target Recognition XXVI, vol. 10200, p. 102001G. International Society for Optics and Photonics, SPIE (2017). https://doi.org/10.1117/12.2263590
5. Golomb, S.: Shift Register Sequences, 3rd edn. World Scientific, Singapore (2017)
6. Imase, M., Itoh, M.: Design to minimize diameter on building-block network. IEEE Trans. Comput. **30**(6), 439–442 (1981). https://doi.org/10.1109/TC.1981.1675809
7. Johnson, D.B.: Finding all the elementary circuits of a directed graph. SIAM J. Comput. **4**(1), 77–84 (1975)
8. Li, Y., Hadjicostis, C.N., Wu, N., Li, Z.: Error- and tamper-tolerant state estimation for discrete event systems under cost constraints (2020). https://doi.org/10.48550/ARXIV.2011.01371
9. Melnikov, S.Y., Samouylov, K.E.: Statistical properties of binary nonautonomous shift registers with internal XOR. Inform. Appl. **14**(2), 80–85 (2020). https://doi.org/10.14357/19922264200211
10. Melnikov, S.Y., Samouylov, K.E.: Polygons characterizing the joint statistical properties of the input and output sequences of the binary shift register. In: The 4th International Conference on Future Networks and Distributed Systems (ICFNDS 2020), pp. 1–6. Association for Computing Machinery, New York, USA (2020). https://doi.org/10.1145/3440749.3442601
11. Melnikov, S.Y., Samouylov, K.E.: Polyhedra of finite state machines and their use in the identification problem. In: NEW2AN/ruSMART -2020. LNCS, vol. 12526, pp. 110–121. Springer, Cham (2020). https://doi.org/10.1007/978-3-030-65729-1_10

12. Potii, O., Poluyanenko, N., Petrenko, A., Pidkhomnyi, O., Florov, S., Kuznetsova, T.: Boolean functions for stream ciphers. In: 2019 IEEE 2nd Ukraine Conference on Electrical and Computer Engineering (UKRCON), pp. 942–946 (2019). https://doi.org/10.1109/UKRCON.2019.8879862

13. Read, R.C.: Graph Theory and Computing. Academic Press, NY (2014)

14. Serov, A.A., Zubkov, A.M.: Testing the NIST statistical test suite on artificial pseudorandom sequences. In: Proceedings of the XII International Conference on Computer Data Analysis and Modeling: Stochastics and Data Science (Minsk, 18–22 September 2019), pp. 291–294. Publishing Center of BSU, Minsk (2019)

15. Sevastyanov, B.A.: The conditional distribution of the output of an automaton without memory for given characteristics of the input. Discret. Math. Appl. 4(1), 1–6 (1994). https://doi.org/10.1515/dma.1994.4.1.1

16. Wolfram, S.: The Mathematica Book. Assembly Automation (1999)

17. Zaytoon, J., Lafortune, S.: Overview of fault diagnosis methods for discrete event systems. Annu. Rev. Control. 37(2), 308–320 (2013). https://doi.org/10.1016/j.arcontrol.2013.09.009

18. Zubkov, A.M., Serov, A.A.: A natural approach to the experimental study of dependence between statistical tests. Matem. Vopr. Kriptogr 12(1), 131–142 (2021). https://doi.org/10.4213/mvk352

Visual Question Answering for Response Synthesis Based on Spatial Actions

Gleb Kiselev[1,2]([✉]) [iD], Daniil Weizenfeld[1,2] [iD], and Yaroslava Gorbunova[1,2] [iD]

[1] Peoples' Friendship University of Russia (RUDN University), 6 Miklukho-Maklaya Street, Moscow 117198, Russian Federation
kiselev@isa.ru
[2] Artificial Intelligence Research Institute, FRC CSC RAS, 44 Vavilova Street, Moscow 119333, Russian Federation

Abstract. The paper considers the automatic analysis problem of a user's natural language query from an image. The mechanism synthesizes a logically correct non-binary response. Synthesis is carried out on the basis of combining the results of convolutional and recurrent networks and projection on a set of valid answers. A three-dimensional data set has been developed to search for an answer in a complex environment using a robotic arm. Similar systems examples and their comparison are given. The experiments results showed that our method is able to achieve indicators comparable with known models.

Keywords: Computer science · Machine learning · Computer vision · Neural networks

1 Introduction

1.1 The VQA Task

Visual Question Answering (VQA) is an artificial intelligence (AI) task from vision-and-language (VL) family that stands on crossing of computer vision and natural language processing. It takes as input an image and a free-form, open-ended, natural-language question about the image and produces a natural-language answer as the output [2].

In early works [2,7,18], a standard structure of VQA system was formed, alongside with free datasets for benchmarking and comparing results with each other research.

All of the [1,2,11,16,17,19] research works (which is the majority of VQA research) utilizes the convolutional and recurrent neural networks.

The standard structure of VQA system mentioned above is built of three main modules: vision, language and cross-modality. Many of VQA solutions have following common features:

– The vast majority of research works utilizes the Bert [5] model as a language module;

- Vision module is made and ran separately from language and cross-modality modules, having in result all images encoded by it as part of dataset, which is used by rest of system (Fig. 1);
- The dataset is pre-tensorized images, 3–6 questions paired with each image and an answer to each pair (for train and val splits)
- Vision module is CNN or RCNN neural networks without classifier layers. Visual data needs to be only encoded prior to cross-modality module, the classifier goes after that.
- Answer synthesis is a classification task, where there is fixed N set of unique text answers provided by the train dataset (*e.g.*, 3000–3200 unique answers in [16] from dataset of 9.18M image-and-sentence pairs)

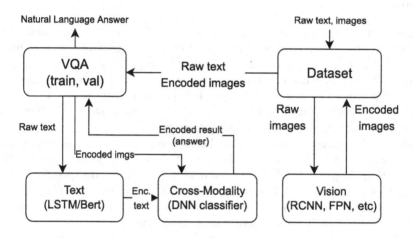

Fig. 1. Typical VQA model architecture

1.2 VQA Models Overview

There are plenty of VQA implementations, including such well-known ones as:

- LXMERT [16] – one of the first cross-modal solutions. Uses BERT as a basis of the language and cross-modal modules. For vision it uses an object detector that provides features of detected objects and its region-of-interest.
- Oscar [11] – the model with an expanded data. The language module is similar to LXMERT's. The vision module provides object labels in order to improve attention to objects, mentioned in the question. The data sample is a Word-Tag-Image triple, where Word and Image are the default text-image pair, and Tag is a set of object labels that are presented both in the question and the image.
- VinVL [19] – Oscar model with an improved vision module and extended dataset. Alongside labels, many object attributes are present. Uses ResNeXt-152 C4 (CNN) with own dataset that is fusion of 4 public ones (*COCO* [12], *OpenImages* [10], *Objects365* [14], *Visual Genome* [9])

- AliceMinD-MMU [17] – a fresh approach to VQA modeling. Three different vision models of object detection in one project (*Region-based, Grid-based, Patch-based*), their outputs then fused into one embedding. Variable cross-modality module: "Mixture-of-Experts" is introduced. These are modules that operates on the image and makes decision: whenever to use default VQA system or any of "expert" submodules including visual text recognition and clock recognition.

1.3   VQA as Part of Bigger Systems

VQA is known to be included as a module in a larger system, if required. Examples are:

- EQA (Embodied Question Answering [3]) – VQA is used for the agent that travels in the scene. To synthesize the answer, the agent can move himself through the scene if it cannot answer from current position. There are module that analyses image from current position and providing decision – whenever the agent can or cannot answer provided question. If it can't – movement sequence is composed then by another module called *planner*, which also takes the current image and question as input.
- IQA (I – Interactive [6]) – system that extends EQA possibilities. Alongside moving, the agent can also interact with some objects in scene. The interactions is pre-defined actions, that planner can use on specific objects in scene (*e.g.*, open the fridge door; turn on the room light switch). Objects are detected by another computer vision module, and the actions mentioned before is paired with one or more classes of objects.
- MQA (M – Manipulation [4]) – system use manipulator for manipulating objects in scene. In the work [4], the scene is a limited, highly loaded area (*a box of objects*). MQA is formulated as extension of IQA (Table 1), however in actual work agent is unable to travel nor use specific actions: it just manipulates objects by pushing these from different sides. You can see an example of MQA work in Table 2.

Table 1. Comparison of EQA, IQA and MQA in paper [4]

	VQA	EQA	IQA	MQA
Understanding	✓	✓	✓	✓
Exploration	–	✓	✓	✓
Interaction	–	–	✓	✓
Manipulation	–	–	–	✓

1.4   End-Product

We develop our own VQA model for use it in a limited, highly loaded area. Such an area can be a fridge, box, cupboard, etc. Example questions and answers are provided in Table 2. In that case, we develop an MQA-like model.

Table 2. Questions-Actions-Answers examples for MQA model

Scene	Question	Actions	Answer
3 small objects	How many objects in the scene?	*No needed*	3
2 big objects, 1 small behind one of big ones	How many objects in the scene?	Moving big objects aside for checking their back	3
1 large object	How many objects in the scene?	Moving object aside	1
High-loaded scene	Is there a pen behind the box in the center?[1]	Moving exact box in center aside	Yes
High-loaded scene	What hided behind that large box?[2]	Moving box aside	Nothing
High-loaded scene	What hided behind that large box?[3]	*No needed*	*No answer*[4]

[1] Assuming that is true
[2] Assuming that box is the most distant object in working area
[3] Assuming that there is more than one large boxes
[4] Answer cannot be provided, clarification required

Before development, we need to choose the model, which we will use as a basis and test it on the diagnostic dataset. After that, we choose the main dataset for final train, validate and fine-tuning.

2   Results

2.1   CLEVR Dataset

CLEVR Dataset [8] consists of three parts: train split and two test splits. Train split includes 70,000 images and 699,989 questions. Every test split includes 15,000 images and 14,988 questions.

CLEVR data structure:

1. Scene objects images set. Objects are a three types of geometric shapes kit: cube, sphere, cylinder. Every figure has color from given set: gray, blue, brown, yellow, red, green, purple, cyan; one of two sizes: big or small; and material: rubber or metal.

2. Scenes description: includes object shape, material, color and 3D position of every object in the scene.
3. Natural language questions about scenes. Question involve relational information, shape, color, material. Most of questions are more difficult than regular VQA datasets because of complex relational information, and its length can reach 30 words. These questions designed especially for agent to be able not only distinguish color and shape of objects, but also be able to understand complex relational information about objects.

There are tho test splits: `testA` and `testB`. The main differences are in the figure colors: in both splits the cube and cylinder color can be only from fixed subset of total color set mentioned above. These subsets are unique and not crosses for cubes and cylinders, while spheres can be any color of main set. Difference of test splits is of that color subsets: there are swapped out in each other split. The train split is only `A` variant it makes `B` variant unavailable during training. That method tests the agent whenever it confuses color with shape of figure (*e.g.*, in bad condition agent can start detect the brown cube as a cylinder, because in train set only cylinder has such color).

The dataset was used by us to test the VQA model – Oscar [11]

2.2 Testing Oscar on CLEVR Dataset

The reasons for choosing this model are:

1. There are another model VinVL [19] that mainly based on Oscar and is famous by its good result on VQA Challenge 2021, ranked 6th and margins from 1st in only $\sim 3\%$[1];
2. Project is open-source and available on public domain;
3. Code is documented and commented better than others [1,16].

As mentioned in Sect. 1.1, Oscar also uses dataset with images already encoded externally. It was decided to prepare CLEVR dataset, used another open-source project – MMF [15], that contains the same pre-trained vision model in its library as used in Oscar dataset.

Already pre-trained Oscar model was extended by CLEVR dataset and fine-tuned on it. Training lasted 20 epochs. Best result had been achieved on 19th – 96.478%.

2.3 Result Analysis

Testing was performed on both variants of `test` split. By visual evaluation of test results, `testA` looks like has 96% accuracy. Example – Fig. 2a.

---

[1] Ranking is based on overall accuracy of answers provided by model. Accuracy is a simple percentage of true answers provided by model on challenge's test dataset. VinVL result – 77.45%; Human performance on same dataset – $\sim 80\%$.

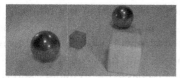

Q: How many gray objects are tiny objects or large balls?
A: 1

Q: There is a small cyan cylinder; are there any gray matte objects behind it?
A: no

Q: Is there a cyan thing made of the same material as the large brown object?
A: no

Q: Do the ball in front of the large brown object and the sphere behind the gray cube have the same material?
A: yes

Q: Is the number of large objects right of the large matte ball greater than the number of green things behind the cyan cube?
A: yes

Q: How many gray rubber objects are the same shape as the gray metal object?
A: 0

(a) Example questions and answers. Answers was synthesized by Oscar model

(b) False answer on first question. Probably because of confusion of shape and color

Q: How many tiny things are the same shape as the large cyan object?
A: 3

Q: How many matte things are either yellow spheres or small cubes?
A: 3

Q: What number of other things are there of the same material as the cyan object?
A: 3

Q: Are there more brown things than metal cubes?
A: yes

Q: There is two small cylinders; does they have the same color?
A: no

Q: How many cylinders is in here?
A: 3

Q: What number of spheres are in here?
A: 1

Q: What number of cubes are in here?
A: 3

Q: What number of cubes are there?
A: 6

(c) Direct input of questions

(d) Example of variation of questions. The 'here' word is frequently causes a false answer

Fig. 2. Test results of Oscar model trained on CLEVR dataset

Testing on variant testB, as expected, was quite worse. In approx. 20% conditions model is confusing cylinders with cubes because of difference of colors from training split. Example – Fig. 2b.

The Oscar code was modified in order to be able ask questions"on fly" (Fig. 2c). Even a small questions form deviation from CLEVR train split produces false answers (Fig. 2d).

The future work is to design our own system. That system will be based on Oscar; that choice is described in Sect. 2.2. The important feature of our model will be unified design: all modules, including vision, will be represented in one code base. That means, the dataset will no longer contain encoded (*tensorized*) images – they will be raw and human-readable. Of course we do understand that reason of pre-tensioning is performance on training – we are experimenting on that.

The system we are developing is called *VQA KWG* (Fig. 3) and is designed against future usage through API. The text module is the newest Bert version, the cross-modality module is based on Oscars' DNN classifier. The vision module architecture is to be decided; we use X152-C4 for early experiments.

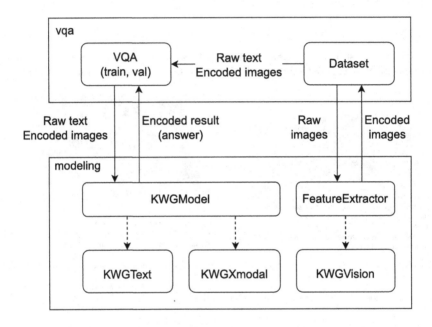

Fig. 3. The VQA system in development

For evaluating quality of that VQA system in environment close to reality, we develop own dataset.

3   Creating Own Dataset

3.1   RL Method

To find answers in the manipulation problem under consideration, RL method is planned to use. Reinforcement Learning (RL) is a machine learning method that mechanism includes: 1) the environment with that the agent interacts; 2) the environment generates "rewards". The following behaviour can be described as equation:

$$V(x_0) = \max_{\{a_t\}_{t=0}^{\infty}} \sum_{t=0}^{\infty} \beta^t F(x_t, a_t) \tag{1}$$

where,

$$a_t \in \Gamma(x_t), \; x_{t+1} = T(x_t, a_t), \; \forall t = 0, 1, 2, \ldots \tag{2}$$

The agent trains by trial and error to optimally interact with the environment. The characteristics of the agent are psychologically plausible, it has memory (creating a hierarchy of the agent's actions), attention (speeding up image processing), curiosity (environment exploring). Examples of RL methods using are: training a car to drive independently on a highway, city, etc.; training an agent to perform simple queries in a house (bring a glass of water, find a car, close a window); interaction with VQA (the closer the prediction is to the truth, the better the model learns to find the necessary features for further answers).

The Table 3 describes the interaction of our method with the RL mechanism.

Table 3. Pseudocode

Algorithm 1 Interaction of our method with the reinforcement learning mechanism
01: $z_{goal} = get\_goal(question)$
02: $z_{actual} = get\_actual(scene)$
03: $r_{actual} = 0$
04: **for** $i = 0, goal\_step$ **do**
05: $s, r_{tmp} = step\_reward\_generation\_TRPO\_alg(z_{actual}, z_{goal}, r_{actual})$
06: $z_{actual} = z_{actual} + step$
07: $r_{actual} = transform\_reward\_TRPO\_alg(r_{actual}, r_{tmp})$
08: **if** $z_{actual} = z_{goal}$
09: **then** go to dot
10: **end if**
11: **end for**
12: dot $a = get\_answer(z_{actual}, z_{goal}, question)$

At the beginning we introduce variables z_{actual} and z_{goal} – these are the current and target states of the environment as tensors of values with information about the environment (number, location, characteristics of objects, etc.), obtained after processing the scene data and natural language question by the method used, r_{actual} is the amount of agent rewards before starting to use the RL mechanism. Then, in a cycle from 0 to $goal\_step$ (the maximum number of steps to achieve the goal), using the reinforcement learning mechanism $TRPO\_alg$, based on Trust Region Policy Optimization[2] (TRPO) [13], we generate a new step s and award r_{tmp} for the agent. To do this, we create a set of valid trajectories applying the policy to the environment, compute advantage estimates, estimate policy gradient, and update the policy. Then we apply the obtained s and r_{tmp} to z_{actual} and r_{actual}. After comparing the current state and the target state of the environment, a decision is made whether to repeat the cycle or to exit the cycle. We get modified environment state tensor as a result of the algorithm. Then the response is generated in natural language using the used method.

It is necessary to thoroughly approach the selection of a dataset for training the model in pattern recognition and the ability to interact directly with objects to performing the agent's activities in close to real conditions.

Currently, there is no suitable 3D dataset for fully training the agent to interact with the scene and the objects located there. For this reason, it was decided to create own 3D dataset for high-quality training of the cognitive agent. The 2D basic CLEVR dataset was chosen as the prototype for creating such a dataset, because it contains minimal biases and includes detailed annotations. Moreover, the dataset is used to analyze various modern systems of visual thinking, providing an understanding of their capabilities and limitations.

3.2 Scene Generation

At the first stages of the work on creating own 3D dataset, CLEVR was studied. A basic V-REP scene was created and the objects of the future scene is placed in random order. To recreate the image of the scene in the modeling environment, initial data was taken from the scene description file, namely: the number of geometric objects in the scene, the shapes of the objects, their scene coordinates.

The Python programming language was chosen for the implementation of scene objects moving. A script is written that gets data about each of the V-REP scene objects, then changes its current coordinates to target coordinates addressing each one.

The result of the work is a scene similar to the target. Comparing the received (Fig. 4a) and the target (Fig. 4b) scenes, we can conclude that the objects are arranged according to the required coordinates. The received scene fully corresponds to the location of geometric objects on the target scene.

3.3 Further Work

The goal is to develop a modern, close to state-of-the-art performance model. The agent will be searching for an answer to a provided question in a 3D space, with a robotic manipulator used if needed to rearrange objects in the scene while finding out the answer to a question (MQA - like model). The question examples for that model are displayed in the Table 2.

---

[2] Trust Region Policy Optimization is a model-free, online, on-policy, policy gradient reinforcement learning method.

Fig. 4. Comparison of a) the received and b) the target scene. The objects are arranged according to the required coordinates. The received scene fully corresponds to the location of geometric objects on the target scene.

4   Conclusion

VQA researches were studied, major systems were formed the comparison and Oscar was selected for testing.

Testing of the Oscar model on the CLEVR dataset is carried out successfully. The accuracy of the answers is high for 20 epochs, correct answers are synthesized to complex questions.

Incorrect answers when testing on variant B exist, as expected, as well as on manually entering different format questions, because a limited range of questions is used in the dataset.

In the future, it is planned to add both other answers and interference for the analysis of the training quality.

Further work on creating own dataset will be aimed at creating code that will automatically create scenes in the V-REP, based on the data about the CLEVR scene. The model will be trained to recognize objects using the VQA module and learn to interact with them using the RL mechanism already with the new dataset.

References

1. StructBERT: visual-linguistic pre-training for visual question answering. ALIbaba (2020). https://github.com/alibaba/AliceMind
2. Antol, S., et al.: VQA: visual question answering. CoRR abs/1505.00468 (2015). http://arxiv.org/abs/1505.00468
3. Das, A., Datta, S., Gkioxari, G., Lee, S., Parikh, D., Batra, D.: Embodied question answering. CoRR abs/1711.11543 (2017). http://arxiv.org/abs/1711.11543
4. Deng, Y., et al.: MQA: answering the question via robotic manipulation. CoRR abs/2003.04641 (2020). https://arxiv.org/abs/2003.04641
5. Devlin, J., Chang, M., Lee, K., Toutanova, K.: BERT: pre-training of deep bidirectional transformers for language understanding. CoRR abs/1810.04805 (2018). http://arxiv.org/abs/1810.04805

6. Gordon, D., Kembhavi, A., Rastegari, M., Redmon, J., Fox, D., Farhadi, A.: IQA: visual question answering in interactive environments. CoRR abs/1712.03316 (2017). http://arxiv.org/abs/1712.03316

7. Goyal, Y., Khot, T., Summers-Stay, D., Batra, D., Parikh, D.: Making the V in VQA matter: elevating the role of image understanding in visual question answering. CoRR abs/1612.00837 (2016). http://arxiv.org/abs/1612.00837

8. Johnson, J., Hariharan, B., van der Maaten, L., Fei-Fei, L., Zitnick, C.L., Girshick, R.B.: CLEVR: a diagnostic dataset for compositional language and elementary visual reasoning. CoRR abs/1612.06890 (2016). http://arxiv.org/abs/1612.06890

9. Krishna, R., et al.: Visual genome: connecting language and vision using crowd-sourced dense image annotations (2016). https://arxiv.org/abs/1602.07332

10. Kuznetsova, A., et al.: The open images dataset V4: unified image classification, object detection, and visual relationship detection at scale. CoRR abs/1811.00982 (2018). http://arxiv.org/abs/1811.00982

11. Li, X., et al.: OSCAR: object-semantics aligned pre-training for vision-language tasks. CoRR abs/2004.06165 (2020). https://arxiv.org/abs/2004.06165

12. Lin, T., et al.: Microsoft COCO: common objects in context. CoRR abs/1405.0312 (2014). http://arxiv.org/abs/1405.0312

13. Schulman, J., Levine, S., Moritz, P., Jordan, M.I., Abbeel, P.: Trust region policy optimization. CoRR abs/1502.05477 (2015). http://arxiv.org/abs/1502.05477

14. Shao, S., et al.: Objects365: a large-scale, high-quality dataset for object detection. In: In Proceedings of the IEEE International Conference on Computer Vision, pp. 8430–8439 (2019)

15. Singh, A., et al.: MMF: a multimodal framework for vision and language research (2020). https://github.com/facebookresearch/mmf

16. Tan, H., Bansal, M.: LXMERT: learning cross-modality encoder representations from transformers. In: Proceedings of the 2019 Conference on Empirical Methods in Natural Language Processing (2019)

17. Yan, M., et al.: Achieving human parity on visual question answering. CoRR abs/2111.08896 (2021). https://arxiv.org/abs/2111.08896

18. Zhang, P., Goyal, Y., Summers-Stay, D., Batra, D., Parikh, D.: Yin and yang: balancing and answering binary visual questions. CoRR abs/1511.05099 (2015). http://arxiv.org/abs/1511.05099

19. Zhang, P., et al.: VinVL: making visual representations matter in vision-language models. CoRR abs/2101.00529 (2021). https://arxiv.org/abs/2101.00529

Application of the Piecewise Linear Approximation Method in a Hardware Accelerators of a Neural Networks Based on a Reconfigurable Computing Environments

Vladislav Shatravin$^{(\boxtimes)}$ and D. V. Shashev

National Research Tomsk State University, 36 Lenin ave., 634050 Tomsk, Russia
shatravin@stud.tsu.ru, dshashev@mail.tsu.ru

Abstract. This paper considers the application of piecewise linear approximation in hardware accelerators of neural networks built according to the concept of reconfigurable computing environments (RCE). A method for providing an uneven approximation step is shown. The results of the application of the proposed approaches for the approximation of exponential and sigmoid functions actively used in machine learning algorithms are presented. The use of RCE allows the development of hardware algorithms for information processing, characterized by high performance characteristics due to the architectural properties of RCE.

Keywords: Neural networks hardware accelerators · Reconfigurable computing environments · Parallel computations

1 Introduction

The outstanding results of applying machine learning algorithms to problems that have long been considered intractable have made them one of the most promising and actively implemented technologies in recent years. Search engines, video surveillance camera analysis systems, predictive analytics, disease diagnosis, autopilots, voice assistants, image processing and recovery tools, text completion tools are a small list of applied applications of machine learning algorithms that modern people encounter in everyday life.

The downside of using machine learning algorithms is their high computational complexity. The problem is especially severe for deep neural networks (DNN), which can have hundreds of billions of parameters. However, even for smaller neural networks, classical computing systems based on central processors are ineffective [1,2]. In this regard, one of the key problems of applying modern neural network architectures is the development of specialized computing devices - hardware accelerators of neural networks.

Supported by by the Russian Science Foundation, grant No. 21-71-00012, https://rscf.ru/project/21-71-00012/.

The problem of the hardware acceleration of neural networks is widely known. Therefore, many different approaches are proposed. All these solutions can be classified according to the hardware platform they use into 3 groups: GPU-based, FPGA-based and ASIC-based [1]. GPU-based accelerators have high performance at the cost of low power efficiency [2]. However, the modern embedded GPU NVIDIA Jetson demonstrates acceptable results. FPGA and ASIC allow to develop a very new implementation of a computing system, well-tuned for specific tasks. As a result, the developed system provides both high performance and high efficiency [3–5]. However, the specialization of these accelerators for concrete machine learning algorithms limits their use in other tasks.

To solve the problem, reconfigurable hardware accelerators have been proposed [6,7]. A key feature of these accelerators is the ability to dynamically change the implemented algorithms to others. This ability allows to implement very different neural network architectures on the same device at different times. Such an accelerator can be implemented using a reconfigurable computing environment (RCE) model. In this paper, we proposed an approach to perform various neuronal activations on the RCE-based hardware accelerator.

2   Reconfigurable Computing Environments

Reconfigurable computing environment (RCE) is a model of a wide class of computing systems, which is a set of simple computing elements of the same type (CE) connected to each other in the form of a regular grid (Fig. 1). In some sources, RCE is referred to as homogeneous structures [8,9]. The key feature of the RCE is tunability - each CE is individually configured for one operation from a predetermined basis, and connections between them can be disabled. The organized behavior of a large group of simple calculators makes it possible to implement complex algorithms for streaming data processing using a computing environment [8,10]. The range of supported algorithms is limited only by a given basis of CE's elementary operations and the number of elements in a computing environment.

3   Hardware Accelerator Architecture

The key stage in the development of any device based on RCE is the design of the CE. We used the principle of the ultimate elementary calculations - all calculations performed in the neural network were divided into simple reused operations that formed the basis of the environment. The analysis of direct distribution networks allowed us to determine the following basis: "source", "transmission", "multiplication with accumulation (MAC)", "linear rectifier (ReLU)", "fixation" (one step delay). With their help, a neuron of arbitrary configuration can be implemented by a chain of CE of sufficient length (Fig. 2) [11]. ReLU is the only activation present in the basis of CE due to its low computational complexity and very wide usage. As we will show further, support for other activations can be introduced by connecting CEs to each other in a certain way.

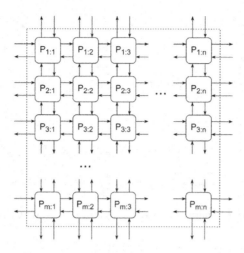

Fig. 1. Reconfigurable computing environment.

Due to the fact that neural networks can have a very large size, it may not be possible to place the entire network in one computing environments. Therefore, our environment model processes the signal by means of its circular movement inside the environment with a step-by-step calculation of the result. This is done by dividing the environment into segments. Each segment implements one of the network layers. As the signal propagates, the segments change their boundaries and are reconfigured to subsequent layers. Thus, the signal moves from one segment to next segment until it is fully processed (Fig. 3). Thanks to this approach, the proposed environment model can implement a neural network of arbitrary depth [7,12]. Other advantages of this approach are high performance due to pipelinening; reduced reconfiguration latency impact; higher utilization of the RCE resources.

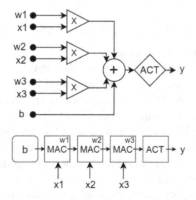

Fig. 2. Implementation of an artificial neuron using CE.

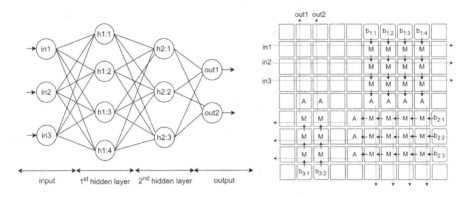

Fig. 3. Fully connected network on RCE. (b - bias, M - MAC, A - activation)

4   Implementation of Piecewise Linear Approximation with Uniform Step Using RCE

The problem of maintaining the diversity and complexity of neural network activation functions using RCE implementation can be solved using approximation. There are various ways to approximate the function, we suggest using piecewise linear approximation due to its computational simplicity.

Piecewise linear approximation is based on the idea of splitting some part of the original function into intervals with replacement by some straight line at each interval (Fig. 4). The distance between two adjacent points of the partition (approximation nodes) is called the approximation step. As the step length decreases, the approximation error may decrease (the amount of discrepancy between the original function and the approximation result), but the computational complexity increases. Since the resulting function is a straight line at each interval, it can be written as:

$$f_i(x) = a_i * x + b_i, \tag{1}$$

where
i - the number of the current interval,
a_i, b_i - the corresponding approximation coefficients.

The block of computational elements presented in the Fig. 5 implements the approximation described in [13] of a widely used sigmoid function in the range $[-5, 5]$ with a step of 1. To implement this part of the environment, in addition to the "signal source" operations already mentioned above (elements with rounded corners) and multiplication with accumulation (MAC), three new operations were added to the basis - minimum (MIN), union (U) and gate (GAT). Dashed arrows passing through the entire height of the medium indicate signals passing through the chain of computing elements without change.

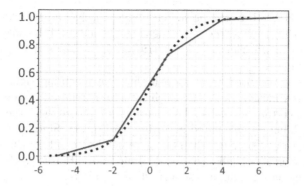

Fig. 4. Piecewise linear approximation of the function.

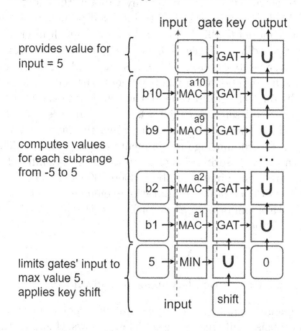

Fig. 5. A part of the RCE that implements the approximation of the sigmoidal function.

The "minimum" operation returns the smallest of two numbers and is used to limit the input signal to the upper bound of the entire approximation section equal to five.

The "union" operation performs a bitwise "OR" on the input signals and serves to combine intermediate results on different lines of the environment.

The "gate" operation controls the propagation of the signal and has two inputs - the main and the key. The computing element configured for this operation compares the value of the signal at the key input with the value stored in the internal memory and, if these values are equal, transmits the signal from the main input forward. In the presented image, the key input is located at the bottom edge of the element, the main one is on the left, and the output is on the right. Regardless of whether the gate is triggered, the key signal is transmitted up to the other gates.

Gate elements are used to check whether the incoming signal belongs to one of the approximation intervals. Each interval corresponds to its own gate, configured for a certain key value. The main gate input receives the value of the approximating function calculated according to Eq. (1). Since all gates are set to different key values, no more than one gate will be open for each input signal. After that, the elements configured for the union operation form the final result from the intermediate results at the output of each gate.

Due to the fact that the gate compares the whole key coming to it, the question arises how to ensure its operation for any value in the range corresponding to the interval. To do this, consider the 16-bit fixed-point number format used in the environment model (Table 1). It can be seen that all intermediate values located between two adjacent integers have the same upper 8 bits. Thus, if we forcibly set the lower 8 bits of both the input signal and the values stored in the gate memory to a fixed value, then the fractional part will no longer affect the comparison result. As such a fixed value of the fractional part, we chose a logical unit, and the operation itself was called a key shift. In order for this correction not to affect the calculation of the function value, the input signal is branched in the lower line of the computing environments (Fig. 5), after which the gate is applied only to the value entering the gate input. The "key shift" element stores the value of the gate – the binary number "0000 0000 1111 1111".

The total number of necessary computational elements for the approximation of the function with k uniform steps according to the proposed algorithm can be calculated by the formula:

$$N = 4k + 8. \tag{2}$$

The proposed implementation has important advantages:

- scalability. Changing the number of intervals does not require reworking the structure and consists only in adding or deleting rows, changing the interpolation step will only require replacing the key gate elements stored in memory;
- parallelism. All intervals are calculated in parallel and independently of each other, which allows you to maintain high performance with an increase in their number;
- simplicity. Despite the fact that in order to maintain the approximation, it was necessary to expand the basis with three new operations, the computational elements retained their internal simplicity. All the added operations have an acceptable hardware implementation and are universal, that is, they can be reused for other tasks.

Table 1. Representation of decimal numbers as 16-bit fixed-point numbers.

Decimal form	16-bit form (8 fraction bits)
−5	1111 1011 0000 0000
−4 - Δ	1111 1011 1111 1111
−4	1111 1100 0000 0000
0 - Δ	1111 1111 1111 1111
0	0000 0000 0000 0000
1 - Δ	0000 0000 1111 1111
1	0000 0001 0000 0000
1.5	0000 0001 1000 0000
1.75	0000 0001 1100 0000
2 - Δ	0000 0001 1111 1111
2	0000 0010 0000 0000
$\Delta = 1/256$	0000 0000 0000 0001

5   Implementation of an Approximation with an Irregular Step

The key disadvantage of piecewise linear approximation is a significant error when applied to functions with strong curvature with a small number of approximation nodes. An example is an attempt to approximate the exponential function (Fig. 6). An increase in accuracy can be achieved by reducing the approximation step, but this will increase the computational complexity of the approximant. In addition, the curvature of the exponential function is uneven and increases significantly with increasing argument. In such cases, a more effective solution is to use an approximation with an uneven step, according to which sections of the function with strong nonlinearity are approximated by a large number of nodes, and more linear sections are approximated by a smaller number.

Consider the approximation of the mentioned exponential function with an irregular step in the interval [−2.5, 2.5]. In the interval [−2.5, 1.5), the approximation step is 0.5, and in the interval [1.5, 2.5] – 0.25. The corresponding approximation coefficients are given in Table 2. In this example, all arguments above the upper limit of the interval equal to 2.5 are equated to it. This is due to the peculiarities of the problem where this approximant was used, and is not essential for the algorithm under consideration for applying the approximation using RCE.

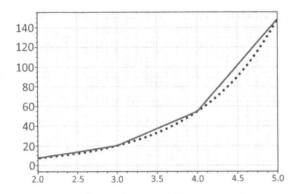

Fig. 6. Piecewise linear approximation of the exponential function using a regular step.

The fragment of the computing environment corresponding to the described approximation is shown in Fig. 7. As you can see, it is in many ways similar to the previous model considered. The key difference is the order of the rows-intervals and the presence of a second shift of the key. Since now the approximation step is 0.5 and 0.25, according to Table 1, their shifts are seven lower bits and six lower bits, respectively. Obviously, if we first set the logical unit to the seven lowest bits of the input signal, then we will not be able to restore the original value of the 7th bit to compare with the shift of 6. For this reason, the shift of the six lowest bits is carried out first, after which the intervals of length 0.25 are calculated. Then the second shift (by the seventh bit) is applied, after which the remaining intervals are calculated. Thus, it is possible to implement an approximation with steps of different sizes. The restriction is the requirement that the lengths of the steps used must be equal to the power of two. The number of required CE can be calculated as:

$$N = 4k + 3l + 5. \tag{3}$$

where
k - the number of intervals,
l - the number of supported step lengths. In this example steps 0.5 and 0.25 are used, thus, $l = 2$.

6   Models Simulation

The described models of approximating blocks of computing environments were implemented in the Verilog hardware description language in the Quartus Prime 20.1 environment. Three key characteristics of the developed models were investigated - approximation error, size and performance. The Altera Cyclone V FPGA (5CGXFC9E7F35C8) was set as the target simulation device.

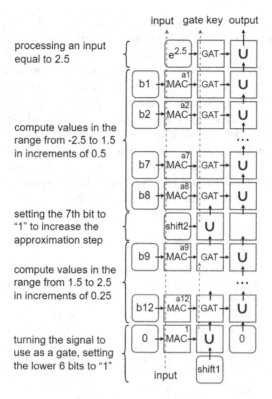

Fig. 7. A fragment of RCE approximating an exponential function with irregular step.

Table 2. Coefficients of approximation of the exponent using irregular step.

Step	Interval	a	b
0	$(-\infty, -2.5)$	0	0
1	$[-2.5, -2.0)$	0.1015625	0.3359375
2	$[-2.0, -1.5)$	0.1796875	0.4921875
3	$[-1.5, -1.0)$	0.2890625	0.65625
4	$[-1.0, -0.5)$	0.4765625	0.84375
5	$[-0.5, 0.0)$	0.7890625	1
6	$[0.0, 0.5)$	1.296875	1
7	$[0.5, 1.0)$	2.1328125	0.58203125
8	$[1.0, 1.5)$	3.53125	-0.81640625
9	$[1.5, 1.75)$	5.09375	-3.16015625
10	$[1.75, 2.0)$	6.53125	-5.67578125
11	$[2.0, 2.25)$	8.390625	-9.39453125
12	$[2.25, 2.5]$	10.78125	-14.7734375
13	$(2.5, +\infty)$	0	$e^{2.5}$

The error was estimated using auxiliary software by calculating the absolute value of the difference between the values of the original and received functions at each point of the approximation interval in accordance with the minimum discretization of a 16-bit number. Despite the fact that the absolute error characterizes not the algorithm proposed by us for implementing the approximating function on the RCE, but the approximating function itself, it allows us to assess the applicability of the developed models for solving applied problems.

The size of the model was estimated as the number of FPGA logic elements necessary for its implementation. It was determined by the results of compiling the Verilog module in the Quartus environment. At the same time, the use of DSP blocks was disabled and the "synthesis keep" submodule preservation directive was used to cancel unintended optimizations.

Table 3. Comparison of simulation results with existing analogues from [14].

Implementation	Absolute error, max	Absolute error, avg	Format	LUT	DSP	Delay, ns
Hajduk(McLaurin)	1.19×10^{-7}	1.45×10^{-8}	32 FP	1916	4	940.8
Hajduk(Pade)	1.19×10^{-7}	3.27×10^{-9}	32 FP	2624	8	494.4
Zaki et al.	1×10^{-2}	N/A	32 FP	363	2	N/A
Tiwari et al.	4.77×10^{-5}	N/A	32 FXP	1388	22	1590–2130
Wei et al.	1.25×10^{-2}	4.2×10^{-3}	12 FXP	140	0	9.856
PLAN	1.89×10^{-2}	5.87×10^{-3}	16 FXP	235	0	30
NRA	5.72×10^{-4}	8.6×10^{-5}	16 FXP	351	6	85
RCE-sigmoid	1×10^{-2}	4×10^{-3}	16 FXP	13175	0	18.5
RCE-exponent	1×10^{-1}	2.7×10^{-2}	16 FXP	15069	0	19.1

N/A – not assessed, FP – floating-point number, FXP – fixed-point number.

The speed was estimated as the longest signal propagation time from the input to the output of the model. At the same time, all computational elements were preconfigured, that is, the time of reconfiguration of elements is not included in the results obtained. To carry out measurements, the Timing Analyzer Tool built into the Quartus environment was used. Simulations were carried out in the "Fast corners 1100 mV 0 °C" mode. The simulation consisted of finding the five hundred slowest paths from the input to the output of the model.

The results obtained in comparison with alternative methods of hardware sigmoid calculation described in [14] are shown in Table 3. It can be seen that the proposed method of implementing the approximation using RCE shows high performance and rather low error. At the same time, the proposed models require a lot of logical elements compared with counterparts. It can be explained by several reasons. First of all, the proposed models are spatially distributed parallel computing systems. This means that high performance is ensured by the execution of different parts of the algorithm in different zones of computing system. Secondly, RCE consists of many identical elements. Each element performs only one operation at a time, although it supports them all. This redundancy is a result of the reconfiguration ability of the environment. Thirdly, due to the large amount of computing elements used and the connections between them, it is difficult for the Quartus compiler to efficiently place and route the entire design for simulation. However, an ASIC implementations of these models can be free from this shortcomings.

7 Conclusion

One of the key tasks of building accelerators based on RCE is to ensure the maintenance of a variety of neuron activation functions, many of which are also complex for hardware computing. As a solution to this problem, general algorithms for implementing piecewise linear approximation of functions on the described models of homogeneous media are proposed in this paper. A method for implementing an approximation with an uneven step for a local increase in accuracy is shown. The application of the proposed algorithms is demonstrated by the example of sigmoid and exponential functions common in machine learning. The results of simulation of the developed models in the form of modules in the Verilog language and their comparison with the results from alternative studies are shown. The described models showed high performance (18–19 ns) and acceptable accuracy (the average absolute error for the sigmoid function is 4e−3), with a large area on the chip (13–15 thousand FPGA logic elements). Such a large size is explained by the overhead costs of ensuring the tunability of each computing element of the environment, as well as parallelization of calculations for a large number of elements. For the vast majority of applications, size will not be a limiting factor for two reasons. Firstly, the described complex activation functions are used only in the output layer of the network, which usually consists of a small number of neurons. Secondly, computing environments often consist of hundreds of thousands of elements, and are quite easily scaled. Thus, the proposed algorithms for using approximations in RCE allow solving the problem of using neuron activation functions in the accelerators being developed.

References

1. Ghimire, D., Kil, D., Kim, S.: A survey on efficient convolutional neural networks and hardware acceleration. Electronics **11**(6), 945 (2022). 1–23. https://doi.org/10.3390/electronics11060945
2. Nabavinejad, S.M., Reda, S., Ebrahimi, M.: Coordinated batching and DVFS for DNN inference on GPU accelerators. IEEE Trans. Parallel Distrib. Syst. **33**(10), 2496–2508 (2022). https://doi.org/10.1109/TPDS.2022.31446140
3. Kyriakos, A., Papatheofanous, E.-A., Bezaitis, C., Reisis, D.: Resources and power efficient FPGA accelerators for real-time image classification. J. Imaging **8**(114), 1–18 (2022). https://doi.org/10.3390/jimaging8040114
4. Chen, Y.H., Yang, T.J., Emer, J., Sze, V.: Eyeriss v2: a flexible accelerator for emerging deep neural networks on mobile devices. IEEE J. Emerg. Sel. Top. Circ. Syst. (JETCAS) **9**, 292–308 (2019). https://doi.org/10.1109/JETCAS.2019.2910232

5. Jouppi, N.P., et al.: In-datacenter performance analysis of a tensor processing unit. In: 2017 Proceedings of the 44th Annual International Symposium on Computer Architecture (ISCA 2017), Toronto, ON, Canada, pp. 1–12 (2017). https://doi.org/10.1145/3079856.3080246

6. Khalil, K., Eldash, O., Dey, B., Kumar, A., Bayoumi, M.: A novel reconfigurable hardware architecture of neural network. In: 2019 Proceedings of the IEEE 62nd International Midwest Symposium on Circuits and Systems (MWSCAS), Dallas, TX, USA, pp. 618–621 (2019). https://doi.org/10.1109/MWSCAS.2019.8884809

7. Shatravin, V., Shashev, D.V., Shidlovskiy, S.V.: Applying the reconfigurable computing environment concept to the deep neural network accelerators development. In: 2021 International Conference on Information Technology (ICIT), pp. 842–845 (2021). https://doi.org/10.1109/ICIT52682.2021.9491771

8. Khoroshevsky, V.G.: Architecture and functioning of large-scale distributed reconfigurable computer systems. In: 2004 International Conference on Parallel Computing in Electrical Engineering, pp. 262–267 (2004). https://doi.org/10.1109/PCEE.2004.14

9. Evreinov, E.V.: Homogeneous computing systems, structures and environments. In: Radio and Communication: Moscow, Russia, p. 208 (1981)

10. Shidlovskii, S.V.: Logical control in automatic systems. J. Comput. Syst. Sci. Int. **45**(2), 282–286 (2006). https://doi.org/10.1134/S1064230706020122

11. Shatravin, V., Shashev, D.V.: Designing high performance, power-efficient, reconfigurable compute structures for specialized applications. In: Journal of Physics: Conference Series, vol. 1611, p. 012071 (2020). pp. 1–6. https://doi.org/10.1088/1742-6596/1611/1/012071

12. Shatravin, V., Shashev, D.V., Shidlovsky, S.V.: Developing of models of dynamically reconfigurable neural network accelerators based on homogeneous computing environments. In: 2021 24th International Conference on Distributed Computer and Communication Networks: Control, Computation, Communications (DCCN), pp. 102–107 (2021). https://doi.org/10.25728/dccn.2021.015

13. Faiedh, H., Gafsi, Z., Besbes, K.: Digital hardware implementation of sigmoid function and its derivative for artificial neural networks. In: ICM 2001 Proceedings. The 13th International Conference on Microelectronics, pp. 189–192 (2001). https://doi.org/10.1109/ICM.2001.997519

14. Pan, Z., Gu, Z., Jiang, X., Zhu, G., Ma, D.: A modular approximation methodology for efficient fixed-point hardware implementation of the sigmoid function. IEEE Trans. Ind. Electron. **69**(10), 10694–10703 (2022). https://doi.org/10.1109/TIE.2022.31465739

Boosting Adversarial Training in Adversarial Machine Learning

Ehsan Saleh⑩, Aleksey A. Poguda⑩, and D. V. Shashev$^{(\boxtimes)}$⑩

Tomsk State University, Tomsk, Russian Federation
dshashev@mail.ru

Abstract. This study aims to investigate Fast Adversarial Training (AT) methods to strengthen models' resilience against aggressive attacks. A software that can train the model on the database in a reasonable length of time is referred to as "fast training". The phrase "adversarial training" describes a code that can survive purposefully altered data that cannot be identified by a human but can have detrimental effects on a Deep Neural Network (DNN) model. Current Fast AT frequently uses a random sample agnostic initialization, which increases efficiency but limits future improvements to robustness. Using a sample-dependent adversarial initialization, Fast AT is enhanced in this study. Experiments on four benchmark datasets show that the proposed technique performs better than state-of-the-art Fast AT methods and is robustly comparable to complex multi-step AT methods. It's a prevalent misconception that networks can't be accurate and robust at the same time, and that improving reliability requires sacrificing precision. Due to its non-smooth character, the commonly used ReLU activation function greatly worsens adversarial training. The second section of this study describes smooth adversarial training (SAT), a method for enhancing adversarial training that substitutes smooth approximations for ReLU. SAT makes advantage of smooth activation functions to find harder adversarial scenarios and compute better gradient updates during adversarial training. In the next part we present the result of combining two presented studies and we witness promising results.

Keywords: Fast · Robust · Adversarial Training · Smooth

1 Introduction

Depending on how many steps are taken to produce adversarial examples, AT can be categorized into two classes: Multi-step AT [1–4], and Fast AT [5–9]. To achieve thorough robustness against various attack methods, multi-step AT employs multi-step adversarial attack methods like projected gradient descent (PGD) [1]. It is suggested that fast AT methods increase training effectiveness by requiring just one gradient calculation and employing a fast gradient

Supported by Tomsk State University Development Programme (Priority-2030).

sign algorithm (FGSM) [10]. Although they can significantly reduce computation and time costs, they do not have the same level of robustness as other Multi-step AT approaches. Some studies [7,8] focus on the initialization issue because it has been shown that using a random initialization in Fast AT greatly enhances robustness. Current Fast AT methods use a variety of random initialization techniques, but most of them are sample-independent, which restricts the development of future robustness improvements. To solve this issue, we introduce FGSM-SDI [11], a sample-dependent adversarial initialization for boosting FGSM-based Fast Adversarial training, which requires gradient calculation both for creating training samples and updating network parameters. Due to ReLU's non-smooth nature, we find that it significantly worsens adversarial training. For example, ReLU's gradient abruptly changes (from 0 to 1) when its input is in the vicinity of zero, as shown in Fig. 1. We look into smooth adversarial training (SAT), which ensures architectural smoothness by replacing smooth ReLU approximations with superior gradient quality in adversarial training (Fig. 1 displays Parametric Softplus as an example of a rough approximation for ReLU). The issue ReLU has created is what SAT seeks to fix. SAT is able to feed the networks with harsher adversarial training samples and calculate better gradient updates for network improvement because smooth activation functions considerably enhance adversarial training. The findings of the experiment demonstrate that SAT enhances adversarial resilience "free", i.e., without requiring additional computations or reducing the precision of conventional calculations.

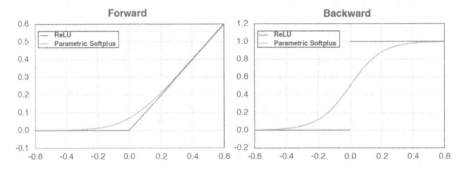

Fig. 1. The demonstration of ReLU and Parametric Softplus. The forward pass for ReLU (blue curve) and Parametric Softplus is shown in the left panel (red curve). The initial derivatives for ReLU (blue curve) and Parametric Softplus (red curve) are shown in the right panel. Parametric Softplus is smooth with continuous derivatives, unlike ReLU [12] (Color figure online).

2 Boosting Fast AT with Learnable Adversarial Initialization

This paper focuses on the image classification issue, where hostile examples can lead a trained image classification model to make a false prediction with a high degree of confidence. A well-trained image classifier $f(\cdot)$ and a clean image x with the corresponding true label y are both used to create the adversarial example x_{adv}, which is used to trick the classifier into producing an incorrect prediction, i.e., $f(x_{\text{adv}}) \neq f(x) = y$, where the distance function satisfies $L_p(x_{\text{adv}}, x) \leq \epsilon$, where ϵ represents the maximum perturbation strength and L_p represents the distance between the adversarial image x_{adv} and the clean image x under the L_p distance metric, where $p \in \{1, 2, \infty\}$. L_∞ is a commonly used distance metric in the recent research of attack methods, which is adopted in this work too.

Attack Methods: Szegedy et al. [10] identify the existence of adversarial examples for DNNs and create adversarial examples using a box-constrained L-BFGS approach. Goodfellow et al. [13] recommend the fast gradient sign method to generate adversarial examples (FGSM). The gradient of the loss function is only calculated once by FGSM to produce adversarial instances, which are then added to clean images. To produce adversarial examples, Madry et al. [1] offer Projected Gradient Descent (PGD). PGD repeatedly iterates to carry out a gradient descent step in the loss function to produce adversarial instances. DeepFool is another approach proposed by Moosavi-Dezfooli et al. [14] for efficiently generating adversarial examples for tricking deep networks. Then, DeepFool creates the least amount of adversarial perturbations that can fool deep neural networks using an iterative linearization of the classifier. Carlini et al. [15] suggest C&W, a more powerful attack model. To address the challenge given by the suboptimal step size and objective function, Croce et al. [16] developed two parameter-free attack techniques, namely auto PGD with cross-entropy (APGD-CE) and auto PGD with the difference of logits ratio (APGD-DLR). Additionally, they build the AutoAttack ensemble (AA) by combining the proposed attack techniques with two already in use attack strategies, namely FAB [17] and Square Attack [18]. Regarding testing model resilience against adversarial examples, AA has outperformed the opposition.

Adversarial Training Methods: Considering a target network $f(\cdot; w)$ with parameters w, a data distribution D including the benign sample x and its corresponding label y, a loss function $L(f(x; w); y)$, and a threat bound Δ, the objective function of AT can be presented as:

$$\min_{w} E_{(x, y)} \sim D \left[\max_{\delta \in \Delta} L(f(x + \delta; w), y) \right] \tag{1}$$

where the threat bound can be defined as $\Delta = \{\delta : \|\delta\| \leq \epsilon\}$ with the maximal perturbation intensity ϵ. The core of the adversarial training is how to find a better adversarial perturbation δ. To create an adversarial perturbation δ, most adversarial training methods use a multi-step adversarial algorithm, such as various steps of projected gradient descent (PGD) [1]. It is presented as:

$$\delta_{t+1} = \prod_{[-\epsilon,\epsilon]^d} [\delta_t + \alpha sign(\nabla_x L(f(x + \delta_t; w), y))] \tag{2}$$

where $\prod_{[-\epsilon,\epsilon]^d}$ represents the projection to $[-\epsilon, \epsilon]^d$ and δ_{t+1} represents the adversarial perturbation after $t + 1$ iteration steps. In general, more iterations can increase the resilience of adversarial training by strengthening the adversarial cases. The prime PGD-based adversarial training framework is proposed in [1]. Using this framework, more PGD-based AT strategies are presented, with an early stopping variant [3] standing out.

Fast Adversarial Training: The one-step fast gradient sign method (FGSM) [13], also known as FGSM-based AT, is adopted to create fast adversarial training variants. It can be defined as:

$$\delta^* = \epsilon sign(\nabla_x L(f(x; w), y)) \tag{3}$$

where ϵ represents the maximal perturbation strength. The target model in FGSM-based AT has a severe overfitting issue where it loses the resilience accuracy of adversarial cases produced by PGD (on the training data) during training. Wong et al. [7] suggest using FGSM-based AT combined with a random initialization to reduce catastrophic overfitting. In terms of performance, it can compete with the best PGD-based AT, according to [1]. Particularly, they perform FGSM with a random initialization $\eta \in U(-\epsilon, \epsilon)$, where U represents a uniform distribution, which can be called FGSM-RS. It can be defined as:

$$\delta^* = \prod_{[-\epsilon,\epsilon]^d} [\eta + \alpha sign(\nabla_x L(f(x + \eta; w), y))] \tag{4}$$

where α represents the step size, which is set to 1.25ϵ in [7]. This study shows that when combined with a good initialization, FGSM-based AT can perform as well as PGD-AT [1].

The Introduced Method: Although it is common and practical to use a random sample-agnostic initialization for Fast AT, this limits the ability to further improve model robustness. This issue was addressed by proposing a sample-dependent adversarial initialization, which would increase the robustness of Fast AT while also addressing the catastrophic overfitting issue [11].

Pipeline of the Introduced Method: A target network and a generative network are used in the proposed technique. Instead of employing a random initialization, the former learns to build a dynamic sample-dependent adversarial initialization for FGSM to produce adversarial instances. The latter increases model resilience by training using the produced adversarial scenarios. As shown in Fig. 2, the generative network receives a benign image and its gradient information from the target network and produces a sample-dependent initialization. The input image is subjected to FGSM along with the produced initialization in order to produce adversarial instances. To make the target network more resistant to hostile attacks, hostile samples are used during training. We use the

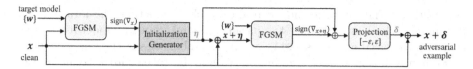

Fig. 2. Adversarial example generation of the proposed Sample-Dependent learnable Initialization algorithm FGSM-SDI [11].

architecture of a conventional image classification network for the target network, which is denoted by the equation $y = f(x; w)$, where x stands for an input picture, y for the predicted label, and w for the target network's parameters.

Three layers make up the generative network. The generative network's inputs are the benign image and its signed gradient. The formula below can be used to calculate the signed gradient:

$$S_x = sign(\nabla_x L(f(x; w), y)) \tag{5}$$

where x is the input image and y is the ground-truth label. The initialization generation process can be defined as:

$$\eta_g = \epsilon g(x, s_x; \theta) \tag{6}$$

where $g(\cdot; \theta)$ represents the generative network with the parameters θ, and η_g represents the generated adversarial initialization. The output pixel value space of $g(\cdot; \theta)$ is $[-1; 1]$. ϵ is a scale factor that maps the value to the range of $[-\epsilon, \epsilon]$.

Formulation of the Introduced Method: The approximate solution of the inner maximization problem, can be represented as:

$$\delta_g = \delta_g(\theta) = \prod_{[-\epsilon, \epsilon]^d} [\eta_g + \alpha sign(\nabla_x L(f(x + \eta_g; w), y))] \tag{7}$$

where η_g is the adversarial initialization defined in Eq. 6, generated by the generative network. It's worth noting that the perturbation involves the generative network's parameters θ via the initialization.

The aim function of jointly learning the generative network and the target network may be stated as follows using the specification of the recommended perturbation. The recommended solution to the inner maximization issue combines the characteristics of the generative network and starts with the objective function of traditional AT in Eq. 1 When the parameters θ are fixed, η_g in Eq. 7 provides a close approximation to the solution. We can further maximize the loss by searching better parameters θ, i.e., $\max_\theta L(f(x + \delta_g(\theta); w); y)$. Hence, the objective function of the joint optimization can be defined as:

$$\min_w \max_\theta E_{(x,y) \sim D} L(f(x + \delta_g(\theta); w); y) \tag{8}$$

The generative network and the target network exhibit game-like behavior, as demonstrated in Eq. 8. While the latter aims to reduce the loss to produce an effective initialization for adversarial example production, the former seeks to minimize the loss to update the parameters and improve model robustness against adversarial examples. Additionally, the robustness of the target model's performance during several training stages may be taken into account by the generative network while generating initializations. By alternately optimizing w and θ, this minimax issue may be resolved. Keep in mind that we repeatedly update *theta* and w. Every k updates of w, we update θ. Additionally, k is a hyper-parameter that requires tuning.

Experiments: Extensive experiments [11] on four benchmark databases (CIFAR10, CIFAR100, Tiny ImageNet, and ImageNet) were done to evaluate the effectiveness of the FGSM-SDI and comparisons with other Fast AT approaches; In these experiments, the only hyper-parameter of the algorithm, k, according to other sets of experiments is set 20. The Table 1 provides comparisons of clean and robust accuracy (%) and Training time (minute) with ResNet18 on the CIFAR10 dataset as presented by [11]. Table 2 provides the results of reimplementations with different hardware.

Table 1. Results of experiments on CIFAR10 dataset as presented in original paper

Method	Clean	PGD-10	PGD-20	PGD-50	C&W	APGD	AA	Time (min)
PGD-AT	82.32	53.76	52.83	52.6	51.08	52.29	48.68	265
FGSM-RS	73.81	42.31	41.55	41.26	39.84	41.02	37.07	51
FGSM-CKPT	90.29	41.96	39.84	39.15	41.13	38.45	37.15	76
FGSM-SDI	84.86	53.73	52.54	52.18	51.00	51.84	48.50	83

Table 2. Results of reimplementation of experiments on CIFAR10 dataset

Method	Clean	PGD-10	PGD-20	PGD-50	C&W	APGD	AA	Time (min)
FGSM-RS	70.95	37.23	38.67	38.22	36.35	38.21	32.00	128
FGSM-CKPT	85.65	40.49	37.41	36.50	40.79	37.42	36.19	190
FGSM-SDI	83.77	52.67	50.39	51.78	49.85	50.32	46.20	210

Results on CIFAR10: On CIFAR10, Resnet18 was chosen as the comparison network for the experiment comparing different defense strategies. Table 1 contains the results. The proposed method, when compared to fast AT methods, achieves the best performance under all attack scenarios and robustness that is comparable to that of advanced PGD-AT [3]. We can verify that the FGSM-SDI performs better than the FGSM-CKPT and FGSM-RS in terms of robust accuracy by looking at Table 2, which is a presentation of the reimplementation of the experiment, aside from the time differences, which are caused by the different hardware used for reimplementation.

3   Smooth Adversarial Training

This study is connected to a theoretical research [19] that suggests smooth alternatives to ReLU can assist networks achieve a manageable constraint while guaranteeing distributional robustness. In this part, a series of controlled experiments [12] were carried out with a focus on the backward pass of gradient calculations to ascertain how ReLU weakens and how its smooth approximation enhances adversarial training.

Experiment setup: The core network is chosen to be ResNet-50 [20] with ReLU as the default function. A PGD attacker [1] was utilized to produce adversarial perturbations, δ. Single-step PGD (PGD-1) was chosen since it is the cheapest PGD attacker in order to lower training costs. According to [6,21], the maximum per-pixel change ϵ and attack step size β were both set to 4 in PGD-1. Standard ResNet training recipes were used to train systems on ImageNet. Models were trained for a sum of 100 epochs using the momentum SGD optimizer, with the learning rate decreasing by $10\times$ at the 30-th, 60-th, and 90-th epochs. The only regularization used was a weight decay of $1e - 4$.

How Does Gradient Quality Affect Adversarial Training? A highly used activation function, ReLU [22,23], is non-smooth, as shown in Fig. 1. The gradient undergoes a sudden change that sharply reduces the gradient quality as the ReLU input approaches zero. Especially when training adversarial models, it was thought that this non-smooth character was causing the training to suffer. Due to the fact that adversarial training necessitates additional calculations for the inner maximization step to produce the perturbation δ, as opposed to regular training, which merely computes gradients for updating network parameters, this is the case. A smooth approximation of ReLU named Parametric Softplus [23] was primarily proposed to address this issue:

$$f(\alpha, x) = \frac{1}{\alpha} \log(1 + \exp(\alpha x)) \tag{9}$$

where the hyper-parameter α is used to control the curve shape. The derivative of this function with respect to the input x is:

$$\frac{d}{dx} f(\alpha, x) = \frac{1}{1 + \exp(-\alpha x)} \tag{10}$$

In order for α to more closely resemble the ReLU curve, it was empirically set to 10. Parametric Softplus ($\alpha = 10$) is smoother than ReLU because it has a continuous derivative, as illustrated in Fig. 1.

The following step was to examine how the standard accuracy and adversarial resilience of ResNet-50 during adversarial training using Parametric Softplus were impacted by the inner maximizing and outer minimization gradient quality. (In order to properly assess the effects, the forward pass was kept unmodified, meaning that ReLU is always used for model inference, and ReLU with Eq. 10 was the only replacement in the backward pass.) As seen in the second row of

Table 3, the resulting model exhibits a performance trade-off when the attacker uses Parametric Softplus' gradient (i.e., Eq. 10) to construct adversarial training examples. Compared to the ReLU baseline, it improves adversarial resilience by 1.5% while lowering standard accuracy by 0.5%.

Table 3. ReLU significantly weakens adversarial training [12].

Improving Gradient Quality for the Adversarial Attacker	Improving Gradient Quality for the Network Optimizer	Accuracy (%)	Robustness (%)
×	×	68.8	33.0
√	×	68.3(−0.5)	34.5(+1.5)
×	√	69.4(+0.6)	35.8(+2.8)
√	√	68.9(+0.1)	36.9(+3.9)

Improving Gradient Quality for Network Parameter Updates. Next, it is examined how gradient quality affects the outer minimization phase of training, which is when network parameters are updated. ReLU is specifically used in the inner step of adversarial training, while Parametric Softplus is used in the outer step. ReLU is used in the forward pass and Parametric Softplus in the backward pass. Interestingly, this strategy increases adversarial robustness for "free" by telling the network optimizer to use Parametric Softplus's gradient (i.e., Eq. 10) to update network parameters. Without requiring additional calculations, adversarial robustness increases by 2.8% and accuracy by 0.6% when compared to the ReLU baseline and the third row of Table 3. The cross-entropy loss on the training set has dropped from 2.71 to 2.59, indicating that the training loss is also declining. Networks may calculate superior gradient updates in adversarial training by applying ReLU's smooth approximation in the backward pass of the outer reduction phase, as shown by the findings of enhanced resilience and accuracy as well as decreased training loss.

Interestingly, we discover that improved gradient updates can benefit both modern and traditional training. For example, using ResNet-50, training with greater gradients increases accuracy from 76.8% to 77.0% while reducing training loss from 1.22% to 1.18%. These findings suggest that altering network parameters with better gradients while leaving the model's inference procedure unaltered (i.e., ReLU is always employed for inference) might be a general method for enhancing performance [12]. This finding applies to both adversarial and normal training.

Improving Gradient Quality for Both the Adversarial Attacker and Network Parameter Updates. Based on the discovery that enhancing ReLU's gradient enhances resilience for either the adversarial attacker or the network optimizer, we can improve adversarial training by substituting Parametric Softplus for ReLU in all backward runs while keeping ReLU in all forward passes.

This level of resilience in a trained model produced the best outcomes to date, as was expected. The final row of Table 3 shows that it outperforms the ReLU baseline in terms of robustness by 3.9%. Surprisingly, this advantage, which reports accuracy that is 0.1% greater than the ReLU baseline, is nevertheless considered "free".

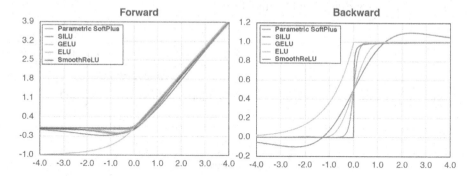

Fig. 3. Visualizations of 5 different smooth activation functions and their derivatives [12].

Smooth Adversarial Training (SAT): Improved ReLU gradient, as seen in the preceding section, may both strengthen the attacker and offer better gradient updates during adversarial training. Although the networks are still trained differently for the forward pass (using ReLU) and the backward pass, this method might not be as successful as it might be (which uses Parametric Softplus). It is suggested that SAT [12], which assures architectural smoothness by utilizing only smooth activation functions (in both the forward and backward passes), be used to fully utilize the potential of training with improved gradients. The majority of the other network elements are smooth and won't result in a gradient issue, so they are all left unchanged. (The non-smooth gradient problem caused by max pooling, which is also present in SAT, is disregarded. This is due to the infrequent use of it in modern architectures; for example, ResNet [20] just uses one max pooling layer, while EfficientNet [24] uses none.)

Adversarial Training with Smooth Activation Functions: The following activation functions are investigated as smooth approximations of ReLU in SAT (Fig. 3 depicts these functions as well as their derivatives):

- Softplus [23]: $\text{Softplus}(x) = \log(1+\exp(x))$. Its parametric version as detailed in Eq. 9 is also considered, where α is set to 10 as in previous section.
- SILU [25–27]: $\text{SILU}(x) = x \cdot \text{sigmoid}(x)$. Compared to other activation functions, SILU has a nonmonotonic "bump" when $x < 0$.
- Gaussian Error Linear Unit (GELU) [28]: $\text{GELU}(x) = x \cdot \phi(x)$, where $\phi(x)$ is the cumulative distribution function of the standard normal distribution.

- Exponential Linear Unit (ELU) [29]:

$$\text{ELU}(x) = \begin{cases} \alpha(\exp(x) - 1) & x < 0 \\ x & x \geq 0 \end{cases} \tag{11}$$

where we set $\alpha = 1$ as default. Note that when $\alpha \neq 1$, the gradient of ELU is not continuously differentiable anymore.

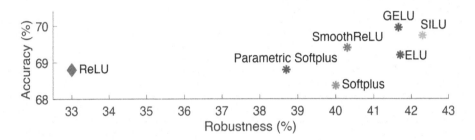

Fig. 4. Adversarial Training is enhanced by smooth activation functions. All smooth activation functions considerably increase robustness in comparison to ReLU while maintaining a nearly unchanged level of accuracy [12].

Main Results: ResNet-50, which has smooth activation functions, was adversarially trained using the conditions stated in the section before. The outcomes are displayed in Figure 11 [12]. All smooth activation functions greatly improve robustness when compared to the ReLU baseline while retaining or almost maintaining standard accuracy. For instance, using Parametric Softplus, smooth activation functions enhance robustness by at least 5.7% (from 33% to 38.7%). The maximum level of robustness is attained by SILU, which enables ResNet-50 to reach 42.3% robustness and 69.7% standard accuracy. These results are likely to be promoted if more sophisticated smooth alternatives (e.g., [30–32]) are used.

We also compare this setup to the one in the previous section, where Parametric Softplus is solely utilized for training on the backward pass. It's interesting to see that switching from ReLU to Parametric Softplus at the forward pass increases robustness by 1.8% (from 36.9% to 38.7%) while keeping roughly the same accuracy. As demonstrated by [12], this outcome emphasizes the value of adopting smooth activation functions in both forward and backward SAT passes.

Dealing with x < 0: The functions mentioned above, in contrast to ReLU, have non-zero responses to negative inputs ($x < 0$), which could have an effect on adversarial training. The SmoothReLU, which flattens the activation function by only changing ReLU after $x \geq 0$ and is inspired by the design of the Rectified Smooth Continuous Unit in [33], is created to rule out this factor.

$$\text{SmoothReLU}(x; \alpha) = \begin{cases} 0 & x < 0 \\ x - \frac{1}{\alpha} \log(\alpha x + 1) & x \geq 0 \end{cases} \tag{12}$$

where α is a learnable variable that is shared by all channels in a layer and must be positive. It's important to note that SmoothReLU is always continuously differentiable, regardless of the choice of α.

$$\frac{d}{dx}\text{SmoothReLU}(x;\alpha) = \begin{cases} 0 & x < 0 \\ \frac{\alpha x}{1+\alpha x} & x \geq 0 \end{cases} \qquad (13)$$

Note when $\alpha \to \infty$, SmoothReLU converges to ReLU. Additionally, the learnable parameter α should be started at a large enough value to prevent the gradient vanishing problem at the beginning of training (e.g., 400 in the provided experiments). Figure 3 illustrates the plot of SmoothReLU and its first derivative. SmoothReLU outperforms ReLU by 7.3% for robustness (from 33.0% to 40.3%) and by 0.6% for accuracy (from 68.8% to 69.4%), showing the significance of smoothness in a function and ruling out the effect of having replies when $x < 0$ [12] (Fig 4).

FGSM-SDI with SAT:
Another potential way to boost AT, is to combine the presented studies. There are many aspects of FGSM-SDI algorithm to consider, and replace ReLU activation functions in its architecture with one of the smooth approximations of ReLU; extensive experiments are needed to firmly conclude that if all or a number of ReLU activation functions being replaced by one special approximations of ReLU it will have the best performance. They have to be tested against different frequently in use datasets and then conclude. Below is the results of two small series of experiments. First experiment series was the reimplementation of one of the experiments in [11]. In second series we replaced all the ReLU activation functions with SmoothReLU ($\alpha = 400$). In all the experiments the target Network was PreActResNet18 and on the Tiny ImageNet dataset.

By comparing the numbers in Table 4, we witness 0.49%, 0.61%, 0.39% and 0.1% increase in the accuracy of different cases, which is quite promising.

Table 4. Results of clean and robust accuracy (%) with PreActResNet18 on the Tiny ImageNet dataset with ReLU activation functions and SmoothReLU activation functions.

Activation functions	Clean	PGD-10	C&W	AA
ReLU	45.72	20.47	18.24	16.55
SmoothReLU	46.21	21.08	18.63	16.65

4 Conclusion

The introduced sample-dependent adversarial initialization to boost Fast AT was first examined in this study. An efficient initialization was produced using a generative network that was conditioned on a benign image and its gradient data from the target network. During the training phase, the generative network and the target network are jointly optimized and engaged in a game. Using a dynamic, sample-dependent initialization, the former gains the ability to produce stronger adversarial instances based on the current target network. The latter employs the generated adversarial cases for training to enhance model robustness.

In the second phase of the study, we investigated another idea known as smooth adversarial training, which imposes architectural smoothness by using adversarial training to swap out non-smooth activation functions for their approximate smooth counterparts. SAT enhances adversarial robustness without degrading accuracy or adding extra computation costs.

And in the last part we witnessed that by using SAT in the FGSM-SDI algorithm, we can further increase the accuracy (as well as the robustness), although this theory needs more testing experiments to firmly state the success of this combination.

References

1. Madry, A., Makelov, A., Schmidt, L., Tsipras, D., Vladu, A.: Towards deep learning models resistant to adversarial attacks. arXiv preprint arXiv:1706.06083 (2017)
2. Zhang, H., Yu, Y., Jiao, J., Xing, E.P., Ghaoui, L.E., Jordan, M.I.: Theoretically principled trade-off between robustness and accuracy. arXiv preprint arXiv:1901.08573 (2019)
3. Rice, L., Wong, E., Kolter, Z.: Overfitting in adversarially robust deep learning. In: ICML, pp. 8093–8104 (2020)
4. Lee, S., Lee, H., Yoon, S.: Adversarial vertex mixup: toward better adversarially robust generalization. In: CVPR, pp. 269–278 (2020)
5. Tramèr, F., Kurakin, A., Papernot, N., Goodfellow, I., Boneh, D., McDaniel, P.: Ensemble adversarial training: attacks and defenses. arXiv preprint arXiv:1705.07204 (2017)
6. Shafahi, A., et al.: Adversarial training for free!. In: NeurIPS, pp. 3353–3364 (2019)
7. Wong, E., Rice, L., Kolter, J.Z.: Fast is better than free: revisiting adversarial training. arXiv preprint arXiv:2001.03994 (2020)
8. Andriushchenko, M., Flammarion, N.: Understanding and improving fast adversarial training. In: Advanced in Neural Information Processing System, vol. 33, pp. 16048–16059 (2020)
9. Kim, H., Lee, W., Lee, J.: Understanding catastrophic overfitting in single-step adversarial training. In: AAAI, pp. 8119–8127 (2021)
10. Szegedy, C., et al.: Intriguing properties of neural networks. arXiv preprint arXiv:1312.6199 (2013)
11. Jia, X., Zhang, Y., Wu, B., Wang, J., Cao, X.: Boosting fast adversarial training with learnable adversarial initialization. IEEE Trans. Image Process. **31**, 4417–4430 (2022)
12. Xie, C., Tan, M., Gong, B., Yuille, A., Le, Q.V.: Smooth adversarial training. arXiv preprint arXiv:2006.14536 (2020)
13. Goodfellow, I.J., Shlens, J., Szegedy, C.: Explaining and harnessing adversarial examples. arXiv preprint arXiv:1412.6572 (2014)

14. Moosavi-Dezfooli, S.-M., Fawzi, A., Frossard, P.: DeepFool: a simple and accurate method to fool deep neural networks. In: Proceedings of the IEEE Conference on Computer Vision and Pattern Recognition, pp. 2574–2582 (2016)
15. Carlini, N., Wagner, D.: Towards evaluating the robustness of neural networks. In: 2017 IEEE Symposium on Security and Privacy (SP), pp. 39–57. IEEE (2017)
16. Croce, F., Hein, M.: Reliable evaluation of adversarial robustness with an ensemble of diverse parameter-free attacks. In: ICML, pp. 2206–2216 (2020)
17. Croce, F., Hein, M.: Minimally distorted adversarial examples with a fast adaptive boundary attack. In: International Conference on Machine Learning, pp. 2196–2205. PMLR (2020)
18. Andriushchenko, M., Croce, F., Flammarion, N., Hein, M.: Square attack: a query-efficient black-box adversarial attack via random search. In: Vedaldi, A., Bischof, H., Brox, T., Frahm, J.-M. (eds.) ECCV 2020. LNCS, vol. 12368, pp. 484–501. Springer, Cham (2020). https://doi.org/10.1007/978-3-030-58592-1_29
19. Sinha, A., Namkoong, H., Volpi, R., Duchi, J.: Certifying some distributional robustness with principled adversarial training (2017). arXiv preprint arXiv:1710.10571
20. He, K., Zhang, X., Ren, S., Sun, J.: Deep residual learning for image recognition. CVPR (2016). arXiv preprint arXiv:1512.03385
21. Wong, E., Rice, L., Kolter, J.Z.: Fast is better than free: revisiting adversarial training (2020). arXiv preprint arXiv:2001.03994
22. Hahnloser, R.H., Sarpeshkar, R., Mahowald, M.A., Douglas, R.J., Seung, H.S.: Digital selection and analogue amplification coexist in a cortex-inspired silicon circuit. Nature **405**(6789), 947–951 (2000)
23. Nair, V., Hinton, G.E.: Rectified linear units improve restricted Boltzmann machines, In: ICML (2010)
24. Yue, Z.: (2021–08–04). On Adversarial Machine Learning and Robust Optimization. ScholarBank@NUS Repository
25. Elfwing, S., Uchibe, E., Doya, K.: Sigmoid-weighted linear units for neural network function approximation in reinforcement learning. Neural Netw. **107**, 3–11 (2018)
26. Hendrycks, D., Gimpel, K.: Gaussian error linear units (GELUs) (2016). arXiv preprint arXiv:1606.08415
27. Ramachandran, P., Zoph, B., Le, Q.V.: Searching for activation functions (2017). arXiv preprint arXiv:1710.05941
28. Finlayson, S.G., Bowers, J.D., Ito, J., Zittrain, J.L., Beam, A.L., Kohane, I.S.: Adversarial attacks on medical machine learning. Science **363**, 1287–1289 (2019)
29. Clevert, D.A., Unterthiner, T., Hochreiter, S.: Fast and accurate deep network learning by exponential linear units (ELUs) (2015). arXiv preprint arXiv:1511.07289
30. Misra, D.: Mish: a self regularized non-monotonic activation function (2019). arXiv preprint arXiv:1908.08681
31. Lokhande, V.S., Tasneeyapant, S., Venkatesh, A., Ravi, S.N., Singh, V.: Generating accurate pseudo-labels in semi-supervised learning and avoiding overconfident predictions via Hermite polynomial activations. In: Proceedings of the IEEE/CVF Conference on Computer Vision and Pattern Recognition, pp. 11435–11443 (2020)
32. Biswas, K., Kumar, S., Banerjee, S., Pandey, A.K.: TanhSoft-a family of activation functions combining Tanh and Softplus (2020) arXiv preprint arXiv:2009.03863
33. Team, G.B.: "TensorFlow" (2015). https://www.tensorflow.org/

Computer and Communication
Networks

On Effectiveness of the Adversarial Attacks on the Computer Systems of Biomedical Images Classification

Eugene Yu. Shchetinin[1]🆔, Anastasia G. Glushkova[2]🆔, and Yury A. Blinkov[3,4(✉)]🆔

[1] Financial University Under the Government of the Russian Federation, Moscow, Russia
[2] Oxford University, Oxford, UK
aglushkova@endeavorco.com
[3] Peoples Friendship University of Russia (RUDN), Moscow, Russia
[4] Saratov State University, Saratov, Russia
BlinkovUA@sgu.ru

Abstract. The problems of vulnerability of the computer systems of biomedical images classification to adversarial attacks on are investigated. The aim of the work is to study the effectiveness of the impact of various models of adversarial attacks on biomedical images and the values of control parameters of algorithms for generating their attacking versions. The effectiveness of attacks prepared using the projected gradient descent algorithm (PGD), Deep Fool (DF) algorithm and Carlini-Wagner algorithm (CW) is investigated. Experimental studies were carried out on the example of solving typical problems of medical images classification using deep neural networks VGG16, EfficientNetB2, DenseNet121, Xception, ResNet50 as well as data containing chest X-rays images and brain MRI-scan images.

Our findings in this work are as follows. Deep models were very susceptible to adversarial attacks, which led to decrease of the accuracy classification of the models for all datasets. Prior to the use of adversarial methods, we achieved a classification accuracy of 93.6% for brain MRI and 99.1% for chest X-rays. During the DF attack the accuracy of the VGG16 model showed a maximum absolute decrease of 49.8% for MRI-scans, and 57.3% for chest X-rays images. The gradient descent (PGD) algorithm with the same values of malicious image disturbances is less effective than the DF and the CW adversarial attacks. VGG16 deep model is more effective in accuracy classification on considered datasets and most vulnerable to adversarial attacks among other deep models. We hope that these results would be useful to design more robust and secure medical deep learning systems.

Y. Blinkov—This paper has been supported by the RUDN University Strategic Academic Leadership Program.

Keywords: adversarial attacks · deep learning · white-box attacks · black-box attack · chest X-ray images · brain tumor MRI-scans

1   Introduction

Deep learning methods demonstrates significant success in solving a wide range of image processing, analysis and recognition tasks. A significant proportion of such tasks are associated with the automation of medical diagnostics processes [9], which include the classification of images in automated systems of computer diagnosis of diseases, detection of suspicious cases requiring the participation of a specialist during large-scale screening of the population, solving such auxiliary tasks as the selection of target areas of medical images, improving their quality, etc.

However, at the moment, the bulk of research and development in the field of deep learning is aimed at achieving the highest possible accuracy of classification and image recognition by Deep Neural Networks (DNN) [8]. At the same time, not enough attention is paid to the problem of safety and reliability of their work. As a result, the pursuit of a single percent increase in accuracy leads to the fact that some serious security problems have not been adequately developed both in the field of research and in the field of practical development.

The problem of DNN robustness is multifaceted and of considerable scientific and practical interest [3–5]. This is due to several reasons. Firstly, deep learning models are unstable to special (potentially malicious) types of modification of input images, which is very unsafe for systems with increased responsibility. Secondly, there is some experimentally confirmed assumption that attacks on the classification system using specially modified attacking images (Adversarial Examples) can be made even if a priori information about both the architecture and the training sample of images used in training the attacked DNN is not available [10,12]. Since there is currently limited information regarding the conditions for successful attacks, there is a great need to accumulate relevant information that can later be used in the development of the necessary means of protection. For obvious reasons, this situation is particularly acute in the field of analysis of biomedical images used in the diagnosis and treatment of various diseases.

Some recent works tested the ability of adversarial attacks to cause serious damage to computer systems of medical images analysis [6,13,19]. In papers [17,20,21] the authors proposed some special models of attacks to boost their influence on computer systems. In this paper we provide understanding of adversarial attacks on biomedical images and detection these attacks. The main purpose of this study is to investigate the dependence of the influence of adversarial attacks on deep neural networks on factors such as the type of attack and the values of the control parameters of the algorithms for generating their attacking versions. Our contributions are: we find that adversarial attacks on medical images computer systems classification are very dangerous and capable of causing serious damage dangerous and capable of causing serious damage. It is necessary to develop special computer software to detect and prevent adversarial attack. We show that vulnerability of DNN designed for large-scale natural images appears due to their overparamerizing. So that most accurate deep model VGG16 it turned out to be more vulnerable compare to other deep models in our investigation.

2 Background Information and Related Work

2.1 Adversarial Attacks Models

An adversarial attack is an action that forces a neural network classifier to make mistakes at the stage of applying a trained deep neural network. Attacks are made using attacking images, which are ordinary input images to be recognized, which have been modified in a special way. These modifications are made so that the modified images are visually practically indistinguishable from the original real images. Despite this, the neural network classifier regards the modified images as completely different and, as a result, makes an unacceptable mistake. Often, the algorithms for generating attacking images are so good that the differences between the attacking and the original images are invisible to the human eye. Nevertheless, the results of attribution to the appropriate class (for example, norm or pathology) are completely erroneous. Actually, the process of attacking a neural network consists in feeding a generated attacking image to its input, which will lead to a recognition error. Meanwhile, when unauthorized access is obtained to a system operating on the basis of DNN, it becomes possible to carry out attacks at the stage of functioning of computer systems, which, of course, is a serious breach in their security. Despite the fact that deep convolutional networks are "well attacked", competitive attacks can also be carried out on shallow networks with a small number of layers. However, in this case, the effect will not be so strong and, in order to make the network make a mistake, the degree of difference between the attacking images from the original ones should be noticeably greater.

Let $x \in R^d$ be an image from the original dataset, where n is the dimension of the image. Let $F : R^d \mapsto (1, ..., p)$ is a function that takes an image as input and returns the class predicted by the DNN, and p is the number of predicted classes. Then an adversarial example is called $x^* \in R^d$, such that the inequality holds

$$F(x^*) \neq F(x) \tag{1}$$

and at the same time for some small constraint is observed

$$\|x^* - x\| \leq \varepsilon \tag{2}$$

Expression (1) shows that the predicted image class x^* differs from the predicted image class x. This is a fairly common phenomenon if x^* is just some image from the original dataset belonging to another class. Therefore, in order to show the essence of the attacking images, namely their minimal difference from the original images, a restriction from formula (2) is introduced. In order to demonstrate the synthetic effect, a condition (2) is introduced for a small value of ε. An acceptable parameter value (such that the attacking image is close enough to the original one) is chosen by the attacking subject. Usually, to make the classifier make a mistake, it is enough to choose a small ε. At different values of ε, the modification of the image is expressed to varying degrees.

2.2   Algorithms for Adversarial Attacking Images Generation

Quite a lot of algorithms for generating attacking images have been developed. Among them, both universal concepts and a variety of different heuristics stand out. To begin with, let's look at the general idea of existing algorithms, and then give a few specific methods. According to the information necessary for the work, the generation algorithms are divided into white box attacks and black box attacks. To carry out a white box attack, it is necessary to know the network configuration, including its architecture and all the parameters obtained as a result of training. In addition, the presence of the original image of the corresponding subject area is required to generate the attacking image itself. To carry out a black box attack, it is enough to have access to the network input where the images are fed and to the prediction results, while the network configuration may remain unknown. Of course, we also need information about the subject area of the recognized images. By the presence or absence of a predefined class that the attacking image needs to be falsified, the algorithms for generating attacks are divided into directed and non-directed. When conducting a directed attack, the class to which the attacking image should be mistakenly attributed is determined in advance. Non-directional attacks, in turn, are designed only to deceive the network, regardless of which erroneous class the image will be assigned to as a result of recognition. Most algorithms for generating attacking images allow you to configure them to carry out an attack of any of the two specified types. In this paper, only non-directional attacks are considered. It is also obvious that in the case of a binary classification of the "norm/pathology" type, directed and non-directed attacks coincide.

The algorithm of Projected Gradient Descent (PGD) [19]. PGD attacks computer system of images classification and optimizes its attack strength by adapting the step size across iterations depending on the overall attack budget and progress of the optimisations. After adapting its steps size Attack Example restarts from the best example found so far. In this white-box attack, the source CNN architecture is trained for a similar task. The gradients from this model are used to produce an adversarial sample which is then transferred to attack the target. Gradients updates are performed in the direction which maximizes the classification loss as per Eq. (1), where x, Adv_x are original and adversarial sample, respectively.

The term ε is the step size that decides the magnitude of the update. The gradient of the loss function is denoted by $\nabla_x J$ and weights corresponding to the CNN is shown as θ. The output label is shown y.

Iterative gradient updates are performed until the loss converges to a higher value. This treatment makes the adversarial image to deviate from the original image, making it unperceivable to humans. Although PGD shows good generalization ability for samples generated on white box source model to be transferred to the black box model, it is limited by the need for the white box source model.

$$Adv_x = x + \varepsilon * sign(\nabla_x J\{\theta, x, y\}). \qquad (3)$$

where $x_0 = x$, $k \in [0, n-1]$, $x^* = x_n$ and the function $clip\_x, \varepsilon$ "cuts" those elements of its argument that differ from the same elements of x by more than ε. The generation of an undirected attacking image using this method depends on the source class m and is defined as follows: $x_{k+1} = clip_{x,\varepsilon}(x_k - \alpha\nabla y_m(x_k))$. Since the function $clip_{x,\varepsilon}$ is used at each iteration, then in the result of the algorithm will be an image x^* that automatically satisfies the constraint (2) for the L_2 norm.

The algorithm DeepFool (DF) [11]. This algorithm is based on the idea of linearization of the neural network output function and iterative calculation of the attacking image as a projection point on some pseudo-plane. One iteration of this algorithm is given by the formula

$$x_{k+1} = x_k - \frac{y_l(x_k) - y_m(x_k)}{||w_l - w_m||^2}(w_l - w_m), \qquad (4)$$

where $w = \Delta y(x_k)$, $x_0 = x$, m is the original class of the object x, and l is the class chosen at each iteration so that the perturbations of the object are minimal. The value $x^* = (1 + \eta) x_p$ is used as the attacking image x_p of the iteration at which the attack was successful.

The Carlini & Wagner(CW) is state-of-the-art optimization-based attack [2]. There are two types of CW attack: L_2 and L_∞ metrics. We concerned on L_∞ version.

3   Adversarial Attacks on Deep Models Effectiveness Exploration

Relying on the transferring hypothesis, which consists in the fact that an attacking image generated to attack one network often successfully attacks another network trained to classify images of the same type [14,15]. Relying on this property, it is possible to formulate a methodology of actions consisting of three main steps:

- on some sample of images from the target network classification domain, we train our own network;
- we carry out a white box attack on a trained tool network and as a result we get an attacking image;
- we submit the generated attacking image to the input of the target network under attack and evaluate the results obtained.

Thus, the target network is attacked using the black box method, since information about the architecture and weights of the target network was not used in any way, only information about the corresponding parameters of the trained tool network playing the role of an auxiliary tool was needed.

To implement the above methodology, our own neural network was trained in a next way. The widely used VGG16, EfficientNetB2, DenseNet121, Xception architectures were chosen as the basic architecture of neural networks. Then the following steps were performed:

1. We trained each of the above networks architectures.
2. We attacked each of the trained networks using the white box method, generating a corresponding attacking image for each source image from the test sample. We save the generated images.
3. In each pair of trained networks, we assign one network to the target, the other to the instrumental. We carry out an attack on the target network by submitting attacking images generated for the instrumental one. We save the results of class predictions made by the target network. Then we swap the target and tool networks and carry out a similar attack.

The PGD, DF, CW adversarial attack models were explored for generating attacking images on deep model classification images. Adam optimizer was chosen as the optimizer. In all cases, less than 100 epochs of training were sufficient to achieve the accuracy acceptable for research. Keras and Tensorflow were used as libraries for training neural networks. During the experiments, the amplitude of the malicious gradient perturbation of the maximum pixel value in increments is $\varepsilon = 0.01$.

Here we considered popular medical datasets images classification with DNN models. First, the dataset of chest X-ray images with three classes [COVID-19, NORMAL, PNEUMONIA] was used to test the models of adversarial attacks described above [1,7,16]. Second, the dataset of brain MRI-scans with two classes [TUMOR, NO TUMOR] was used [18]. Examples of successful attacks on X-ray images used in the diagnosis of COVID-19 are shown in Fig. 1. Examples of successful attacks on brain tumor MRI-scans are shown in Fig. 2. For training and testing our deep models under attacks we produced dataset $Train$ for pretraining and dataset $Test$ for evaluating the DNN models and attacking experiments. In the attack testing we split $Test$ dataset into $AdvTrain$ and $AdvTest$ datasets.

For efficiency evaluating of success of adversarial attacks the rate of success of the adversarial attack metric was chosen [11,22]. The main factors that influenced on this metric we proposed the number of iterations of training the DNN on $AdvTrain$ dataset and amplitude perturbation parameter. The results of the study of the influence of the studied factors on the success of adversarial attacks on DNN models are presented in Fig. 3, 4, 5, 6. On Fig. 3 the dependence of attack success rate on amplitude perturbation for chest X-ray images is shown. As could be seen, DF algorithm is most robust to amplitude of the image perturbation compare to the other algorithms. On Fig. 4 the dependence of attack success rate on amplitude perturbation for MRI-scans of brain images is shown.

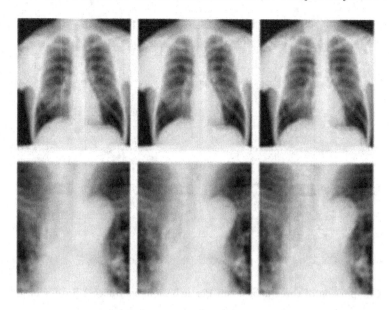

Fig. 1. Examples of real and adversarial attack chest X-ray images

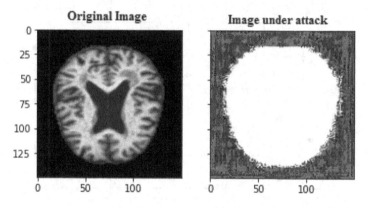

Fig. 2. Examples of real and under adversarial attack brain tumor X-ray images

Also it could be seen, that DF algorithm is most robust to amplitude of the image perturbations compare to the other algorithms.

On Fig. 5 the dependence of attack success rate vs number of iterations for Chest X-ray images is shown. As could be seen, DF algorithm is most resistant to amplitude of the image perturbation compare to the other algorithms. On Fig. 6 the dependence of attack success rate on amplitude perturbation for MRI-scans of brain images is shown. Also it could be seen, that DF algorithm is most robust to amplitude of the image perturbations compare to the other algorithms. It can be seen that for classification tasks the ratio of successful attacks using the black box method is significant. For this problem, the influence of such attacks

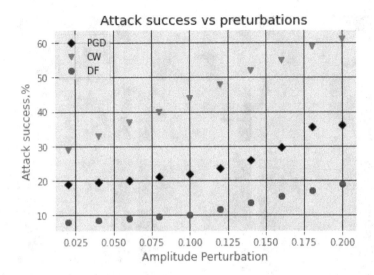

Fig. 3. Comparison of three attacks on Perturbations vs Success Rate. Chest X-ray

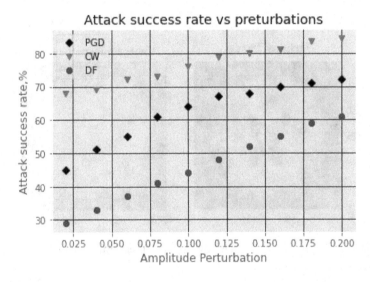

Fig. 4. Comparison of three attacks on Perturbations vs Success rate. MRI-scans of brain

is noticeable, and for the brain tumor MRI-scans classification problem, their strength is comparable to the strength of attacks using the white box method (although, of course, still less). It is also worth noting that the percentage of successful attacks using the black box method is determined by the network input with which specific architecture the generated images were submitted.

At the same time, there is no dependence on which network the images were generated for.

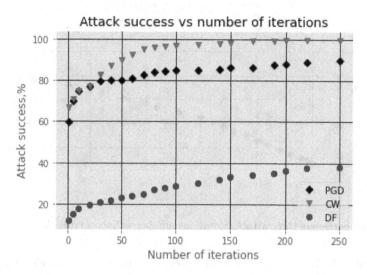

Fig. 5. Comparison of three attacks on Number of iterations vs Success rate. Chest X-ray

The results of three black-box attacks on investigated datasets with deep models mentioned above were presented in Table 1, Table 2. As it could be seen from Table 1, most vulnerable deep model is VGG16 and robust to adversarial attacks is ResNet50 model. In Table 2 as it could be seen, most vulnerable deep model is Xception and more robust turned to be EfficientNetB2 deep model.

Table 1. The ratio of successful attacks on Chest X-rays images dataset

model under attack	Basic deep model				
	VGG16	Xception	ResNet50	DenseNet121	EfficientNetB2
VGG16	66	29	19	17	15
Xception	14	47	9	8	11
ResNet50	11	12	31	18	9
DenseNet121	28	23	32	47	26
EfficientNetB2	11	14	14	14	33

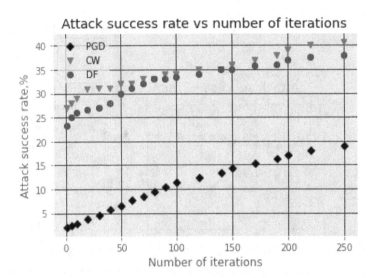

Fig. 6. Comparison of three attacks iterations number vs Success rate. MRI-scans brain

Table 2. The ratio of successful attacks on MRI-scans images of brain

model under attack	Basic deep model				
	VGG16	Xception	ResNet50	DenseNetl21	EfficientNetB2
VGG16	25	3	3	2	1
Xception	4	25	2	2	1
ResNet50	2	2	19	3	1
DenseNetl21	2	2	2	15	1
EfficientNetB2	2	2	1	2	11

4   Conclusion

The results obtained in this work allow us to draw the following conclusions. The problem of adversarial attacks is relevant for biomedical image recognition tasks, since the tested algorithms successfully attack trained neural networks so that their accuracy in some cases falls below 15%. With an increase in the amplitude of the malicious disturbance, the probability of an error in predicting an adversarial example increases. Increasing the number of iterations of the predicted gradient descent method gradually increases the number of errors allowed by the network, with asymptotic convergence to a certain maximum value. Images that are classified by the network with a confidence of more than 95% are much more resistant to adversarial attacks. The predicted gradient descent (PGD) algorithm with the same values of malicious image distortion is less effective than the DeepFool (DF) algorithm and the Carlini-Wagner algorithm (CW), and the DF algorithm turned out to be the most effective in mostly all cases.

Deep models were very susceptible to adversarial attacks, which led to decrease of the accuracy classification of the models for all datasets. Prior to the use of adversarial methods, we achieved a classification accuracy of 93.6% for brain MRI and 99.1% for chest X-rays. During the DF attack, the accuracy of the VGG16 model showed a maximum absolute decrease of 49.8% for MRI-scans, and 57.3% for chest X-rays images. As we have seen from the results of the study the VGG16, Xception model, which are the most effective at classifying medical images on both studied datasets, also turned out to be the most vulnerable to attacks among other deep learning models. This may indicate that state-of-art deep models designed for large scaled image analysis may be overfitted and as the resultare highly vulnerable to strong adversarial attacks. The results show that a clinical diagnosis based on DNN is easier to deceive thanks to adversarial attacks. Attackers can cause erroneous diagnoses, moreover, they can affect the diagnosis. Our results emphasize that more careful consideration is required when developing DNNs for medical imaging and their practical application.

In future work, we are going to extend our work to more complex model architectures and adversarial attacks generated using various algorithms. Also in the future, it is planned to develop effective methods and programs to increase the resistance of computer systems to adversarial attacks.

References

1. Asgari Taghanaki, S., Das, A., Hamarneh, G.: Vulnerability analysis of chest X-ray image classification against adversarial attacks. In: Stoyanov, D., et al. (eds.) MLCN/DLF/IMIMIC 2018. LNCS, vol. 11038, pp. 87–94. Springer, Cham (2018). https://doi.org/10.1007/978-3-030-02628-8_10
2. Carlini, N., Wagner, D.: Towards evaluating the robustness of neural networks. In: 2017 IEEE Symposium on Security and Privacy (SP), pp. 39–57 (2017). https://doi.org/10.1109/SP.2017.49
3. Chakraborty, A., Alam, M., Dey, V., Chattopadhyay, A., Mukhopadhyay, D.: Adversarial attacks and defences: a survey (2018). https://doi.org/10.48550/arxiv.1810.00069
4. Fawzi, A., Fawzi, O., Frossard, P.: Analysis of classifiers' robustness to adversarial perturbations. Mach. Learn. **107**(3), 481–508 (2017). https://doi.org/10.1007/s10994-017-5663-3
5. Finlayson, S.G., Bowers, J.D., Ito, J., Zittrain, J.L., Beam, A.L., Kohane, I.S.: Adversarial attacks on medical machine learning. Science **363**(6433), 1287–1289 (2019). https://doi.org/10.1126/science.aaw4399. https://www.science.org/doi/abs/10.1126/science.aaw4399
6. Ilyas, A., Santurkar, S., Tsipras, D., Engstrom, L., Tran, B., Madry, A.: Adversarial examples are not bugs, they are features. In: Wallach, H., Larochelle, H., Beygelzimer, A., d'Alché-Buc, F., Fox, E., Garnett, R. (eds.) Advances in Neural Information Processing Systems, vol. 32. Curran Associates, Inc. (2019). https://proceedings.neurips.cc/paper/2019/file/e2c420d928d4bf8ce0ff2ec19b371514-Paper.pdf

7. Jain, G., Mittal, D., Thakur, D., Mittal, M.K.: A deep learning approach to detect Covid-19 coronavirus with X-ray images. Biocybern. Biomed. Eng. **40**(4), 1391–1405 (2020). https://doi.org/10.1016/j.bbe.2020.08.008. https://www.sciencedire ct.com/science/article/pii/S0208521620301005

8. Kermany, D.S., et al.: Identifying medical diagnoses and treatable diseases by image-based deep learning. Cell **172**(5), 1122–1131.e9 (2018). https://doi.org/ 10.1016/j.cell.2018.02.010. https://www.sciencedirect.com/science/article/pii/S0 092867418301545

9. Litjens, G., et al.: A survey on deep learning in medical image analysis. Med. Image Anal. **42**, 60–88 (2017). https://doi.org/10.1016/j.media.2017.07.005. https:// www.sciencedirect.com/science/article/pii/S1361841517301135

10. Madry, A., Makelov, A., Schmidt, L., Tsipras, D., Vladu, A.: Towards deep learning models resistant to adversarial attacks (2017). https://doi.org/10.48550/ARXIV. 1706.06083

11. Moosavi-Dezfooli, S.M., Fawzi, A., Frossard, P.: DeepFool: a simple and accurate method to fool deep neural networks. In: 2016 IEEE Conference on Computer Vision and Pattern Recognition (CVPR), pp. 2574–2582 (2016). https://doi.org/ 10.1109/CVPR.2016.282

12. Ozdag, M.: Adversarial attacks and defenses against deep neural networks: a survey. Procedia Comput. Sci. **140**, 152–161 (2018). https://doi.org/10.1016/j.procs. 2018.10.315

13. Papernot, N., McDaniel, P., Goodfellow, I., Jha, S., Celik, Z.B., Swami, A.: Practical black-box attacks against machine learning. In: Proceedings of the 2017 ACM on Asia Conference on Computer and Communications Security, ASIA CCS 2017, pp. 506–519. Association for Computing Machinery, New York (2017). https://doi. org/10.1145/3052973.3053009

14. Papernot, N., McDaniel, P.D., Goodfellow, I.J.: Transferability in machine learning: from phenomena to black-box attacks using adversarial samples. CoRR abs/1605.07277 (2016). http://arxiv.org/abs/1605.07277

15. Shchetinin, E.Y., Sevastianov, L.A.: On transfer learning methods in biomedical image classification tasks. Informatika i ee Primeneniya **15**(4), 59–64 (2021). https://doi.org/10.14357/19922264210408

16. Shchetinin, E.Y., Sevastianov, L.A.: Automated detection of Covid-19 coronavirus infection based on analysis of chest X-ray images by deep learning methods. Tomsk State Univ. J. Control Comput. Sci. **58**, 98–105 (2022). https://doi.org/10.17223/ 19988605/58/9

17. Shi, Y., Wang, S., Han, Y.: Curls & Whey: boosting black-box adversarial attacks. In: 2019 IEEE/CVF Conference on Computer Vision and Pattern Recognition (CVPR), pp. 6512–6520 (2019). https://doi.org/10.1109/CVPR.2019.00668

18. Siar, M., Teshnehlab, M.: Brain tumor detection using deep neural network and machine learning algorithm. In: 2019 9th International Conference on Computer and Knowledge Engineering (ICCKE), pp. 363–368 (2019). https://doi.org/10. 1109/ICCKE48569.2019.8964846

19. Wang, H., Yu, C.: A direct approach to robust deep learning using adversarial networks. CoRR abs/1905.09591 (2019). http://arxiv.org/abs/1905.09591

20. Wang, J., Yin, Z., Tang, J., Jiang, J., Luo, B.: PICA: a pixel correlation-based attentional black-box adversarial attack. CoRR abs/2101.07538 (2021). https:// arxiv.org/abs/2101.07538

21. Yang, J., Jiang, Y., Huang, X., Ni, B., Zhao, C.: Learning black-box attackers with transferable priors and query feedback. In: Proceedings of the 34th International Conference on Neural Information Processing Systems, NIPS 2020, vol. 33, pp. 12288–12299. Curran Associates Inc., Red Hook (2020)
22. Yuan, J., Zhou, S., Lin, L., Wang, F., Cui, J.: Black-box adversarial attacks against deep learning based malware binaries detection with GAN. In: Giacomo, G.D., et al. (eds.) 24th European Conference on Artificial Intelligence, ECAI 2020, Santiago de Compostela, Spain, 29 August–8 September 2020, Including 10th Conference on Prestigious Applications of Artificial Intelligence (PAIS 2020). Frontiers in Artificial Intelligence and Applications, vol. 325, pp. 2536–2542. IOS Press (2020). https://doi.org/10.3233/FAIA200388

Multipath Transmission of Heterogeneous Traffic in Acceptable Delays with Packet Replication and Destruction of Expired Replicas in the Nodes that Make Up the Path

V. A. Bogatyrev[1,2], A. V. Bogatyrev[3]($\boxtimes$), and S. V. Bogatyrev[2,3]

[1] Department Information Systems Security, Saint-Petersburg State University of Aerospace Instrumentation, Saint Petersburg, Russia
[2] ITMO University, Saint Petersburg, Russia
vabogatyrev@itmo.ru
[3] Yadro Cloud Storage Development Center, Saint Petersburg, Russia
gangleon@gmail.com

Abstract. The possibilities of increasing the efficiency of multi-path transmissions of traffic that is heterogeneous in terms of acceptable delays in the delivery of traffic packets are analyzed. The efficiency of servicing individual flows is estimated by the probability of timely delivery of flow packets to the addressee. The complexity of providing and evaluating the efficiency of non-uniform traffic transmission reservation is due to the fact that an increase in the probability of timely reserved packet delivery of some flows leads to a decrease in this probability for other flows.

Models are proposed that make it possible to justify the choice of organizing redundant transmissions of heterogeneous traffic, taking into account the risks associated with untimely delivery of packets of different flows to the addressee.

The efficiency of solutions for organizing redundant transmissions was evaluated by a multiplicative criterion summing up the probabilities of timely delivery of packets of various streams to the addressee, taking into account the frequency of their arrival. Calculations of the efficiency of options for organizing redundant transmissions are carried out according to a generalized criterion that takes into account the influence of the probability of timely delivery of packets of different streams on the overall economic efficiency of servicing a non-uniform total stream.

The significance of the effect of redundant transmissions of packets with the shortest allowable delivery time to their addressee is shown. It has been established that there is a limit to the intensity of receipt of requests of the total flow, if it is exceeded, the negative impact of the growth of delays in the delivery of packets to the addressee on the overall economic efficiency of multi-path transmissions increases. If this limit is exceeded by the incoming thread, it is proposed to block the servicing of part of the thread's requests. It is shown that the efficiency of multi-path transmissions is enhanced by the implementation in communication nodes of the destruction of packets that have exceeded the allowable total waiting time in queues by the node involved in their transmission along the established route.

Keywords: Packet delivery path · Request intensity · Timeliness · Multiplicative criterion · Waiting time in node queues · Probability of timely delivery · Reservation multiplicity

1 Introduction

Currently, there is an intensive development of the methodology for designing industrial automation systems based on distributed fault-tolerant highly reliable computer systems and networks. The construction of such systems involves the consolidation of artificial intelligence technologies, big data and the industrial Internet of things.

Computer systems of industrial automation provide for their multi-level construction. At the lower level, data is collected from numerous different types of sensors and the formation of control actions directly on physical objects. The development of control actions can be carried out autonomously from embedded computer tools or when they interact with data processing and storage centers of the upper level of the system. At the top level, there is a distributed processing of information from a variety of sensors, including those based on clustering technologies, cloud computing, machine learning and big data. Communication between the computer facilities of the lower and upper levels is carried out by the telecommunication level. The control of complex industrial facilities operating in real time requires high reliability and fault tolerance at all levels of the industrial automation system [1–4], subject to restrictions on the allowable service time of requests generated in it [5–7]. An important influence on ensuring the timeliness of the implementation of applied tasks is made by the organization of telecommunications. Ensuring the reliability of the interaction of computer facilities through the network can be achieved with link aggregation and multipath routing [8–10].

The construction of multi-path routing networks requires justification of decisions to ensure the reliability, fault tolerance and timeliness of multi-path transmissions, including in real time. The efficiency of multi-path transmissions largely depends on the choice and implementation of transport, network and link layer protocols. Protocols intended for use in networks with multi-path transmissions should provide mechanisms for ensuring high reliability of real-time transmissions [9–11], including those based on the information redundancy of protocol data units [12, 13].

Known network multipath routing, implemented on the basis of the protocol Multipath TCP. Note that the use of this protocol in real-time networks, due to the formation of confirmations of packet delivery to the addressee, can lead to disruption of transmissions at the scheduled time. This shortcoming of the Multipath TCP protocol has led to interest in the use of protocols without acknowledgment generation in multipath real-time networks, including the QUIC protocol [14,15]. The QUIC protocol provides for encoding the frames of the transmitted message, which makes it possible to restore the entire message in case of transmission errors of a part of the frames, eliminating the need for retransmissions of lost frames. Multipath redundant transmission is focused on increasing the reliability and probability of timely delivery of packets to the addressee [16–18].

Reservation of transmissions is accompanied by replication of packets with a multiplicity depending on the restrictions on the allowable time of their delivery to the addressee. The generated replicas are transmitted along different paths connecting the source and destination of the transmitted packet. The success of the transfer consists in the delivery to the addressee within the maximum allowable time of at least one transferred replica. Models of networks with multipath, including redundant, transmissions are proposed in [16–20], in which a study was made of the effectiveness of solutions for organizing redundant transmissions with the definition of areas of their appropriate use. The probability of delivering a packet to the addressee with a delay less than the maximum allowable one is determined taking into account the accumulation of waiting time in the queues of all path nodes involved in the sequential transmission of the packet (replica) to the addressee. It is possible to additionally reduce the total delays in the nodes that make up the path of data delivery to the addressee by destroying irrelevant packets in the nodes of the path, for which the delay in the queues of passed nodes has already exceeded the allowable limit.

The complexity of increasing the efficiency of transmission reservation and its evaluation in case of traffic heterogeneity is due to the fact that an increase in the probability of timely reserved packet delivery of some flows leads to a decrease in this probability for other flows. Ensuring the required probability of timely delivery of packets of all streams to the addressee is possible if the distribution of request intensities of different streams, such that the transmission of the most delay-critical packets does not lead to violation of the time limits for delivery of packets less critical to delays. When traffic is heterogeneous in terms of admissible packet delays to the addressee (including redundant transmission), in [20] the efficiency of servicing individual flows is estimated by the probabilities of not exceeding the maximum admissible delays set for them. In view of the conflict in ensuring the timely service of all non-uniform traffic flows, the overall efficiency of its service should be based on a multi-criteria vector assessment. If the Pareto area of efficient solutions for servicing all non-uniform traffic flows includes a set of solutions, then the choice of options for organizing multipath transmissions should be carried out according to a scalar criterion.

The search for the most efficient solution, taking into account the servicing of all non-uniform traffic flows, can be carried out according to additive and multiplicative criteria. The additive criterion is formed as the product of the probabilities of timely delivery of packets of all streams to the addressee. The multiplicative criterion summarizes the probabilities of timely transmission of packets from different streams, taking into account the frequency (probability) of their arrival.

Thus, the multiplicative criterion can be represented as the mathematical expectation of the probabilities of delivering packets of all streams within the allowable time. The additive criterion makes it possible to estimate the probability that packets of different flows will be executed within the maximum allowable time specified for them. The probability of requests for the transmission of packets from different streams, as well as the importance of their delivery to the addressee, is not taken into account by the additive criterion. With a multiplicative criterion, the importance of delivering packets from different streams is compared only with the frequency (probability) of their arrival. The considered criteria do not take into account the criticality of the timely delivery of packets from different streams and the risks associated with violation of the required deadlines for the delivery of packets to recipients.

A generalized criterion for the efficiency of multipath transmissions can be formed taking into account the influence of individual flows on the economic efficiency of packet transmission through the network, including the profit from the timely delivery of packets of different flows and penalties for their non-delivery or late delivery.

Justification of decisions on the organization of multipath transmissions based on a generalized criterion involves the construction of analytical network models that take into account the variance in the organization of redundant transmissions and the possibility of destroying expired packets in network nodes.

Thus, the purpose of the work is to build models that make it possible to justify the choice of organizing redundant transmissions of heterogeneous traffic, including taking into account the risks associated with untimely delivery of packets of different streams to the addressee.

2   Variants of Communication Between Sources and Data Receiver in Multipath Transmissions

Let's consider some options for combining the existing communication paths of the source and receiver of requests for organizing multipath redundant transmissions. A data processing and storage center, including one with a cluster architecture, can act as a request receiver. The communication paths of the source and receiver of requests include serially connected switching nodes.

In the first variant, there is one source (I) of a non-uniform request flow connected to the destination (receiver) F via n paths, including m serially connected switching nodes. If there is a group of nearby sources, all communication paths with the addressee are available for each of them.

In the second variant, two (or more) geographically distributed sources of requests (or their groups) are singled out. The first and second sources can be directly connected to the request receiver (Fig. 1b) via n_1 and n_2 paths combined for redundant transmissions. The number of communication nodes included in the path for delivering packets to the addressee from different sources can be different (m_1 and m_2).

In the system under research, switching nodes-path switches R can be used to combine and switch paths (path segments). The number of paths emanating from the first and second sources n_1 and n_2 before they are connected to the path switching node may vary. It is possible to combine paths from sources through one or two path switches R (Fig. 1c and 1d). The number of communication paths of the switching node R with the destination may differ from the total number of paths included in it. The number of communication nodes that make up the communication path with the switch node from the first m_1 and second m_2 sources may be different.

The switch nodes R, in addition to combining flows from different sources, can be additionally used to improve the reliability of the system during its reconfiguration based on the switching of path segments that remained operational after failures, located before and after the switching node.

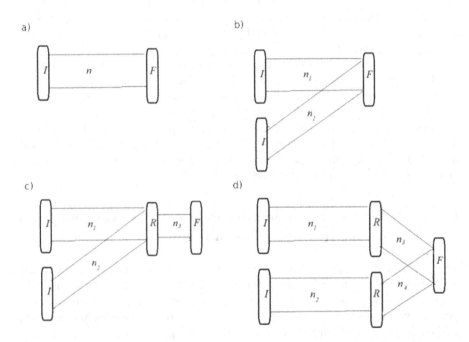

Fig. 1. Options for combining paths for delivering packets to a destination.

3   Model of Redundant Transmissions of Heterogeneous Traffic Through Series-Connected Switching Nodes with the Destruction of Expired Packets

The study of the influence of the organization of multipath transmissions on the delays in the delivery of transmitted packets to the addressee will be carried out when the switching nodes are represented by the simplest queuing models [21, 22].

Consider the option of combining paths for redundant transmissions in accordance with the structure in Fig. 1a.

With the intensity of heterogeneous traffic, Λ_S, the intensity of the flow of packets distributed to one of the n delivery paths to the addressee is equal to $\Lambda = \Lambda_S / n$.

When combined for redundant transmissions of z streams, the intensity of sending packets to one of the n communication paths with the destination will be:

$$\Lambda_1 = \Lambda \sum_{b=1}^{z} \beta_b k_b, \tag{1}$$

At the same time, β_b share of packets of the b-th flow, k_b - multiplicity of their reservation, and t_b are allowed for them delays in the queues of the nodes involved in the transmission of the packet to the destination.

$$\sum_{b=1}^{z} \beta_b = 1.$$

The probability of passing without destruction due to the loss of relevance of the packet of the b-th flow through the i-th node of the path (i>1) is calculated as:

$$P_{ib} = (1 - \Lambda_i v_i e^{(\Lambda_i - \frac{1}{v_i})(t_b - \sum_{j=1}^{i-1} w_j)}),$$

where v_j is the average packet transmission time through the i-node of the path, Λ_i - intensity of the incoming flow to the i-th node for transmission, for the first node of the path this intensity is determined by the formula (1).

For $i > 1$:

$$\Lambda_i = \Lambda \sum_{b=1}^{z} k_b \beta_b \prod_{j=1}^{i-1} P_{jb},$$

average delay in the queue of the j-th node

$$w_j = \frac{\Lambda_j v_j^2}{1 - \Lambda_j v_j}.$$

The probability P_b of successful delivery of a packet of the b-th flow to the addressee in a time less than tb is equal to

$$P_b = \prod_{i=1}^{m} P_{ib},$$

and the probability of success of the redundant transmission of this stream will be

$$R_b = 1 - (1 - P_b)^{k_b}.$$

Let us estimate the efficiency of heterogeneous traffic transfers by a multiplicative indicator as

$$P_c = \sum_{b=1}^{z} \beta_b R_b.$$

Consider the evaluation of the efficiency of redundant transmissions for the structure according to Fig. 1b. We will assume that the first source generates packets of the first flow, the share of which is β, and the allowable waiting delay at the nodes of the path is t_1. The second source generates requests of the second thread, the share of which is $1 - \beta$, and the allowable delay is t_2. The intensity of packets arriving for transmission through one of n1 paths connecting the first source with the destination is defined as

$$\Lambda_1 = \Lambda k_1,$$

where

$$\Lambda = \beta \Lambda_S / n_1$$

When irrelevant query replicas are destroyed along the way, the intensity of the flow of packets arriving at the i-th node of the path for $i > 1$ will be

$$\Lambda_i = \Lambda k_1 \prod_{j=1}^{i-1} P_j,$$

where

$$P_i = \left(1 - \Lambda_i v_i e^{(\Lambda_i - \frac{1}{v_i})(t_1 - \sum_{j=1}^{i-1} w_j)}\right).$$

With redundant transmissions, the probability of successful delivery to the addressee within the allowable time of at least one replica of packets generated by the first source of requests will be

$$R_1 = 1 - (1 - \prod_{i=1}^{m_1} P_i)^{k_1}.$$

Similarly, the intensity of the flow of packets arriving for transmission through one of the n_1 paths from the second source is defined as

$$\Lambda_1 = \Lambda k_2,$$

with $\Lambda = (1 - \beta) \Lambda_S / n_2$.

The intensity of the flow of packets sequentially arriving at the nodes that make up the communication path of the second source with the addresser will be

$$\Lambda_i = \Lambda k_2 \prod_{j=1}^{i-1} P_j,$$

wherein

$$P_i = \left(1 - \Lambda_i v_i e^{(\Lambda_i - \frac{1}{v_i})(t_2 - \sum_{j=1}^{i-1} w_j)}\right).$$

The probability of successful delivery to the addressee within a reasonable time of at least one replica of packets from the second source as

$$R_2 = 1 - (1 - \prod_{i=1}^{m_2} P_i)_2^k.$$

We estimate the overall efficiency of redundant transmissions from the first and second sources by a multiplicative indicator as:

$$P_c = \beta R_1 + (1 - \beta) R_2.$$

4 Examples of Evaluating the Timeliness of Transmissions

Let us assume that the heterogeneous stream being transmitted includes two streams that differ in acceptable delays in the delivery of packets to the addressee. Delivery of packets to the destination involves sequential transmission through three communication nodes.

We will carry out the calculation at $v_1 = v_2 = v_3 = 0.1$ s. In the case when the share of the first stream $\beta = 0.7$, the results of calculating the efficiency by the multiplicative criterion of non-reserved transmissions with and without destruction of requests that have lost their relevance are shown in Fig. 2 by curves 1.2 at $t_1 = 0.2$ s and $t_2 = 2$ s and curves 3.4 $t_1 = 0.2$ s and $t_2 = 1$ s.

The results of evaluating the efficiency by the multiplicative criterion of redundant transmissions with and without destruction of requests that have lost their relevance at $\beta = 0.8$ and $t_1 = 0.4$ s, $t_2 = 0.8$ s are shown in Fig. 3. When destroying packages that have lost their relevance, options without redundancy, with duplication and tripling of all transmissions, are represented by graphs 1, 2, 3, and without their destruction, graphs 4, 5, 6.

The results of the efficiency analysis according to the multiplicative criterion of transmissions with reservation of only packets of the first stream, for which $\beta = 0.5$ and $t_1 = 0.2$ s at $t_2 = 1$ s, are shown in Fig. 4. Curves 1, 2, 3 represent options without redundancy, with duplication and tripling of packet transmissions of the first stream, if the destruction of outdated packets is implemented, and if it is not implemented, then curves 4, 5, 6 correspond to the indicated redundancy options.

The analysis performed showed the effectiveness of increasing the probability of timely delivery of packets of traffic that is heterogeneous in terms of acceptable delays when reserving packet transmissions with the smallest acceptable delays along the way. It is shown that an additional effect of increasing the probability

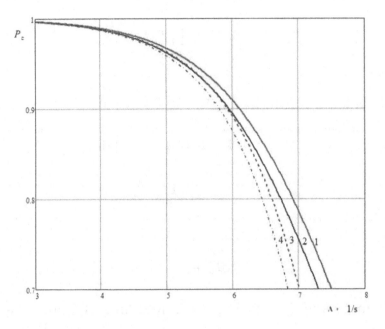

Fig. 2. Efficiency by the multiplicative criterion of non-redundant transmissions of two flows.

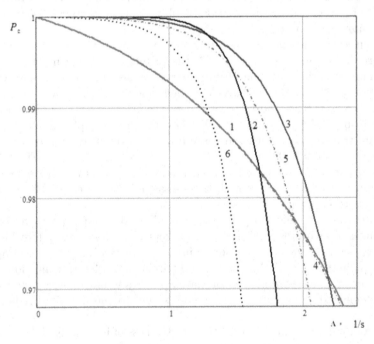

Fig. 3. Efficiency by the multiplicative criterion of redundant transmissions of two flows.

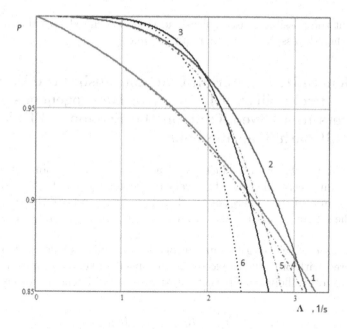

Fig. 4. Efficiency by the multiplicative criterion for redundant transmissions of only packets of one flow.

of delivering packets to the addressee of all streams in the time allowed for them is achieved when the replicas of packets that have lost their relevance are destroyed at the path nodes. The loss of relevance of the replicas can be caused by the distortion of packets or the accumulation of their delays in the queues of previously traversed nodes included in the communication path with the destination. Destroying irrelevant replicas of packets can reduce the load on communication nodes and, accordingly, reduce waiting delays in them.

The conducted studies have shown the feasibility of adaptive regulation of the transmission frequency depending on the intensity of non-uniform traffic. So Fig. 4 confirms that at low traffic intensity, it is advisable to triple packet transmissions, with the lowest allowable delays in the network. With an increase in the intensity of the incoming stream, it is advisable to switch to duplicate transmission of packets with a shorter allowable time for their delivery to the addressee. With a significant intensity of the total flow, transmission without reservation of all packets becomes the most effective.

The studies carried out in this section are based on the use of a multiplicative criterion that combines (sums) the probabilities of timely delivery of packets of various streams to the addressee, taking into account the frequency of their arrival. In fact, this multiplicative criterion represents the mathematical expectation of the probability of delivering packets to the addressee of all streams in the time allowed for them. The criterion does not reflect the criticality of the timely delivery of packets of different streams to the addressee, including taking into account the possible consequences and risks associated with their failure to deliver to the addressee within the required time.

5 A Generalized Criterion that Establishes the Influence of the Probability of Delivery of Heterogeneous Traffic Packets in a Given Time on the Economic Efficiency of Multipath Transmissions

We will establish the overall efficiency of multipath transmissions of heterogeneous up to an acceptable time of delivery to the addressee of traffic, taking into account the impact on the economic efficiency of the probabilities of transmissions for the set time and the risks associated with their non-delivery or untimely delivery.

For the case of selecting two gradations of transmission success (the packet was delivered with or without exceeding the specified time t_i) for each of the z streams, the total efficiency of packet transmission of different streams is given as

$$C = \sum_{i=1}^{z} \beta_i (R_i c_i - (1 - R_i) s_i).$$

where R_i is the probability of delivering packets to the addressee of the i-th flow within the time t_i set for it, and c_i and s_i are the income and the penalty for the delivery or non-delivery of the packet for the required time.

This criterion, in fact, determines the average income from the transmission of packets of different flows.

The average income (profit) from the transmission of packets of the total flow per unit of time can be found as:

$$D = \Lambda \sum_{i=1}^{z} \beta_i (R_i c_i - (1 - R_i) s_i).$$

For the case of selecting three gradations of transmission success (the packet of the i-th flow was delivered without exceeding the specified time t_i, the packet was delivered with an excess of time t_i and the packet was not delivered, that is, it was lost or corrupted)

$$D = \Lambda \sum_{i=1}^{z} \beta_i (R_i c_i - (1 - R_i) s_i - G_i a_i),$$

where a_i is the penalty for the loss (non-delivery) of packets of the i-th stream. G_i - the probability of losing this packet,

$$R_i + G_i = 1.$$

For the case of allocation of four gradations of transmission success (the packet of the i-th stream was delivered not exceeding the specified time t_i, the packet was delivered exceeding the time t_i, the packet was not delivered (lost or corrupted), the packet was refused to be transmitted)

$$D = \Lambda \sum_{i=1}^{z} \beta_i (R_i c_i - (1 - R_i)s_i - G_i a_i - H_i h_i),$$

where h_i is the penalty for denial of service for the transmission of a packet of the i-th stream, H_i is the probability of losing this packet

$$R_i + G_i + H_i = 1.$$

It is possible to form a generalizing criterion for the efficiency of heterogeneous traffic transfers with the allocation of gi gradations of the allowable delivery time of packets of the i-th stream and ui gradations of its untimely delivery

$$D = \Lambda \sum_{i=1}^{z} \beta_i \left(\sum_{j=1}^{g_i} R_{ij} c_{ij} - \sum_{j=1}^{u_i} B_{ij} s_{ij} \right),$$

where c_{ij} is the income for transmitting a packet of the i-th flow in accordance with the j-th gradation, the allowable time of its delivery to the addressee, R_{ij} is the probability of such a transfer, s_{ij} is the penalty for delivering the packet of the i-th stream in accordance with the j-th gradation of the invalid delivery time, B_{ij} is the probability of this event.

The last criterion can be modified to take into account the loss and denial of service of packets from different flows.

When calculating the efficiency of redundant multipath transmissions, we consider non-uniform traffic, which includes two streams that differ in acceptable delays in packet delivery to the addressee.

We will evaluate the efficiency of heterogeneous traffic transfers according to the criterion

$$D = \Lambda(\beta(R_1 c_1 - (1 - R_i)s_i) + (1 - \beta)(R_2 c_2 - (1 - R_2)s_2)). \tag{2}$$

Consider the impact on the overall performance of non-redundant transmissions of the destruction of packets, delayed in the queues of the nodes transmitting them, overtime on criterion D. When calculating, we will assume that $c_1 = 10$ c.u., $c_2 = 3$ c.u. $s_1 = 50$ c.u., $s_2 = 4$ c.u., $v_1 = v_2 = v_3 = 0.1$ s $t_1 = 0.4$ s, and $t_2 = 2$ s. The results of the calculation are shown in Fig. 5. Curves 1. 2, 3 correspond to transmissions without destroying overdue packets in transit, and curves 4.5, 6 with their destruction at $\beta = 0.7, 0.5, 0.2$.

In Fig. 6, curves 1 and 2 correspond to transmissions with the destruction of unreserved and duplicated packets that have lost their relevance in the delivery path to the addressee, curves 3 and 4 correspond to transmissions without the destruction of unreserved and duplicated packets that have expired along the

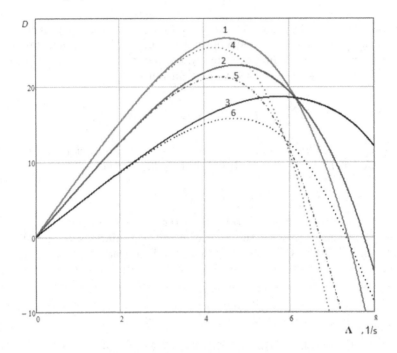

Fig. 5. Profit from non-reserved transfers.

way. Calculations were performed at $c_1 = 10$ c.u., $c_2 = 5$ c.u. $s_1 = s_2 = 60$ c.u. $\beta = 0.7$, $v_1 = v_2 = v_3 = 0.01$ s and $t_1 = 0.2$ s, $t_2 = 2$ s.

The results of evaluating the efficiency of redundant packet transmissions of only the first flow with a shorter allowable packet delivery time to the addressee are shown in Fig. 7. On Fig. 7, options with no replication of packets of the first stream, with their duplication and tripling, are represented by curves 1, 2 and 3 in the case of the destruction of overdue packets along the way, and by curves 4, 5 and 6 if their destruction is not implemented. The calculation was performed at $c_1 = 10$ c.u., $c_2 = 5$ c.u.; $s_1 = s_2 = 40$ c.u. ; $t_1 = 1$ s, $t_2 = 2$ s and $\beta = 0.7$.

When destroying packets delayed beyond the allowable limit, curves 1, 2 and 3 in Fig. 8 at $\beta = 0.5$ show the cases of transmission of packets of the first stream without redundancy, with duplication and with tripling. The same redundancy options at $\beta = 0.2$ are shown by curves 4, 5, 6 in Fig. 8.

The calculations carried out according to the generalized criterion, which takes into account the probabilistic influence of the timeliness of packet delivery of individual flows on the overall economic efficiency of servicing a non-uniform total flow, showed the significant effect of redundant packet transmissions with the shortest allowable delivery time to their addressee. At the same time, it is shown that the efficiency of multipath transmissions is enhanced by the implementation in communication nodes of the destruction of packets that have exceeded the allowable total waiting time in the queues of nodes that transmit them along the path established for them.

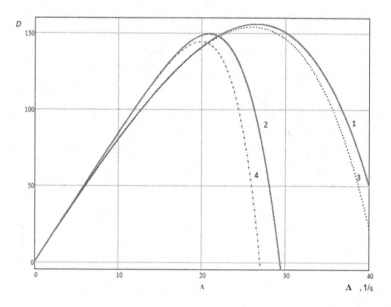

Fig. 6. Efficiency of non-redundant and duplicate transmissions of two flows.

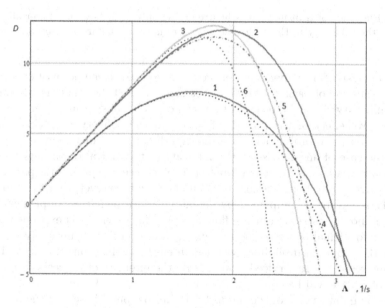

Fig. 7. Efficiency of multipath transfers when reserving only packets with a shorter allowable time for their delivery to the destination.

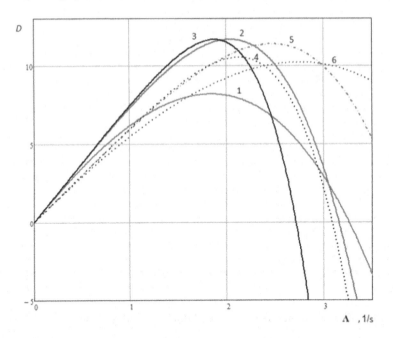

Fig. 8. Efficiency of transmissions when reserving only packets with a lower allowable time for their delivery to the addressee with the shares of the first stream 0.2 and 0.5.

An analysis of the dependence of the total profit (received per unit of time) on the transmission of packets of different streams showed that there is a boundary of the intensity Λ_g of the receipt of packets of the total stream, when exceeded, the negative effect of the growth of delays in the delivery of packets to the addressee on the overall efficiency of multipath transmissions increases.

In the case of an increase in the intensity of the arrival of packets over the established boundary Λ_g, it is proposed to block the reception of a part of the packets arriving for transmission. It should be noted, however, that this blocking of packets may be accompanied by denial-of-service penalties. Implementation of the proposed restriction of the flow accepted for servicing is expedient if the penalties for denial of service do not exceed the penalties for untimely service.

For the proposed discipline, with the intensity of the input flow $\Lambda \leq \Lambda_g$, all incoming requests are serviced, and $\Lambda > \Lambda_g$ the intensity of requests receiving a denial of service will be equal to $(\Lambda - \Lambda_g)$.

The boundary value of the intensity Λ_g can be found based on the fact that the profit calculated by formula (2) at $\Lambda = \Lambda_g$ becomes equal to

$$D_g = D(\Lambda_g) - (\Lambda - \Lambda_g)(\beta h_1 + (1 - \beta)h_2), \tag{3}$$

where h_1 and h_2 are penalties for denial of service requests for the transmission of packets of the first and second flows. The calculation $D(\Lambda_g)$ is carried out according to the formula (2) at $\Lambda = \Lambda_g$.

The profit received from the transmission of packets of a non-uniform flow in the proposed discipline if $\Lambda \leq \Lambda_g$ is calculated by the formula (2) and if $\Lambda > \Lambda_g$ then by the formula (3).

The results of assessing the profit from servicing heterogeneous traffic according to the proposed discipline, which provides for the possibility of refusing to service requests, depending on the intensity of their receipt, are shown in Fig. 9., where curves 1 and 2 correspond to the discipline without request service failures at $\beta = 0.5$ and $\beta = 0.2$. Curves 3 and 4 correspond to a discipline with the possibility of denials in servicing requests at $\beta = 0.5$ and $\beta = 0.2$. The calculation was carried out at $c_1 = 10$ c.u., $c_2 = 3$ c.u.; $s_1 = 50$, $s_2 = 4$ c.u.; $t_1 = 0.8$ s, $t_2 = 2$ s.

The presented graphs confirm the effectiveness of the proposed discipline for serving heterogeneous traffic with the possibility of refusing to service part of the requests at a high intensity of their receipt.

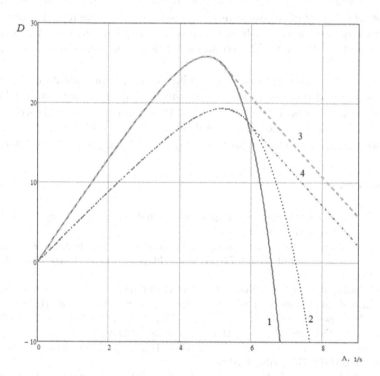

Fig. 9. The effectiveness of the discipline that provides for the possibility of denial of service requests.

The conducted research can be used in the development of distributed computer systems and networks, which are subject to high requirements for fault tolerance and reliability of operation under conditions of restrictions on permissible delays in packet transmission, including in real time [5, 6].

6   Conclusion

A model and a generalized efficiency criterion are proposed to justify the choice of organizing redundant transmissions of heterogeneous traffic, including taking into account the risks associated with untimely delivery of packets of different streams to the addressee. The proposed generalized criterion takes into account the impact of individual streams on the economic efficiency of packet transmission of different streams, including the profit from their timely delivery and penalties for their non-delivery or late delivery.

The analysis performed showed the effectiveness of increasing the probability of timely delivery of packets of traffic that is heterogeneous in terms of acceptable delays when reserving packet transmissions with the smallest acceptable delays along the way. The expediency of adaptive regulation of the multiplicity of transmissions depending on the intensity of heterogeneous traffic is shown.

The existence of the boundary of the intensity of receipt of requests of the total flow is established, above which, in order to reduce the negative impact of the increase in transmission delays, it is proposed to block the servicing of a part of the requests of the flow. The economic efficiency of such a service discipline is shown.

It is shown that an additional effect of increasing the probability of delivering packets to the addressee of all streams in the time allowed for them is achieved when the replicas of packets that have lost their relevance are destroyed at the path nodes. The loss of relevance of replicas can be caused by packet corruption or the accumulation of their delays in the queues of previously passed nodes.

References

1. Aysan, H.: Fault-tolerance strategies and probabilistic guarantees for real-time systems Mälardalen University, p. 190. Västerås, Sweden (2012)
2. Kim, S., Choi, Y.: Constraint-aware VM placement in heterogeneous computing clusters. Clust. Comput. **23**(1), 71–85 (2019). https://doi.org/10.1007/s10586-019-02966-6
3. Tatarnikova, T.M., Poymanova, E.D.: Differentiated capacity extension method for system of data storage with multilevel structure. Sci. Tech. J. Inf. Technol. Mech. Opt. **20**(1), 66–73 (2020). https://doi.org/10.17586/2226-1494-2020-20-1-66-73
4. Bennis, M., Debbah, M., Poor, H.V.: Ultrareliable and low-latency wireless communication: tail, risk and scale. Proc. IEEE **106**, 1834–1853 (2018). https://doi.org/10.1109/JPROC.2018.2867029
5. Ji, H., Park, S., Yeo, J., Kim, Y., Lee, J., Shim, B.: Ultra-reliable and low-latency communications in 5g downlink: physical layer aspects. IEEE Wirel. Commun. **25**, 124–130 (2018). https://doi.org/10.1109/MWC.2018.1700294
6. Sachs, J., Wikström, G., Dudda, T., Baldemair, R., Kittichokechai, K.: 5G radio network design for ultra-reliable low-latency communication. IEEE Netw. **32**, 24–31 (2018). https://doi.org/10.1109/MNET.2018.1700232
7. Pokhrel, S., Panda, M., Vu, H.: Analytical modeling of multipath TCP over last-mile wireless. IEEE/ACM Trans. Netw. **25**(3), 1876–1891 (2017). https://doi.org/10.1109/TNET.2017.266352

8. Peng, Q., Walid, A., Hwang, J., Low, S.H.: Multipath TCP: analysis, design, and implementation. IEEE/ACM Trans. Netw. **24**(1), 596–609 (2016)
9. Perkins, C.: RTP, p. 414. Addison-Wesley (2003). ISBN 9780672322495
10. Zurawski, R.: RTP, RTCP and RTSP Protocols. The Industrial Information Technology Handbook, pp. 28–70. CRC Press (2004). ISBN 9780849319
11. Krouk, E., Semenov, S.: Application of coding at the network transport level to decrease the message delay. In: Proceedings of 3rd International Symposium on Communication Systems Networks and Digital Signal Processing, Staffordshire University, UK, pp. 109–112 (2002)
12. Kabatiansky, G., Krouk, E., Semenov, S.: Error Correcting Coding and Security for Data Networks. Analysis of the Superchannel Concept, p. 288. Wiley (2005)
13. De Coninck, Q., Bonaventure, O.: Multipath QUIC: design and evaluation. In: Proceedings of the Conext 2017, Seoul, Korea (2010)
14. Roskind, J.: QUIC(Quick UDP Internet Connections): Multiplexed Stream Transport Over UDP. Technical report, Google (2013)
15. Bogatyrev, V.A., Bogatyrev, A.V., Bogatyrev, S.V.: Redundant servicing of a flow of heterogeneous requests critical to the total waiting time during the multipath passage of a sequence of info-communication nodes. In: Vishnevskiy, V.M., Samouylov, K.E., Kozyrev, D.V. (eds.) DCCN 2020. LNCS, vol. 12563, pp. 100–112. Springer, Cham (2020). https://doi.org/10.1007/978-3-030-66471-8_9
16. Bogatyrev, V.A., Bogatyrev, A.V., Bogatyrev, S.V.: The probability of timeliness of a fully connected exchange in a redundant real-time communication system. In: Wave Electronics and its Application in Information and Telecommunication Systems (WECONF 2020) (2020). https://doi.org/10.1109/WECONF48837.2020.9131517. https://ieeexplore.ieee.org/document/9131517
17. Bogatyrev, V.A., Bogatyrev, S.V., Bogatyrev, A.V.: Multipath redundant real-time transmission with the destruction of expired packets in intermediate nodes. In: CEUR Workshop Proceedings, vol. 3027, pp. 971–979 (2021)
18. Bogatyrev, V.A., Bogatyrev, S.V., Bogatyrev, A.V.: Inter-machine exchange of real time in distributed computer systems. In: CEUR Workshop Proceedings, vol. 2744 (2020)
19. Merindol, P., Pansiot, J., Cateloin, S.: Improving load balancing with multipath routing. In: Proceedings of the 17th International Conference on Computer Communications and Networks, ICCCN 2008, pp. 54–61. IEEE (2008)
20. Prasenjit, C., Tuhina, S., Indrajit, B.: Fault-tolerant multipath routing scheme for energy efficient wireless sensor networks. Int. J. Wirel. Mob. Netw. (IJWMN) **5**(2), 33–45 (2013)
21. Kleinrock, L.: Queueing Systems: Volume I. Theory, p. 417. Wiley Interscience, New York (1975). ISBN 978-0471491101
22. Kleinrock, L.: Queueing Systems: Volume II. Computer Applications, p. 576. Wiley Interscience, New York (1976). ISBN 978-0471491118

The Probability of Timely Redundant Service in a Two-Level Cluster of a Flow of Requests that is Heterogeneous in Functionality and Allowable Delays

V. A. Bogatyrev[1,3]($\boxtimes$) (iD), S. V. Bogatyrev[2,3] (iD), and A. V. Bogatyrev[2] (iD)

[1] Department Information Systems Security, Saint-Petersburg State University of Aerospace Instrumentation, Saint Petersburg, Russia
[2] Yadro Cloud Storage Development Center, Saint Petersburg, Russia
[3] ITMO University, Saint Petersburg, Russia
vabogatyrev@itmo.ru

Abstract. Currently, the development of methods for the synthesis and analysis of fault-tolerant computer clusters designed to service a flow of requests that is heterogeneous in functionality and acceptable delays in queues is relevant. The clusters under consideration assume a hierarchical structure with a multi-stage request service.

At the first stage of servicing a request, its primary processing is performed in a general-purpose server, and at the second stage, it is processed on functionally oriented servers. The options for building clusters with and without combining servers for various purposes into groups (cluster nodes) are considered.

The set of two-level cluster nodes includes a general-purpose server, as well as a complete or incomplete set of functional servers. Incomplete functional configuration of cluster nodes can be laid down during design or formed as a result of system degradation with the accumulation of failures in it. When designing a system, the incomplete functionality of two-level nodes can be caused by the heterogeneity of the cluster load for the execution of various functional requests, taking into account the restrictions on the allowable delays in their service. For the considered options for building clusters, models are proposed for estimating the probability of timely servicing of requests of various functionality and the allowable waiting time, taking into account the multiplicity of their reservation depending on the traffic intensity and the accumulation of delays in the queues of servers involved in various stages of service.

Keywords: replication · two-level cluster · functional configuration of nodes · degradation · accumulation of failures · multi-functionality · redundant maintenance

V. M. Vishnevskiy et al. (Eds.): DCCN 2022, CCIS 1748, pp. 122–134, 2023.
https://doi.org/10.1007/978-3-031-30648-8_10

1   Introduction

At present, there is great interest in the synthesis and analysis of high-performance [1–5] highly reliable fault-tolerant [6–8] computer systems with the consolidation of computing resources when they are combined in clusters [9,10]. For cluster systems that execute requests in real time, in addition to providing the required level of structural reliability, the fundamental organization of functioning with the execution of requests that are critical to waiting for the maximum allowable time. The solution of these problems relates to ensuring the functional reliability of the real-time system. The tasks of assessing and ensuring the functional reliability of distributed systems that are critical to the timeliness of servicing requests have not yet been fully researched.

This article discusses models and methods for ensuring the functional reliability of distributed systems when servicing a functionally heterogeneous flow of requests in multilevel clusters with functional heterogeneity of cluster nodes, formed during system design or formed as a result of system degradation due to accumulation of failures. At the same time, the possibility of increasing functional reliability based on redundant real-time service of a request flow that is heterogeneous in functionality is considered.

Increasing the functional reliability of info-communication systems and networks is largely based on the concepts of multi-path routing [11,12], broadcast service [16–18] and dynamic distribution of requests in clusters, taking into account the accumulation of failures in nodes [19].

For cluster architecture computer systems with a heterogeneous flow of requests in terms of allowable waiting time, the probability of timely servicing of the most delay-critical requests can be achieved as a result of traffic prioritization and redundant request execution, as well as their combination [12,13]. Such a combination is especially effective at a high probability of transmission errors, including under the influence of destabilizing factors [12,13].

If the flow is heterogeneous, the frequency of request replication during redundant service can be set depending on the maximum allowable waiting time in queues. With redundant service, the generated query replicas are sent to different nodes of the cluster, and it is possible to set different or identical priorities for them.

When designing fault-tolerant clusters designed to serve a stream of requests that is heterogeneous in functionality and acceptable delays in queues, it is currently required to justify the structural and functional solutions for building clusters that combine the resources of general-purpose servers and functional-oriented servers. Combining into clusters designed to service a functionally heterogeneous flow determines the multi-stage nature of request servicing. The synthesis of servers of the class under consideration should be based on a model-oriented approach. To support this approach, it is necessary to develop models, assess the probability of timely servicing of requests of various functionality and the allowable waiting time, taking into account the multiplicity of their reservation depending on the traffic intensity and the accumulation of delays in the queues of the servers involved in their servicing. The construction of models for assessing the timeliness of service in fault-tolerant clusters becomes more complicated taking into account the degradation of the system as a result of the accumulation of failures of its functional resources.

The aim of the work is to build models to justify the structural and functional solutions for organizing fault-tolerant two-level clusters that combine general-purpose servers and functional servers, taking into account the possibility of system degradation due to failures of functional resources.

2   The Structure of a Cluster of Two-Level Computer Nodes Equipped with Functionally Oriented Servers

Let's consider info-communication systems with functional heterogeneity of traffic, in which servicing requests may require the use of functionally oriented computer facilities (functionally oriented servers). Consolidation of computer resources in such systems is aimed at increasing reliability and reducing delays in servicing requests and is implemented when both general-purpose and functionally oriented servers are combined into a cluster.

The choice of the number of general-purpose and functionally oriented servers to be clustered depends on the system load for performing various functional tasks and the allowable delays in their maintenance.

Let us consider configurations of a cluster of m two-level nodes that combine general-purpose servers (basic servers) at the top level and servers designed to perform functional tasks at the bottom level.

The structure of such a two-level cluster node is shown in Fig. 1a.

The configuration of cluster nodes may include a complete or incomplete set of functional servers.

Incomplete functional configuration of cluster nodes can be laid down during design or formed during system degradation with the accumulation of failures in it.

When designing a system, the incomplete functionality of two-level nodes can be caused by the heterogeneity of the cluster load for the execution of various functional requests or restrictions on the allowable delays in their service. Incomplete functionality of cluster nodes may also be due to various economic, energy or design constraints.

Let us also consider the option of building a two-level cluster without combining servers for various purposes into two-level nodes. Such a variant of building a cluster, shown in Fig. 1b, general-purpose servers belong to the upper level, and functionally oriented servers to the lower level. The structure is characterized by fully connected servers of different levels. The number of connections between servers can be reduced when they are combined through switches.

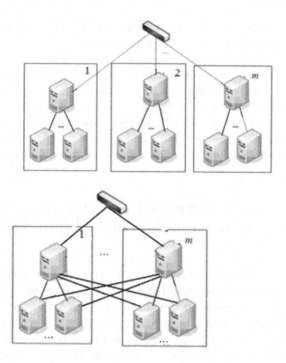

Fig. 1. The structure of a two-level cluster with clustered nodes of general-purpose servers and functionally oriented servers.

Requests arriving for service in the cluster are distributed to one of the cluster nodes, taking into account its functionality and workload. Redundant query servicing is allowed when query replicas are created and sent for servicing to different cluster nodes. The computational process in the structure of Fig. 1a is organized in such a way that, first, a request (or its replica) arriving at a node is processed at its upper level of the node, and then at its lower level. For the structure shown in Fig. 1b, the request is first executed in one of the general-purpose servers, and then distributed to any lower-level server, depending on its functionality, readiness (operability) and workload.

Let's represent the two-level nodes of the cluster as multi-functional elements. The distribution of functions $f_1, f_2, ..., f_n$ performed in the cluster over m of its two-level nodes will be displayed by a binary matrix $\|\phi_{ij}\|$ of dimensions $n \times m$ [19].

Wherein

$$\phi_{ij} = \begin{cases} 1 & \text{if if the request of the } i\text{-th functionality is executed by the } j\text{-th node,} \\ 0 & \text{if the request of the } i\text{-th functionality is not executed by the } j\text{-th node.} \end{cases}$$

With the full functionality of the cluster nodes, all elements of the matrix are equal to "1".

The failure of the server of the j-th node, which performs the function f_i, is accompanied by the transition of the element ϕ_{ij} from the "1" state to the "0" state.

Two-level multi-functional cluster nodes, due to the sequential execution of requests by servers of two levels, are characterized by the overlap of resources involved in the implementation of various requests. As a result of this overlap, the failure of the top-level server of the cluster node leads to its complete failure with the loss of all functionality. Thus, the failure of the top-level server leads to the transition of all elements of the corresponding column of the matrix $\|\phi_{ij}\|$ to zero states.

3   Timeliness of Servicing a Heterogeneous Flow in Clusters with Servers Combined into Two-Level Nodes

Consider cluster systems when cluster nodes are equipped with a general-purpose server with functional servers connected to it (Fig. 1 a). In the beginning, we will consider the option of a fully functional configuration of cluster nodes.

Let us analyze the case of a balanced distribution of functionally inhomogeneous total traffic arriving with intensity Λ between m two-level cluster nodes. With a full functional configuration of the nodes, the intensity coming to each (j-th) general-purpose server will be

$$\Lambda_j = \sum_{i=1}^{n}(b_i\Lambda/m),$$

where b_i is the share in the total flow of requests for the execution of functions f_i.

$$\sum_{i=1}^{n} b_i = 1.$$

Average request service time in general purpose clusters will be

$$v_0 = \sum_{i=1}^{n} b_i v_{0i}.$$

Delays in the queues of top-level servers when they are represented as a queuing system of the $M/M/1$ type [21–23] can be calculated as:

$$w = (\sum_{i=1}^{n} b_i\Lambda v_{0i}^2/m)/(1 - \sum_{i=1}^{n} b_i\Lambda v_{0i}/m). \tag{1}$$

In a general-purpose server, the probability of timely execution of i-functionality requests f_i with their maximum allowable delay in queues t_i is estimated as

$$r_i = (1 - \frac{\Lambda v_0}{m}e^{(\frac{\Lambda}{m}-\frac{1}{v_0})t_i}). \tag{2}$$

Probability of timely two-stage servicing of requests of the i-th functionality

$$R_i = r_i(1 - \frac{\Lambda b_i v_i}{m} e^{(\frac{\Lambda b_i}{m} - \frac{1}{v_i})(t_i - w - v_{0i})}),$$

where v_i is the average execution time of a request in a server dedicated to executing functional requests f_i. With full functionality of cluster nodes, the total number of servers capable of performing the function of each type is m.

With incomplete functionality of the configuration of two-level cluster nodes, their number capable of fulfilling requests for the i-th functionality is equal to

$$m_i = \sum_{j=1}^{m} \phi_{ij}.$$

As generalized indicators that allow us to resolve the contradiction in ensuring the timeliness of servicing requests of different functionality and criticality to the maximum allowable waiting delays, we use probabilistic indicators of the form:

$$R = \prod_{i=1}^{n} R_i,$$

$$A = \sum_{i=1}^{n} b_i R_i.$$

4   Redundant Servicing of a Functionally Heterogeneous Flow

Ensuring the probability of serving functionally heterogeneous requests in the required time under certain conditions can be achieved with replication and redundant servicing of functional requests [21], the most critical to the total delays of the queues serving their servers

For a complete functional configuration of two-level nodes in the case of replication of requests for the execution of the i-th function with multiplicity k_i, the average waiting time for requests in general-purpose servers is calculated as

$$w = (\sum_{i=1}^{n} b_i k_i \Lambda v_{0i}^2 / m) / (1 - \sum_{i=1}^{n} b_i k_i \Lambda v_{0i} / m). \tag{3}$$

In a general-purpose server, the probability of executing with an acceptable delay less than t_i some replica of the request of the i-th functionality is equal to

$$r_i = (1 - \frac{\Lambda \sum_{i=1}^{n} k_i b_i v_{0i}}{m} e^{(\frac{\Lambda \sum_{i=1}^{n} k_i b_i}{m} - \frac{1}{\sum_{i=1}^{n} k_i b_i v_{0i}})t_i}). \tag{4}$$

The probability of a timely two-stage service for at least one of the k_i replicas of the request of the i-th functionality will be equal to:

$$R_i = \{1 - [1 - r_i(1 - \frac{k_i \Lambda b_i v_i}{m_i} e^{(\frac{k_i \Lambda b_i}{m_i} - \frac{1}{v_i})(t_i - w - v_{0i})]^{k_i}\}.$$

Consider the case of incomplete functional configuration of cluster nodes represented by matrix $\|\phi_{ij}\|_{n \times m}$.

Consider the distribution of requests, in which requests for i-functionality are evenly distributed among the cluster nodes capable of servicing them.

In this case, the intensity of incoming requests to the j-th node of the cluster for executing requests for i-functionality will be equal to

$$\Lambda_{ij} = \phi_{ij} b_i \Lambda / m_i,$$

$$m_i = \sum_{j=1}^{m} \phi_{ij}.$$

The total intensity of requests of various functionality sent to the j-th node of the cluster is defined as

$$\Lambda_j = (\sum_{i=1}^{n} \phi_{ij} b_i \Lambda / m_i). \tag{5}$$

Average delay in the queue of the general purpose server of the j-th node

$$w_j = (\sum_{i=1}^{n} \phi_{ij} b_i \Lambda v_{0i}^2 / m_i)/(1 - \rho_j),$$

where the load of the general purpose server of the j-th node of the cluster is defined as

$$\rho_j = \sum_{i=1}^{n} \phi_{ij} b_i \Lambda v_{0i} / m_i).$$

Probability of timely execution of requests for the i-th functionality by a general-purpose server in the j-th cluster node

$$r_{ij} = 1 - \rho_j e^{(\Lambda_j - \frac{1}{v_j})t_i},$$

where is the average execution time for requests of various functionality in the general purpose server of the j-th cluster node

$$v_j = \sum_{i=1}^{n} (\phi_{ij} b_i v_{0i} / m_i).$$

With a uniform distribution of requests over all nodes of functional orientation capable of fulfilling it, the average residence time of requests of the i-th functionality in one of the m_i servers allocated to perform the corresponding function will be

$$T_i = v_i/(1 - \Lambda_i v_i),$$

$$\Lambda_i = \Lambda b_i/m_i.$$

The probability of timely execution of requests for the i-th functionality in the j-th cluster node, provided that it is executed in a timely manner in the general purpose server of this node

$$R_{ij} = (1 - \Lambda_i v_i e^{(\Lambda_i - \frac{1}{v_i})(t_i - v_{0i} - w_j)})$$

For the case under consideration, when the request of the i-th functionality with a probability of $1/m_i$ is distributed to any cluster node capable of executing it, the average probability of timely execution of the request of the i-th functionality at two service stages will be

$$R_i = \sum_{j=1}^{m} \frac{1}{m_i} r_{ij} (1 - \Lambda_i v_i e^{(\Lambda_i - \frac{1}{v_i})(t_i - v_{0i} - w_j)}).$$

5   Timely Maintenance in Two-Tier Clusters Without Merging Servers into Cluster Nodes

Let us consider a two-level computer system of cluster architecture, which does not imply the integration of servers into multi-functional cluster nodes. In the structure of the analyzed system (corresponds to Fig. 1b), two levels are distinguished. At the top level of the cluster there are servers used to serve all types of requests, and at the bottom there are servers specialized for performing certain functions. Requests are first made to the top servers and then the bottom servers. When balancing the load of top-level servers without replicating requests, the average waiting time in them is determined by formula (1), and the probability of servicing requests for the execution of functions f_i with admissible delays for them is determined by formula (2).

The probability of timeliness of the two-stage service of the request for the execution of the function f_i is calculated by the formula

$$R_i = r_i (1 - \frac{\Lambda b_i v_i}{m_i} e^{(\frac{\Lambda b_i}{m_i} - \frac{1}{v_i})(t_i - w - v_{0i})}).$$

When replicating requests, the average waiting time in top-level servers is calculated by formula (3), and the probability r_i of timely service of a certain replica for the execution of function f_i is calculated by formula (4).

The probability of timely service by top-level servers of at least one of the replicas of the request for the execution of functions f_i is defined as

$$p_i = 1 - (1 - r_i)^{k_i}.$$

The average latency for the first completed replica of the i-th functionality by one of the k_i top-level servers involved in the redundant maintenance of this replica is defined as

$$g_i = \int_0^\infty (1 - \frac{\Lambda \sum_{i=1}^n k_i b_i v_{0i}}{m} e^{(\frac{\Lambda \sum_{i=1}^n k_i b_i}{m} - \frac{1}{\sum_{i=1}^n k_i b_i v_{0i}})t})^{k_i} dt.$$

Consider a two-stage service discipline in which the top-level server, which is the first to serve the replica of the i-th functionality, re-implements the request replication for its further service in the lower-level servers. Replicas of the i-th functionality that are not first served by the top servers are ignored.

For the discipline of redundant two-stage service under consideration, we estimate the probability of timely execution of requests for the i-th functionality as

$$R_i = p_i(1 - [\frac{\Lambda b_i v_i}{m_i} e^{(\frac{\Lambda b_i}{m_i} - \frac{1}{v_i})(t_i - g_i - v_{0i})}]^{k_i}),$$

The considered organization of the cluster potentially makes it possible to increase the probability of timely service of traffic. But at the same time, it is associated with the complication of the interaction of servers to identify replicas served first and destroy irrelevant replicas.

6   An Example of Estimating the Probability of Timely Service of Functional Requests

In a cluster of two-level nodes, we will consider the maintenance of a functionally heterogeneous flow with the allocation of two types of functional requests that differ in the permissible total waiting delays in the queues of the servers executing them. We will assume that the cluster nodes are fully functional. For requests of the first and second functionality, the allowed total waiting time at the two stages of maintenance is t_1 and t_2. The reserved service is carried out with a multiplicity of k only for the most critical waiting requests.

The average delay in the queue of the top-level server with an average time v_0 of execution of all types of requests is calculated as

$$T_1 = v_0/(1 - \Lambda(1 + b(k - 1))v_0/m).$$

The probabilities of timely execution of requests of two types in the top-level server are defined as

$$r_1 = (1 - \frac{\Lambda(1 + b(k - 1))v_0}{m} e^{(\frac{\Lambda(1 + b(k-1))}{m} - \frac{1}{v_0})t_1}),$$

$$r_2 = (1 - \frac{\Lambda(1 + b(k - 1))v_0}{m} e^{(\frac{\Lambda(1 + b(k-1))}{m} - \frac{1}{v_0})t_2}).$$

We estimate the probability of timely servicing of a functionally heterogeneous flow by the generalized indicators R and A and calculate as:

$$R = (1 - \{1 - r_1(1 - [[k\Lambda bv_1/m]e^{(\frac{k\Lambda b}{m} - \frac{1}{v_1})(t_1 - T_1)}])\}^k)$$
$$\times \{r_2(1 - [(\Lambda(1-b)v_2)/m]e^{(\frac{\Lambda(1-b)}{m} - \frac{1}{v_2})(t_2 - T_1)})\}.$$

$$A = b\{1 - \{1 - r_1(1 - [[k\Lambda bv_1/m]e^{(\frac{k\Lambda b}{m} - \frac{1}{v_1})(t_1 - T_1)}])\}^k\}$$
$$+(1 - b)\{r_2(1 - [(\Lambda(1-b)v_2)/m]e^{(\frac{\Lambda(1-b)}{m} - \frac{1}{v_2})(t_2 - T_1)})\}.$$

The calculation will be carried out at $b = 0.5$ and $t_2 = 1$ s, $m = 5$ nodes, $v_0 = 0.01$ s and $v = 0.1$ s. The dependence of the timeliness indicators R and A in the case of duplication of requests that are critical to delays is shown in Fig. 2a and b.

Curves 1–3 correspond to $t_1 = 0.1$; 0.15; 0.2 s when duplicating more critical requests, and curves 4–6 without redundancy.

The dependencies of the probabilities R and A on the multiplicity of reservations for waiting-critical requests are shown in Fig. 2a and b.

Curves 1–3 correspond to the intensities of the input flow $\Lambda = 10, 15, 20$ $1/s$ at fractions $b = 0.5$, and curves 4–6 at $b = 0.7$.

The performed calculations show the possibility of increasing the probability of timely servicing of a flow that is heterogeneous in functionality as a result of reserving requests that are critical to delays. The dependence of the possibilities of increasing the probability of servicing the total flow on its intensity and the shares of flows of different functionality has been established. The conducted studies confirm the existence of an optimal multiplicity of redundant servicing of waiting-critical functional requests (Figs. 3 and 4).

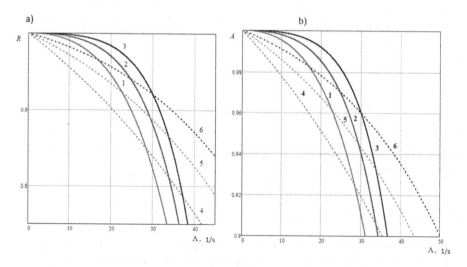

Fig. 2. The probability of timely servicing of a functionally heterogeneous flow R (a) and A (b) depending on its intensity.

a)   b)

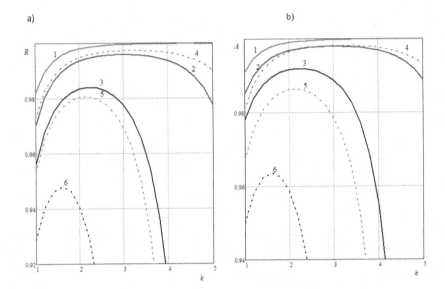

Fig. 3. Dependences of the probabilities R (a) and A (b) of timely servicing of a flow that is heterogeneous in functionality on the multiplicity of reservations for the most critical functional requests to wait.

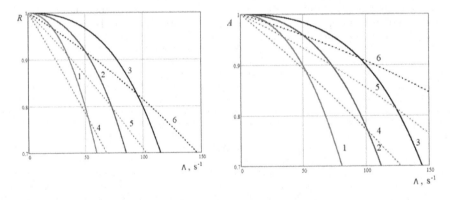

Fig. 4. Probabilities of timely servicing of R and A flow, heterogeneous in functionality, in case of duplication of delay-critical requests.

7 Conclusion

Options for building fault-tolerant clusters designed to serve the flow of requests, heterogeneous in functionality and acceptable delays in queues, are considered.

For the considered variants of cluster construction, models are proposed for estimating the probability of timely two-stage servicing of requests of various functionality and acceptable waiting time, taking into account the multiplicity of their redundancy depending on traffic intensity and the accumulation of delays in the queues of servers involved in their maintenance. The efficiency of the reserved service of the most critical to waiting requests is shown.

References

1. Aab, A.V., Galushin, P.V., Popova, A.V., Terskov, V.A.: Mathematical model of reliability of information processing computer appliances for real-time control systems. Sib. J. Sci. Technol. **21**(3), 296–302 (2020). https://doi.org/10.31772/2587-6066-2020-21-3-296-302
2. Sorin, D.: Fault Tolerant Computer Architecture, p. 103. Morgan & Claypool, San Rafael (2009)
3. Pavsky, V.A., Pavsky, K.V.: Mathematical model with three-parameters for calculating reliability indices of scalable computer systems. In: 2020 International Multi-Conference on Industrial Engineering and Modern Technologies, FarEastCon 2020, p. 9271332 (2020)
4. Malik, V., Barde, C.R.: Live migration of virtual machines in cloud environment using prediction of CPU usage. Int. J. Comput. Appl. **117**(23), 1–5 (2015). https://doi.org/10.5120/20691-3604
5. Houankpo, H.G.K., Kozyrev, D.V., Nibasumba, E., Mouale, M.N.B.: Mathematical model for reliability analysis of a heterogeneous redundant data transmission system. In: 2020 12th International Congress on Ultra Modern Telecommunications and Control Systems and Workshops (ICUMT), Brno, Czech Republic, pp. 189–194 (2020). https://doi.org/10.1109/ICUMT51630.2020.9222431
6. Tatarnikova, T.M., Poymanova, E.D.: Differentiated capacity extension method for system of data storage with multilevel structure. Sci. Tech. J. Inf. Technol. Mech. Opt. **20**(1), 66–73 (2020). https://doi.org/10.17586/2226-1494-2020-20-1-66-73
7. Astakhova, T.N., Verzun, N.A., Kasatkin, V.V., Kolbanev, M.O., Shamin, A.A.: Sensor network connectivity models. Informatsionno-upravliaiushchie sistemy [Information and Control Systems] (5), 38–50 (2019). https://doi.org/10.31799/16848853-2019-5-38-50
8. Zakoldaev, D.A., Korobeynikov, A.G., Shukalov, A.V., Zharinov, I.O.: Workstations Industry 4.0 for instrument manufacturing. In: IOP Conference Series: Materials Science and Engineering, vol. 665, p. 012015. IOP Publishing (2019). https://doi.org/10.1088/1757-899X/665/1/012015
9. Kutuzov, O.I., Tatarnikova, T.M.: Model of a self-similar traffic generator and evaluation of buffer storage for classical and fractal queuing system. In: Moscow Workshop on Electronic and Networking Technologies, MWENT 2018 - Proceedings, no. 1, pp. 1–3 (2018)
10. Jiang, Z., Wang, G., He, Y., Huang, Z., Xue, R.: Reliability prediction for computer numerical control machine servo systems based on an ipso-based rbf neural network Shock and Vibration, vol. 2022, p. 2684942 (2022)
11. Gómez, C.E., Chavarriaga, J., Castro, H.E.: Reliability analysis in desktop cloud systems. Communications in Computer and Information Science. 2020. vol. 1277 CCIS. pp. 165–180 (2020)
12. Mukhin, V., Loutskii, H., Kornaga, Y., Steshyn, V., Barabash, O.: Models for analysis and prognostication of the indicators of the distributed computer systems' characteristics. Int. Rev. Comput. Softw. **10**(12), 1216–1224 (2015)
13. Kim, S., Choi, Y.: Constraint-aware VM placement in heterogeneous computing clusters. Clust. Comput. **23**(1), 71–85 (2019). https://doi.org/10.1007/s10586-019-02966-6

14. Newman, M., de Castro, L.A., Brown, K.R.: Generating fault-tolerant cluster states from crystal structures. Quantum **4** (2020)
15. Tourouta, E., Gorodnichev, M., Polyantseva, K., Moseva, M.: Providing fault tolerance of cluster computing systems based on fault-tolerant dynamic computation planning. In: Zaramenskikh, E., Fedorova, A. (eds.) Digitalization of Society, Economics and Management. LNISO, vol. 53, pp. 143–150. Springer, Cham (2022). https://doi.org/10.1007/978-3-030-94252-6_10
16. Bukhsh, M., Abdullah, S., Arshad, H., Rahman, A., Asghar, M.N., Alabdulatif, A.: An energy-aware, highly available, and fault-tolerant method for reliable IoT systems. IEEE Access **9**, 145363–14538 (2021)
17. Shcherba, E.V., Litvinov, G.A., Shcherba, M.V.: Securing the multipath extension of the OLSR routing protocol. In: 13th International IEEE Scientific and Technical Conference Dynamics of Systems, Mechanisms and Machines, Dynamics 2019 - Proceedings. 2019, p. 8944743 (2019)
18. Lee, M.H., Dudin, A.N., Klimenok, V.I.: The SM/V/N queueing system with broadcasting service. Math. Probl. Eng. **2006**, 18 Article ID 98171 (2006)
19. Bogatyrev, S.V., Bogatyrev, A.V., Bogatyrev, V.A.: Priority maintenance with replication of wait-critical requests. In: Wave Electronics and its Application in Information and Telecommunication Systems (WECONF 2021), p. 9470640 (2021)
20. Bogatyrev, V.A., Bogatyrev, A.V., Bogatyrev, S.V.: Redundant servicing of a flow of heterogeneous requests critical to the total waiting time during the multipath passage of a sequence of info-communication nodes. In: Vishnevskiy, V.M., Samouylov, K.E., Kozyrev, D.V. (eds.) DCCN 2020. LNCS, vol. 12563, pp. 100–112. Springer, Cham (2020). https://doi.org/10.1007/978-3-030-66471-8_9
21. Bogatyrev, V.A., Bogatyrev, A.V., Bogatyrev, S.V.: Reliability and probability of timely servicing in a cluster of heterogeneous flow of query functionality. In: Wave Electronics and its Application in Information and Telecommunication Systems (WECONF 2020) - 2020, p. 9131165 (2020). https://doi.org/10.1109/WECONF48837.2020.9131165
22. Kleinrock, L.: Queueing Systems: Volume I. Theory, p. 417. Wiley Interscience, New York (1975). ISBN 978-0471491101
23. Vishnevsky, V.M.: Theoretical foundations of computer network design (2003). Technosphere ISBN 5-94836-011-3

Characterizing the Effects of Base Station Variable Capacity on 5G Network Slicing Performance

Nikita Polyakov[ID] and Anna Platonova[✉]

Peoples' Friendship University of Russia (RUDN University),
6 Miklukho-Maklaya St, Moscow 117198, Russian Federation
platonova_aa@pfur.ru

Abstract. Network slicing is a technique based on virtualization of base station resources that allows to divide the network into logical parts (slices). In 5G, traffic with very heterogeneous transmission requirements is expected, which will need to be transmitted efficiently over a common infrastructure. Data transmission using the millimeter range is planned, which is quite sensitive to blockages. Due to the fact that the channel can be unstable, it is important to investigate resource allocation policies taking into account the variable capacity of the base station. In this work, we focus on to such an aspect of network modeling as the abstraction of the base station resource, which depends on the communication channel state. Quite often, it is presented by researchers in the form of a constant value of the data transfer rate in some units. However, in reality, the behavior of the channel is complex and this value is changing. We explore our resource allocation scheme among slices through simulation. We characterize the effect of the variable capacity of the base station on the stability of the network slicing technology and present numerical results in the second half of the article. We conclude that the variable capacity requires a careful choice of the re-slicing frequency. This can be achieved by invoking re-slicing in timely manner, upon arrival of sessions, or by choosing an appropriate interval.

Keywords: 5G · network slicing · variable capacity · radio channel · simulation

1 Introduction

5G networks are very complex both in terms of supported communication technologies, traffic and workload diversity. To efficiently manage infrastructure and resources, the network slicing technique has been proposed, which also allows third-party companies to act as network operators using the provider's infrastructure.

This publication has been supported by the RUDN University Scientific Projects Grant System, project No. 021929-2-000.

Developers of communication network models and algorithms have to make assumptions about the behavior of the communication channel. This is an important point, especially since the access radio technologies envisioned in 5G, such as New Radio and mmWave, are susceptible to blocking due to the use of millimeter-wave frequencies.

Quite often the communication channel is abstracted by some data speed or amount of resources. For example, in works on the queuing theory, resource abstraction is approached differently. The authors of [8] model a 5G multi-service network and abstract the communication channel with a constant, expressing the capacity of the base station in resource units. This allows them to use the advanced method of queuing theory. The paper considers the possibility of dynamic traffic prioritization for mission-critical sessions using the proposed strategies. In [16] the authors also express capacity in resource units, focusing on the study of the join transmission of heterogeneous traffic, which is modeled using the Markovian arrival process (MAP).

However, the communication channel is far from stable. Millimeter waves are sensitive to both weather [5,6] and blockages by human [12] and static structures [15,18] such as buildings. Therefore, for example, in the paper [3] the admission control scheme model for LTE Wireless Network is being studied, which is presented as unreliable devices operating in a random environment. Thus, the authors form the state space of the queuing system, taking into account the communication channel. And in [14] the authors approach the channel abstraction in an interesting way: they take into account its volatile behavior with the help of random coefficients that affect the session service rate.

Therefore, accounting for the state of the communication channel, especially in the case of mmWave, is an important aspect when designing network resource management policies. This work is aimed at evaluating the influence of the communication channel time variability on the efficiency of the resource allocation scheme among network slices that we proposed and studied in [10,17]. The results are obtained using a 5G network simulator developed by the authors [9].

The paper is organized as follows. Section 2 introduces the basic model assumptions. Section 3 describes our simulation setup. Section 4 offers numerical results. Section 5 concludes the paper.

2   System Model

We assume a base station (BS) equipped with several mmWave radio access technologies (RAT) and supporting resource virtualization and network slicing. Therefore, the total capacity of the BS, denoted by C, can be expressed in bits per second. Previously we considered the capacity C constant. In this work, we take into account that the communication channel is not stable and the BS capacity, $C(t)$, changes in time.

2.1   Radio Channel Assumptions

Propagation Losses. We use the Urban-Micro (UMi) Street-Canyon model for representing propagation losses and express the path loss as

$$\text{PL}_{[dB]}(d) = 10\zeta \log_{10} d + \beta + 20 \log_{10} f_c, \tag{1}$$

where $d_{[m]}$ is the 3D-distance from the BS to the UE, $f_{c[GHz]}$ is the carrier frequency, α is a coefficient being 2.1 under the line-of-sight (LoS) and 3.19 under non-line-of-sight (nLoS) conditions, β is 32.4 dB when the UE is not blocked by human and 52.4 dB otherwise.

The received signal-to-interference-plus-noise ratio (SINR) for $X \in \{\text{LoS}, \text{nLoS}\}$ using (1) can be obtained as

$$\text{SINR}_X(d) = \frac{P_{\text{BS}}G_{\text{BS}}G_{\text{UE}}}{\text{PL}(d)N_0 R_{\text{PRB}}\chi_\sigma A_X}, \tag{2}$$

where P_{BS} is the emitted power, G_{BS} and G_{UE} are respectively the BS and UE antenna gains, N_0 is the thermal noise, χ_σ is the shadow fading which is normally distributed in dB with zero mean and standard deviation σ, R_{PRB} is the size of the physical resource block (PRB), and A_X represents aggregated losses due to interference, cable losses, etc.

Signal Blockages. We take into account two types of blockages:

– Human body blockage probability [4]

$$p_B(y) = 1 - e^{-2\lambda_B r_B \left[\sqrt{y^2 + (h_{M,B} - h_U)^2}\frac{h_B - h_U}{h_{M,B} - h_U} + r_B\right]}, \tag{3}$$

where $\lambda = 0.1$ bl./m^2 is the intensity of blockers, $r_B = 0.3$ m and h_B are the blockers' radius and height, h_U is the height of UE, $h_{M,B}$ is the height of the mmWave BS.
– Blockage by buildings [1], with the LoS probability for a user at a 2D distance r from the BS being

$$p_{\text{LoS}}(r) = \begin{cases} 1, & r \leq 18, \\ 18r^{-1} + e^{-\frac{r}{36}}(1 - 18r^{-1}), & r > 18. \end{cases} \tag{4}$$

The LoS/nLoS state of the user determines the values of ζ_X and A_X in (2), whereas human blockage determines the value of β.

2.2   Assumptions About Slices

As in our previous works, we assume that each slice deployed at the BS serves only one type of traffic, which allows us to more accurately determine its QoS requirements. Thus, for each slice $s \in \mathcal{S}$, we define the following parameters:

– $R_s^{\text{min}} > 0$ and $R_s^{\text{max}} \geq R_s^{\text{min}}$ are the data rate bounds,

- $N_s^{\text{cont}} > 0$ is the contractual maximum number of users for which QoS will be guaranteed (in each slice, we treat each session as a unique user),
- $0 < R_s^{\text{deg}} \leq C$ is the data rate threshold under which the slice is considered degraded (degradation of a slice). Serving users at a rate below this threshold results in a QoS guarantee violation.

It is assumed that the parameters $N_s^{\text{cont}}, R_s^{\text{min}}, R_s^{\text{max}}$ are specified in the SLA before the initialization of each slice.

Let the number of users in slice s, N_s, and its capacity C_s form the state of slice s. The average service rate of one user in each slice state is $R_s = C_s/N_s$. Since now the capacity of the base station $C(t)$ is time-varying, we proportionally change the capacity of each slice s every modelling second $t = 1, 2, \ldots, T$ according to the formula

$$C_s(t) = \frac{C_s(t-1)}{C(t-1)} C(t), s \in \mathcal{S}, \tag{5}$$

and then omit t for brevity.

2.3   Resource Allocation Among Slices

The assumed slicing scheme relies on the solution of the following optimization problem [9,17]:

$$\text{maximize } U(R_1, \ldots, R_S) = \sum_{i=1}^{S} W_i(N_i) N_i \ln(R_i), \tag{6}$$

$$\text{subject to } \sum_{i=1}^{S} N_i R_i = C, \tag{7}$$

$$\text{over } R_1, \ldots, R_S : R_i^{\text{min}} \leq R_i \leq R_i^{\text{max}}, \tag{8}$$

where $W_i(N_i)$ is given by

$$W_i(N_i) = \begin{cases} 1, & N_i \leq N_i^{\text{cont}} \\ N_i^{\text{cont}}/N_i, & N_i > N_i^{\text{cont}} \end{cases} \tag{9}$$

Now,

- if $\sum_{i=1}^{S} N_i R_i^{\text{max}} \leq C$ then we set $C_s = N_s R_s^{\text{max}}$ for all $s \in \mathcal{S}$;
- if $\sum_{i=1}^{S} N_i R_i^{\text{max}} > C$ and $\sum_{i=1}^{S} N_i R_i^{\text{min}} \leq C$ then we set $C_s = N_s R_s$, where R_s, $s \in \mathcal{S}$, are the solution to the above optimization problem;
- if $\sum_{i=1}^{S} N_i R_i^{\text{min}} > C$ then
  - if $\sum_{i=1}^{S} \min\{N_i, N_i^{\text{cont}}\} R_i^{\text{min}} \geq C$, we set

$$C_s = \frac{N_s^{\text{min}} R_s^{\text{min}}}{\sum_{i=1}^{S} \min\{N_i, N_i^{\text{cont}}\} R_i^{\text{min}}} C \tag{10}$$

- else

$$C_s = N_s^{\min} R_s^{\min} + \frac{(N_s - N_s^{\min}) R_s^{\min} (C - \sum_{i=1}^{S} \min\{N_i, N_i^{\text{cont}}\} R_i^{\min})}{\sum_{i=1}^{S} (N_i - \min\{N_i, N_i^{\text{cont}}\}) R_i^{\min}}.$$
(11)

3 Simulation Setup

Before modeling the process of resource allocation among slices, we compute the BS capacity $C(t)$, $t = 1, 2, \ldots, T$, for each second of simulation time. Using the random direction mobility (RDM) model [7], we simulate mobility for a total average number of users in slices U_{avg} per cell and compute the aggregated data rate under the previously stated assumptions [1]. For calculating $C(t)$ we use simulation model (see [11]) with implementation of users RDM and parameters of SINR from [13]. Parameters for capacity calculation are shown in the Table 1.

Table 1. Capacity simulation parameters

Symbol	Value	Description
B	400 MHz	Bandwidth
f_c	28 GHz	Operating frequency
F_N	7 dB	Noise figure
G_{BS}	11.57 dBm	BS antenna gain
G_{UE}	5.57 dBm	UE antenna gain
h_{BS}	10 m	BS height
h_{UE}	1.7 m	UE height
L_C	2 dB	Cable losses
M_I	3 dB	Interference margin
N_0	–174 dBm/Hz	Noise power
P_{BS}	33 dBm	Transmitting BS power
p_{out}	0.01	Outage probability
$\text{SINR}_{\text{thre}}$	–9.478 dB	SINR threshold
T	10000	Simulation time
v_i	1.5 m/s	User speed
Δt	1 s	Time slot length
$\sigma_{\text{SF,nLoS}}$	8.2 dB	nLoS shadow fading STD
$\sigma_{\text{SF,LoS}}$	4 dB	LoS shadow fading STD
τ_i	30 s	Mean run time in the RDM model

In the simulator, the slice capacities $C_s, s \in \mathcal{S}$, are computed and allocated to slices by a slicing manager (Slicer). During initialization, the capacity of the base station is distributed proportionally to $\gamma_s = N_s^{\text{cont}} R_s^{\min}/C$. Subsequent re-slicing is called on triggers:

- Timer is once every specified number of seconds (set 5 s);
- Arrivals is a user arrival in any slice;
- Degradation is when any of the slices is degraded ($R_s < R_s^{\text{deg}}$).

When an event occurs, the slicer calls the slicing scheme, to which the parameters C and $N_s, R_s^{\min}, R_s^{\max}, \gamma_s, s \in \mathcal{S}$, are passed. After calculating new capacities, the Slicer sets them to the slices.

Based on our previous research, we consider a scenario of slices' traffic characteristics presented in Table 2. Voice traffic represents classical cellular telephony services. We assume that the best effort slice includes such services as social networking, web browsing, messaging, file sharing, cloud services, VPN, and security. For these services, there are no strict requirements or rate guarantees, and the rates themselves are rather moderate. Video streaming represents all types of movie streaming applications, like Netflix, YouTube, etc. Online gaming, virtual reality and augmented reality are the most demanding slices with strict speed guarantees and latency requirements [2]. Corporate customers use a mix of traffic.

We assume that user sessions in slice $s \in \mathcal{S}$ arrive according to a Poisson process of rate λ_s, while session durations are exponentially distributed with parameter μ_s. The set of slices $\mathcal{S}$ is defined once, and based on it and the SLA parameters, we calculate the guaranteed share of capacity γ_s for each slice s using the formula

$$\gamma_s = \frac{p_{0.9}^s R_s^{\min}}{\sum_{i=1}^S p_{0.9}^i R_i^{\min}}, \tag{12}$$

where $p_{0.9}^s$ is the 90-percentile of the Poisson distribution with parameter λ_s/μ_s. For Voice slice γ_1 is predefined and fixed at 2%, which must be taken into account when calculating this parameter for the remaining slices.

3.1   Scenario

Table 2. Numerical evaluation parameters

Slice type	s	$R_{s[Mbps]}^{\min}$	$R_{s[Mbps]}^{\max}$	γ_s	$1/\mu_s$	$1/\lambda_s$
Voice	1	0.1	1	2% fixed	210	–
Best Effort	2	1	C	13.87%	10	0.05
Movie Streaming	3	3	25	22.14%	3600	35
Online Gaming	4	10	50	16.54%	1800	45
VR/AR	5	20	100	29.21%	1800	100
Corporate	6	5	50	16.24%	3600	85

As it can be seen from Table 2, we consider a scenario, which describes a situation with increased load in the Online Gaming slice. For example, when a long-expected online game or add-on is released, the traffic in the corresponding slice may temporarily more than double, because players are massively connecting to game servers. The scenario does not specify workload parameters for Voice slice, because we assume that it is hard-isolated from the rest of the slices.

3.2 Performance Criteria

To evaluate the efficiency of resource allocation, we use the following metrics:

– Slicing frequency;
– Slice degradation probability,

$$P_s^{\text{deg}} = P\{R_s < R_s^d\} = \lim_{T \to \infty} \frac{1}{T} \sum_{i=1}^{D_s(T)} (d_{s,i} - d_{s,i-1})\mathcal{H}\{R_{s,i} < R_s^d\}, \qquad (13)$$

where $s \in \mathcal{S}$, T is model time, $D_s(T)$ is counter of slice s degradation threshold R_s^d crossing (in any direction), $R_{s,i}$ is rate in i-th moment, $d_{s,i}$ is time of the i-th degradation threshold R_s^d crossing, and $\mathcal{H}$ is Heaviside step function;
– Averaged degradation probability,

$$P^{\text{deg}} = \frac{\sum_{i=1}^{S} P_i^{\text{deg}}}{S}, \qquad (14)$$

– Capacity utilization,

$$\text{UTIL} = \frac{T}{\sum_{t \in T} C(t)} \sum_{s \in \mathcal{S}} \lim_{T \to \infty} \frac{1}{T} \sum_{i=1}^{Y_s(T)} (y_{s,i} - y_{s,i-1})N_{s,i}R_{s,i}, \quad s \in \mathcal{S}, \qquad (15)$$

where $Y_s(T)$ is counter of slice s service rate R_s and number of jobs N_s changes, $y_{s,i}$ is moment i of changing R_s or N_s.
– User satisfaction,

$$R_s^{\text{sat}} = \lim_{T \to \infty} \frac{1}{T} \sum_{i=1}^{K_s(T)} \frac{(r_{s,i} - r_{s,i-1})R_{s,i}}{R_s^{\min}(1 - P_s^{empty})}, \qquad (16)$$

where $s \in \mathcal{S}$, T is model time, $K_s(T)$ is counter of slice s rate R_s change, $r_{s,i}$ is time of the i-th rate change, P_s^{empty} is the probability of an empty slice.

4 Numerical Results

As part of a numerical experiment, we intend to determine the degree of the base station variable capacity influence on the processes of re-slicing, and in particular on the frequency of calls and the efficiency of the scheme. We compare the system with variable capacity of base station and a fixed one.

For evaluation we set $U_{avg} = 500$ for calculation vector $C(t)$ with t from 0 to 7200 s. The simulation duration of each scenario was 500000 s of model time and averaged by 10 runs with cyclic use of data for the $C(t)$ vector. These parameters are chosen so that the slice degradation probability is within 95% confidence intervals, the length of which does not exceed 0.005. Every second of model time, the capacity C of the base station changes to the next value from the $C(t)$ vector. The capacity of each slice is recalculated according to formula (5). This event is not a re-slicing. For cases with fixed capacity we set averaged value $C = 1902$ Mbps.

Thus, in terms of base station capacity, we consider two options/abstractions: when it varies and changes once per second of model time ($varC$) and when it is averaged and fixed throughout the entire simulation period ($avgC$).

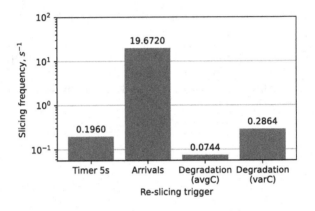

Fig. 1. Slicing frequency vs. trigger and capacity abstraction.

In this experiment, we consider re-slicing by Arrivals as an approximation to online slicing. Figure 1 shows the frequency of re-slicing calls. Re-slicing by Arrivals happens 19.672 times per second and coincides with the average number of sessions arriving in the Best Effort slice, whose arrival rate is the largest among all slices and equals 20. Re-slicing by Degradation is called much less frequently (0.0744/0.286 times per second) and is comparable to re-slicing by Timer (0.196 times per second). It is a reactive re-slicing, which happens only when it is needed. In the case $varC$ with variable capacity, the frequency is approx. 3.8 times higher than in the case $avgC$, which indicates a more frequent occurrence of degradation in slices.

Figure 2 shows that the averaged system degradation is noticeably higher if variable capacity ($varC$) is taken into account. However, an interesting effect is revealed in the case of re-slicing by Degradation. With a variable capacity ($varC$), it occurs more often and the average slice degradation probability is slightly lower than with a fixed capacity ($avgC$). Re-slicing by Timer when taking into account variable capacity shows a poor result of 27%. However, with a "smart" call for re-slicing, i.e. calls on events that pose a potential seriously reduce to the QoS, the degradation probability does not exceed 2.5%.

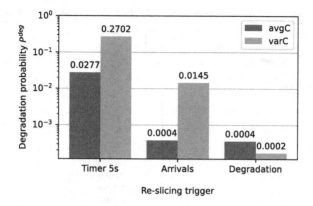

Fig. 2. Averaged degradation probability vs. trigger and capacity abstraction.

Let us consider in more detail the reason for such a difference in the degradation probability by referring to Fig. 3 (Table 3). Variable capacity has the least effect on Degradation re-slicing calls, this is because if the SLA in the slice is violated, as a result of a decrease in the total capacity of the base station, it immediately causes re-slicing. For the rest of the triggers, everything directly depends on how often the re-slicing will occur, the calls of which do not depend on the capacity change events.

Table 3. Slice degradation probability with confidence interval (95% confidence level) vs. trigger and capacity abstraction.

Slice	Timer 5s		Arrivals		Degradation	
Averaged capacity ($avgC$)						
Best Effort	1.004×10^{-5}	$\pm 3.546 \times 10^{-6}$	6.399×10^{-6}	$\pm 4.863 \times 10^{-6}$	9.836×10^{-6}	$\pm 5.831 \times 10^{-6}$
Movie Streaming	0.036	$\pm 1.632 \times 10^{-3}$	1.796×10^{-7}	$\pm 4.006 \times 10^{-7}$	6.144×10^{-6}	$\pm 1.076 \times 10^{-5}$
Online Gaming	0.05	$\pm 4.626 \times 10^{-4}$	0.001	$\pm 5.189 \times 10^{-4}$	9.476×10^{-4}	$\pm 9.29 \times 10^{-4}$
VR/AR	0.024	$\pm 4.464 \times 10^{-4}$	8.496×10^{-4}	$\pm 5.128 \times 10^{-4}$	7.827×10^{-4}	$\pm 7.762 \times 10^{-4}$
Corporate	0.028	$\pm 1.941 \times 10^{-4}$	3.818×10^{-5}	$\pm 3.137 \times 10^{-5}$	4.41×10^{-5}	$\pm 9.733 \times 10^{-5}$
Variable capacity ($varC$)						
Best Effort	1.804×10^{-9}	$\pm 4.08 \times 10^{-9}$	1.593×10^{-7}	$\pm 3.502 \times 10^{-7}$	6.588×10^{-7}	$\pm 9.473 \times 10^{-7}$
Movie Streaming	0.111	$\pm 7.58 \times 10^{-3}$	0.005	$\pm 5.067 \times 10^{-4}$	9.197×10^{-5}	$\pm 1.866 \times 10^{-5}$
Online Gaming	0.446	$\pm 1.605 \times 10^{-4}$	0.025	$\pm 4.095 \times 10^{-4}$	4.55×10^{-4}	$\pm 7.498 \times 10^{-5}$
VR/AR	0.4	$\pm 1,072 \times 10^{-3}$	0.021	$\pm 4.354 \times 10^{-4}$	1.332×10^{-4}	$\pm 7.322 \times 10^{-5}$
Corporate	0.394	$\pm 1.296 \times 10^{-3}$	0.021	$\pm 7.559 \times 10^{-5}$	1.59×10^{-4}	$\pm 1.952 \times 10^{-5}$

The system resource utilization (UTIL) for selected arrival rates fluctuates around 84.8–87% regardless of the capacity abstraction and the re-slicing trigger. Also, user satisfaction is greater than or equal to 1 in all cases. This suggests that even if SLA violations occur, this is not reflected in the average value even with a high system load.

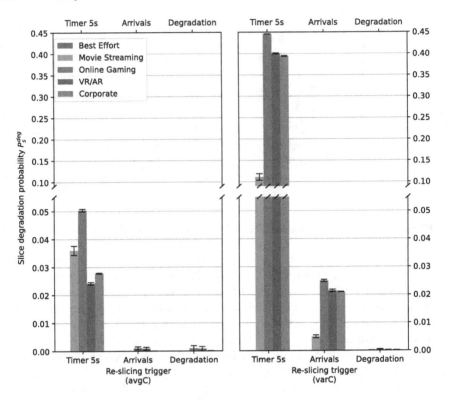

Fig. 3. Slice degradation probability vs. trigger and capacity abstraction.

5   Conclusion

In this work, we characterize the influence of the channel state fluctuations on the operation of the 5G wireless network resource slicing scheme. We demonstrate that changes in the base station capacity may have significant impact on the quality of service under high load conditions resulting in up to tenfold degradation.

Re-slicing by Arrivals, if they occur very often, may be redundant. And calling the scheme by such a trigger does not show itself in the best way in the case of a variable capacity. Degradation trigger allows to do reactive re-slicing, but at the same time it is difficult to implement on real equipment, because it introduces a functional dependence on the state of the system. It will be necessary to listening this state and performing calculations with a certain frequency, which also must be selected. And finally, choosing a period for Timer empirically is not an easy task.

To reduce the slice degradation probability, re-slicing needs to be performed at much higher frequency. We show that triggering re-slicing at session arrivals is sufficient for keeping performance degradation at the acceptable limit while keeping the complexity low.

References

1. 3GPP: Study on channel model for frequencies from 0.5 to 100 GHz (Release 14). 3GPP TR 38.901 V14.1.1, July 2017
2. 3GPP: Virtual reality (VR) media services over 3GPP. 3GPP TR 26.918 V16.0.0, December 2018
3. Adou, Y., Markova, E., Gudkova, I.: Performance measures analysis of admission control scheme model for wireless network, described by a queuing system operating in random environment. In: 2018 10th International Congress on Ultra Modern Telecommunications and Control Systems and Workshops (ICUMT), pp. 1–5 (2018). https://doi.org/10.1109/ICUMT.2018.8631263
4. Gapeyenko, M., et al.: Analysis of human-body blockage in urban millimeter-wave cellular communications. In: Communications (ICC), 2016 IEEE International Conference on, pp. 1–7. IEEE (2016)
5. Hosseini, S.A., Zarepour, E., Hassan, M., Chou, C.T.: Analyzing diurnal variations of millimeter wave channels. In: 2016 IEEE Conference on Computer Communications Workshops (INFOCOM WKSHPS), pp. 377–382 (2016). https://doi.org/10.1109/INFCOMW.2016.7562105
6. Kántor, P., Bító, J.: Influence of climate variability on performance of millimeter wave cellular 5th generation wireless systems in consequence of precipitation. In: Proceedings of the 8th European Conference on Antennas and Propagation (EuCAP 2014), pp. 303–307 (2014). https://doi.org/10.1109/EuCAP.2014.6901752
7. Nain, P., Towsley, D., Liu, B., Liu, Z.: Properties of random direction models **3**, 1897–1907 (2005)
8. Petrov, V., Yarkina, N., Moltchanov, D., Andreev, S., Samouylov, K.: Session-level reliability analysis for multi-service communication in autonomous vehicular fleets. IEEE Access **8**, 174629–174642 (2020). https://doi.org/10.1109/ACCESS.2020.3024790
9. Polyakov, N.A., Yarkina, N.V., Samouylov, K.E.: A simulator for analyzing a network slicing policy with SLA-based performance isolation of slices. Discret. Contin. Models Appl. Comput. Sci. **29**(1), 36–52 (2021). https://doi.org/10.22363/2658-4670-2021-29-1-36-52
10. Polyakov, N., Yarkina, N., Samouylov, K., Koucheryavy, Y.: Network slice degradation probability as a metric for defining slice performance isolation. In: Koucheryavy, Y., Balandin, S., Andreev, S. (eds.) NEW2AN/ruSMART -2021. LNCS, vol. 13158, pp. 481–492. Springer, Cham (2022). https://doi.org/10.1007/978-3-030-97777-1_40
11. Prosvirov, V., Khayrov, E., Polyakov, N.: A model for 5G millimeter wave service rate abstraction. In: Distributed Computer and Communication Networks - 25rd International Conference, DCCN 2022. LNCS, Springer, submitted (2022)
12. Romero-Peña, J.S., Cardona, N.: Applicability limits of simplified human blockage models at 5G mm-wave frequencies. In: 2019 13th European Conference on Antennas and Propagation (EuCAP), pp. 1–5 (2019)
13. Samuylov, A., et al.: Characterizing resource allocation trade-offs in 5G NR serving multicast and unicast traffic. IEEE Trans. Wirel. Commun. **19**(5), 3421–3434 (2020). https://doi.org/10.1109/TWC.2020.2973375
14. Sopin, E., Daraseliya, A., Shorgin, S.: Probabilistic model for the analysis of elastic session duration in wireless networks. J. Phys. Conf. Ser. **2134**(1), 012029 (2021). https://doi.org/10.1088/1742-6596/2134/1/012029

15. Tafintsev, N., et al.: Handling spontaneous traffic variations in 5G+ via offloading onto mmwave-capable UAV bridges. IEEE Trans. Veh. Technol. **69**(9), 10070–10084 (2020). https://doi.org/10.1109/TVT.2020.3005253

16. Vishnevskiy, V.M., Samouylov, K.E., Yarkina, N.V.: Mathematical model of LTE cells with machine-type and broadband communications. Autom. Remote Control **81**(4), 622–636 (2020). https://doi.org/10.1134/S0005117920040050

17. Yarkina, N., Correia, L., Moltchanov, D., Gaidamaka, Y., Samouylov, K.: Multi-tenant resource sharing with equitable-priority-based performance isolation of slices for 5G cellular systems. Comput. Commun. **188** (2022). https://doi.org/10.1016/j.comcom.2022.02.019

18. Yoon, Y., Park, S., Kim, J.H.: Empirical model including the statistics of location variability for the over-rooftop path in the 32 GHz band. IEEE Access **8**, 77263–77271 (2020). https://doi.org/10.1109/ACCESS.2020.2988919

On Waiting Time Maxima in Queues with Exponential-Pareto Service Times

Irina Peshkova[1,2]([✉]) [iD], Alexander Golovin[2] [iD], and Maria Maltseva[2] [iD]

[1] Petrozavodsk State University, Lenin str. 33, Petrozavodsk 185910, Russia
iaminova@petrsu.ru
[2] Institute of Applied Mathematical Research of the Karelian Research Centre
of RAS, Pushkinskaya str. 11, Petrozavodsk 185910, Russia
golovin@krc.karelia.ru

Abstract. In this paper we study the extreme behavior of waiting time in steady-state $GI/G/1$ systems with Exponential-Pareto service times. We show that this distribution belongs to the subclass of subexponential distributions, that allows to apply known tail asymptotic for stationary waiting times via the equilibrium distribution of service times. We propose the expressions for the normalizing constants and derive the limiting distribution of waiting time maxima (Frechet-type distribution). Simulation results show that approximation by Frechet-type distribution works well for $GI/G/2$ systems as well.

Keywords: Regenerative Simulation · Extreme Values · Exponential-Pareto Mixture · Extremal Index

1 Introduction

Among finite mixture distributions a special importance belongs to the Exponential-Pareto mixture distributions [16]. Their application is motivated by the fact that the first component of mixture relates to so-called *light-tailed distributions*, and the second relates to *heavy-tailed* ones. Heterogeneous data is typical for communication systems. For example, data processing time (associated with service time in queueing models) may be actually generated by such distributions. Some devices might create huge amounts of data traffic, like a media content On the contrast, others can generate "small and bursty" dara like a temperature, pressure or light reading from a sensor. *Mixture model* may work quite well in simulation of such systems.

Modeling and forecasting of extreme values in the performance analysis of queueing systems is an important problem which can be performed using the *extreme value theory* [7,9,12]. The classical extreme-value limit theorems for

The publication has been prepared with the support of Russian Science Foundation according to the research project No. 21-71-10135, https://rscf.ru/en/project/21-71-10135/..

independent and identically distributed (iid) random variables (rv's) don't work because the successive variables in the queueing processes are typically strongly dependent [4]. Nevertheless, the limit theorems for queueing processes (maximum waiting time, maximum virtual waiting time and the maximum queue size) has been proved in [6,11,19]. In our research, we rely on limit theorems for strictly stationary sequences, and moreover, we evaluate the quality of the approximation. Based on perfect simulation [5,14] of the system.

Extremes for strictly stationary systems are provided by the limit theorems via the so-called extremal index. The extremal index θ reflects a cluster structure of an underlying sequence or its local dependence. For example, for iid sequence $\theta = 1$ holds. The value of extremal index equal to zero, $\theta = 0$, implies a total dependence that corresponds to "very wide" clusters of exceedances when observations of the stochastic process likely never exceed a sufficiently high threshold u_n tending to infinity [17,20,21]. It is important for practice that $1/\theta$ approximates the mean cluster size. This allows one to estimate the extremal index by non-parametric methods like blocks, runs and intervals estimators [22]. Distributions of the cluster and inter-cluster sizes of the stationary waiting times are obtained for a given value of the threshold in [13]. The threshold selection method for non-parametric estimation of the extremal index of a stationary sequence proposed in [18]. The algorithm of computing the extremal index of the stationary waiting time of a stable $G/G/1$ system with distribution belonging to the domain of attraction of Gumbel distribution is given in [10].

In this research we consider $GI/G/1$ queuing system with Exponential-Pareto mixture distribution of services. On the base of well-known criteria [1] we show that this mixture distribution belongs to the subclass S^* of subexponential distributions. In this case tail waiting time asymptotics is defined by the equilibrium service time distribution [2,3]. This property can be applied to derive the limiting distribution of waiting time maxima based on the extreme value theory. To implement this idea, we derived an expression for normalizing constants, which eventually allowed us to prove that the limiting distribution of maxima of steady-state waiting times has Frechet-type distribution (see Lemma 2). This analysis can be used to the extremal index evaluation via perfect simulation and regenerative approach.

Large waiting times asymptotics for multiserver queues with heavy tails are studied in [8], where the authors present upper and lower bounds for the tail distribution of the stationary waiting time in the stable $GI/G/c$ FCFS queue for $c \geq 2$.

In [15] the mixture of Weibull distribution (and therefore it is in the maximum domain of attraction of the Gumbel law) and the regularly varying distribution (maximum domain of attraction of the Fréchet law) is considered. It is shown that if mixing proportion varies as the number of observations grows the set of possible limit distributions includes Gumbel and Fréchet distributions and also one discontinuous law.

The main theoretical contribution of this research is formulated in Lemma 2, where it is proved that the limiting distribution of waiting time maxima in

$GI/G/1$ steady-state system with Exponential-Pareto mixture service time distribution has Frechet-type form. Simulation results via perfect simulation confirm this theoretical conclusion.

The paper is structured as follows. In Sect. 2, we obtain the equilibrium distribution B_e and the tail asymptotics of the stationary waiting time distribution (in a $GI/Exponential - Pareto\ mixture/1$ system). The limiting distribution of maxima of waiting times is derived in Sect. 3 (see Lemma 2). In Sect. 4, we describe two methods of the extremal index estimation: block method and regeneration approach. Numerical results of simulation $M/G/1$ and $M/G/2$ systems are presented in Sect. 5.

2   Tail Asymptotic of Waiting Times

We consider $GI/G/1$ queueing system with iid interarrivals and iid service times and FCFS service discipline. Suppose that service time S has the distribution function B and belongs to the class of so-called subexponential distributions with a finite mean $1/\mu$. Then it is known that [2,3] the tail of stationary waiting time W is related to the service time distribution tail B via the equivalence:

$$\mathbf{P}(W > x) \sim \frac{\rho}{1-\rho}\mathbf{P}(S_e > x), \quad x \to \infty, \tag{1}$$

where the stationary remaining renewal time S_e has the equilibrium density $\overline{B}(x)/\mathbf{E}S$ and belongs to subexponential distributions, $\rho = \lambda_\tau/\mu < 1$ is traffic intensity, $\overline{B}(x) = 1 - B(x)$. (Relation $a \sim b$ in (1) means the asymptotic equivalence, that is $a/b \to 1$.)

Let service times S have an *Exponential-Pareto mixture distribution* with distribution function (df) [16]

$$B(x) = 1 - pe^{-\lambda x} + (1-p)\left(\frac{x_0}{x_0 + x}\right)^\alpha, \quad \lambda > 0, \alpha > 0, x_0 > 0, x \geq 0, \tag{2}$$

where $0 < p < 1$ is *mixing proportion*. Equation (2) shows that rv S coincides with Exponential distribution with the probability p, and with Pareto distribution with the probability $1 - p$.

The expressions for the mean and the equilibrium distribution B_e are given respectively by:

$$1/\mu = \frac{p}{\lambda} + \frac{(1-p)x_0}{\alpha - 1}, \tag{3}$$

$$B_e(x) = \mu \int_0^x \overline{B}(t)dt = 1 - \mu\left(\frac{pe^{-\lambda x}}{\lambda} + \frac{(1-p)x_0^\alpha}{(\alpha - 1)(x_0 + x)^{\alpha-1}}\right). \tag{4}$$

We emphasize that both expressions (3) and (4) exist if parameter $\alpha > 1$.

We calculate the failure rate function of the Exponential-Pareto distribution

$$r(x) := \frac{f_B(x)}{\overline{B}(x)} = \frac{p\,\lambda a(x) + (1-p)\,\alpha/(x_0 + x)}{p\,a(x) + (1-p)}, \tag{5}$$

where

$$a(x) = e^{-\lambda x} \left(1 + \frac{x}{x_0}\right)^\alpha.$$

is auxiliary function and f_B is density function. To verify that the Exponential-Pareto mixture distribution B (of the service time S) and the corresponding equilibrium distribution B_e (with parameter $\alpha > 1$) both belong to the class of the subexponential distributions, it is enough [1] to verify that B belongs to a special subclass $\mathcal{S}^*$ of the subexponential distributions.

Lemma 1. *The Exponential-Pareto mixture distribution with df defined by expression (2) (with parameter $\alpha > 1$) belongs to a special subclass $\mathcal{S}^*$.*

Proof. One of the following criteria for df to belong $\mathcal{S}^*$ can be applied [1].

I. If

$$\limsup_{x \to \infty} x r(x) < \infty, \tag{6}$$

then $B \in \mathcal{S}^*$.

II. Suppose that

$$\lim_{x \to \infty} r(x) = 0. \tag{7}$$

Then

$$B \in \mathcal{S}^* \iff \lim_{x \to \infty} \int_0^x e^{yr(x)} \overline{B}(y) dy = 1/\mu. \tag{8}$$

It is easy to check that for Exponential-Pareto mixture distribution

$$r(x) \longrightarrow 0 \quad \text{as} \quad x \to \infty \quad \text{and} \quad x r(x) \longrightarrow \alpha \quad \text{as} \quad x \to \infty.$$

Moreover,

$$\int_0^x e^{yr(x)} \overline{B}(y) dy = -\frac{pe^{-y(\lambda - r(x))}}{\lambda - r(x)} \bigg|_0^x + (1-p)\frac{(x_0 r(x))^\alpha}{r(x)} e^{-x_0 r(x)}$$

$$\cdot [\gamma(-\alpha + 1, x_0 r(x) + x r(x)) - \gamma(-\alpha + 1, x_0 r(x))]$$

$$\longrightarrow \frac{p}{\lambda} + \frac{(1-p)x_0}{\alpha - 1} = \frac{1}{\mu} \quad \text{as} \quad x \to \infty$$

since

$$\gamma(-\alpha + 1, x_0 r(x) + x r(x)) \longrightarrow \gamma(-\alpha, \alpha) \quad \text{as} \quad x \to \infty$$

and

$$\gamma(-\alpha + 1, x_0 r(x)) \sim \frac{(x r(x))^{-\alpha + 1}}{-\alpha + 1} \quad \text{as} \quad x \to \infty,$$

where $\gamma(\alpha, x) = \int_0^x y^{\alpha - 1} e^{-y} dy$ is the lower incomplete gamma function.

Hence conditions (6)–(8) are satisfied and Exponential-Pareto distribution belongs to subclass $\mathcal{S}^*$, which allows to apply the tail asymptotics (1) for stationary waiting times.   □

For the Exponential-Pareto mixture distribution the traffic intensity is determined by the relation

$$\rho = \lambda_\tau \left[\frac{p}{\lambda} + \frac{(1-p)x_0}{\alpha - 1} \right], \tag{9}$$

where λ_τ is the input rate. It now follows from relations (1), (4) and Lemma 1 that the waiting time tail distribution satisfies

$$\mathbf{P}(W > x) \sim \frac{\mu \lambda_\tau}{\mu - \lambda_\tau} \left(\frac{pe^{-\lambda x}}{\lambda} + \frac{(1-p)x_0^\alpha}{(\alpha-1)(x_0+x)^{\alpha-1}} \right) \text{ as } x \to \infty. \tag{10}$$

3 The Extreme Behavior of Stationary Waiting Times

Let $\{X_n,\ n \geq 1\}$ be a family of the iid rv's with a distribution function F. Then the distribution of $M_n = \max(X_1, \dots, X_n)$ satisfies $\mathbf{P}(M_n \leq x) = F^n(x)$.

Suppose there exists a sequence of real constants b_n, $a_n > 0$, $n \geq 1$ such that

$$\lim_{n \to \infty} \mathbf{P}((M_n - b_n)/a_n \leq x) = G(x), \quad n \to \infty, \tag{11}$$

for every continuity point x of G, and G a nondegenerate distribution function. Then $G(x)$ is one of the three types of *extreme value distributions*: Gumbel, Frechet or reversed Weibull [12]. The class of extreme value distributions (which combines all three types) is $G_\eta(cx + d)$ with real $c > 0$, d, where

$$G_\eta(x) = \begin{cases} e^{-(1+\eta x)^{-1/\eta}}, & \eta \neq 0,\ 1 + \eta x > 0; \\ e^{-e^{-x}}, & \eta = 0. \end{cases}$$

It is easy to check that the first component of Exponential-Pareto mixture is in the maximum domain of attraction (MDA) of the Gumbel law, while the second is in the maximum domain of attraction of the Fréchet law.

For given $0 \leq \tau \leq \infty$ and a sequence $\{u_n,\ n \geq 1\}$ of real numbers the following are equivalent [7]

$$n\overline{F}(u_n) \to \tau \text{ as } n \to \infty \tag{12}$$

and

$$\mathbf{P}(M_n \leq u_n) \to e^{-\tau} \text{ as } n \to \infty. \tag{13}$$

If condition (11) is satisfied, then convergence (13) is preserved for any linear normalizing sequence $u_n(x) = a_n x + b_n$, $n \geq 1$ and expression (13) becomes

$$\mathbf{P}(M_n \leq u_n(x)) \to \tau(x),$$

where a concrete form of the function $\tau(x)$ depends on the type of the limiting distribution.

Denote the maximum waiting time by $W_n^* = \max(W_1, \dots, W_n)$ for each $n \geq 1$. In the following Lemma 2 the normalizing sequences and the limiting distribution for stationary waiting times are derived.

Lemma 2. *If the service time in a steady-state $GI/G/1$ queueing system has Exponential-Pareto mixture distribution (2) with parameter $\alpha > 1$, then the limiting distribution of W_n^* has the following Frechet-type form:*

$$\mathbf{P}(W_n^* \leq u_n(x)) \to e^{-\dfrac{(1-p)\mu\lambda_\tau}{(\alpha-1)(\mu-\lambda_\tau)}x^{1-\alpha}} \qquad as\ n \to \infty, \qquad (14)$$

with the normalizing sequence

$$u_n(x) = a_n x + b_n = x_0^{\alpha/(\alpha-1)} x n^{1/(\alpha-1)} - x_0. \qquad (15)$$

Proof. To prove the lemma it is enough to substitute in the expression (12) the tail distribution (10) and the normalizing sequence (15).

$$
\begin{aligned}
n\overline{F}(u_n(x)) &= n\frac{\mu\lambda_\tau}{\mu-\lambda_\tau}\left[\frac{p}{\lambda}e^{-\lambda u_n(x)} + \frac{1-p}{\alpha-1}\frac{x_0^\alpha}{(x_0+u_n(x))^{\alpha-1}}\right] \\
&= n\frac{\mu\lambda_\tau}{\mu-\lambda_\tau}\frac{1-p}{\alpha-1}x_0^\alpha(x_0 + x_0^{\alpha/(\alpha-1)}xn^{1/(\alpha-1)} - x_0)^{-\alpha+1} \\
&\quad \cdot\left[1 + \frac{p(\alpha-1)(x_0+u_n(x))^{\alpha-1}e^{-\lambda u_n(x)}}{\lambda(1-p)x_0^\alpha}\right] \\
&= \frac{\mu\lambda_\tau}{\mu-\lambda_\tau}\frac{1-p}{\alpha-1}x^{-\alpha+1}\left[1 + \frac{p(\alpha-1)(x_0+u_n(x))^{\alpha-1}e^{-\lambda u_n(x)}}{\lambda(1-p)x_0^\alpha}\right] \\
&\longrightarrow \frac{\mu\lambda_\tau}{\mu-\lambda_\tau}\frac{1-p}{\alpha-1}x^{-\alpha+1}
\end{aligned}
$$

since $u_n(x) \to \infty$ as $n \to \infty$. It is known [7], that if $B_e \in MDA(\Phi_{\alpha-1})$ and

$$\lim_{n\to\infty} B_e^n(u_n(x)) = \Phi_{\alpha-1}(x), \quad x > 0.$$

Then

$$\lim_{n\to\infty} F^n(u_n(x)) = \Phi_{\alpha-1}(cx), \quad x > 0, \qquad (16)$$

for some constant $c > 0$ if and only if d.f.'s are tail equivalent,

$$\lim_{x\to\infty} \frac{\overline{B}_e(x)}{\overline{F}(x)} = c^{\alpha-1}. \qquad (17)$$

From last relation (17) we can find

$$c = \frac{(\mu-\lambda_\tau)^{1/(\alpha-1)}}{\lambda_\tau^{1/(\alpha-1)}}$$

and substituting this constant in (16) we get (14).

$\square$

In what follows that the asymptotic behaviour of waiting time maxima is determined by the second component of service time distribution (Pareto distribution). We propose to check this statement via perfect simulation technique in Sect. 5.

4 Extremal Index

The strictly stationary sequence $\{X_n, \ n \geq 1\}$ with common df F is said to have the *extremal index* θ [12], if for each positive τ there exists a sequence of real numbers $u_n = u_n(\tau)$ such that

$$\lim_{n\to\infty} n(1 - F(u_n)) = \tau \tag{18}$$

and

$$\lim_{n\to\infty} \mathbf{P}(M_n \leq u_n) = e^{-\theta\tau}, \tag{19}$$

where $M_n = \max(X_1 \ldots, X_n)$ and $\theta \in [0, 1]$. This means that for large n, one can use the following approximation for the distribution of M_n

$$\mathbf{P}(M_n \leq u_n) = e^{-\theta\tau} \approx F^{n\theta}(u_n).$$

Hence, the extremal index θ is a key parameter for the distribution of sample extremes for strictly stationary sequences.

The well-established block method [7,12] can be used to estimate θ. Convergence (19), together with (18), imply the basic relation

$$\theta_b = \lim_{n\to\infty} \frac{\log \mathbf{P}(M_n \leq u_n)}{n \log F(u_n)}. \tag{20}$$

Now let's divide the sequence $X_1, \ldots, X_n$ into m blocks of the size k, assuming that $n = m\, k$. Firstly, we need to calculate the maximum of the sequence over each block, that is consider the sequence

$$M_{l,k} = \max(X_{(l-1)k+1}, \ldots, X_{lk}), \ l = 1, \ldots, m.$$

Denote $N(u_n) = \#(i \leq n : X_i > u_n)$, the number of the exceedances of the level u_n by the sequence $X_1, \ldots, X_n$, and let

$$m(u_n) = \#(l \leq m, M_{l,k} > u_n)$$

be the number of blocks with at least one exceedance of the level u_n. Then we can estimate the extremal index θ with the use of the following formula [7]

$$\hat{\theta}_b(n) = \frac{1}{k} \frac{\log\left(1 - \frac{m(u_n)}{m}\right)}{\log\left(1 - \frac{N(u_n)}{n}\right)}. \tag{21}$$

Block method is not the only way to estimate the extremal index. Regenerative approach can also be applied to estimate θ [11,19]. More precisely, consider a stationary regenerative sequence $\{Z_n\}$ with the regeneration instants β_k, and denote by M_{Y_i} its maximum in the ith regeneration cycle, that is,

$$M_{Y_i} = \sup_n \{Z_n, \ \beta_{i-1} \leq n < \beta_i\},$$

and also denote $\mathbf{E}\delta$ as the mean cycle length. Then, as it is stated in [19], the stationary regenerative sequence $\{Z_n\}$ has the extremal index θ_δ if and only if there exists the limit

$$\theta_\delta = \lim_{n\to\infty} \frac{\mathbf{P}(M_{Y_1} > u_n)/\mathbf{E}\delta}{\mathbf{P}(Z_1 > u_n)}, \qquad (22)$$

for some sequence u_n satisfying relation (12).

By (22), the regenerative approach allows to estimate the extremal index in the following way (for n large enough):

$$\hat{\theta}_\delta(n) = \frac{1}{\hat{\delta}} \frac{\log\left(1 - \frac{\hat{m}(u_n)}{m_\delta}\right)}{\log\left(1 - \frac{\hat{N}(u_n)}{n}\right)}, \qquad (23)$$

where

$$\hat{N}(u_n) = \#(i \le n : Z_i > u_n), \quad \hat{m}(u_n) = \#(i \le m_\delta : M_{Y_i} > u_n),$$

$\hat{m}(u_n)$ is the number of the exceedances within regeneration cycle, m_δ is the number of regeneration points which are detected during simulation procedure, and $\hat{\delta}$ is the sample mean of the cycle length.

5   Numerical Experiments

5.1   Convergence to Frechet Distribution

We use the *perfect simulation technique* that allows exact sampling from the (unknown) steady-state distribution of the waiting time W. The algorithm for $M/G/c$ perfect simulation is outlined in [14].

We simulate $M/G/1$ system with Exponential-Pareto mixture service times with parameters $x_0 = 1$, $\alpha = 4.5$, $\lambda = 2$, input rate $\lambda_\tau = 0.2$, traffic intensity $\rho = 0.1$, mixing proportion $p = 0.95$. According to our simulation study the perfect simulation demonstrates convergence to Frechet-type distribution for sample size $n = 10^6$ which corresponds to our theoretical conclusions obtained in Lemma 1 for maximum waiting time W_n^*. Figure 1a) demonstrates the convergence of empirical to theoretical density, Fig. 1c) shows the proximity of cdf's (the theoretical functions are shown in red). The linearity of the points of Q-Q plot (Fig. 1b) and P-P plot (Fig. 1d) suggests that the data have Frechet-type distribution in the form (14).

Figures 2 show the simulation results of $M/G/1$ system with parameters $x_0 = 1$, $\alpha = 2.1$, $\lambda = 2.2$, input rate $\lambda_\tau = 0.88$, traffic intensity $\rho = 0.6$ and mixing proportion $p = 0.5$. The goodness-of-fit for the maxima of stationary waiting times distribution is checked by the Kolmogorov-Smirnov criterion. Test does not reject the null hypothesis of the corresponding theoretical Frechet distribution.

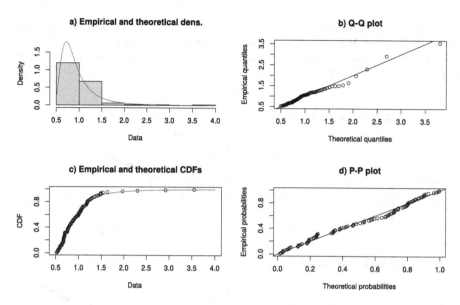

Fig. 1. Convergence of empirical a) density c) CDF to theoretical Frechet-type density and CDF; b) Q-Q-plot, d) P-P plot for maximum waiting time W_n^* in $M/G/1$ with Exponential-Pareto mixture services with parameters $x_0 = 1$, $\alpha = 4.5$, $\lambda = 2$, input rate $\lambda_\tau = 0.2$, traffic intensity $\rho = 0.1$, mixing proportion $p = 0.95$, sample size $= 10^6$.

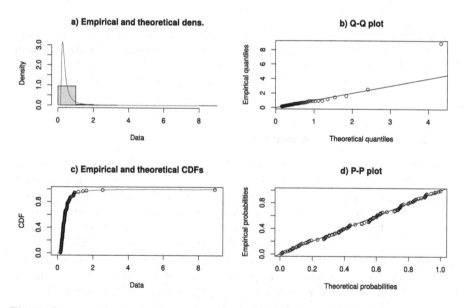

Fig. 2. Convergence of empirical a) density c) CDF to theoretical Frechet-type density and CDF; b) Q-Q-plot, d) P-P plot for maximum waiting time W_n^* in $M/G/1$ with Exponential-Pareto mixture services with parameters $x_0 = 1$, $\alpha = 2.1$, $\lambda = 2.2$, input rate $\lambda_\tau = 0.88$, traffic intensity $\rho = 0.6$, mixing proportion $p = 0.5$.

Figures 3 demonstrate simulation results of $M/G/2$ system with the same parameters as for $M/G/1$ with the exception of the input rate $\lambda_\tau = 0.41$. There results demonstrate that approximation by Frechet-type distribution works well in $GI/G/2$ case as well.

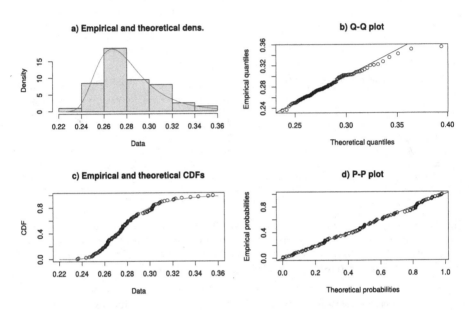

Fig. 3. Convergence of empirical a) density c) CDF to theoretical Frechet-type density and CDF; b) Q-Q-plot, d) P-P plot for maximum waiting time W_n^* in $M/G/2$ with Exponential-Pareto mixture services with parameters $x_0 = 1$, $\alpha = 4.5$, $\lambda = 2$, input rate $\lambda_\tau = 0.41$, traffic intensity $\rho = 0.1$, mixing proportion $p = 0.95$, sample size $= 10^6$.

5.2   Extremal Index Estimation

Figures 4 show the simulation results of $M/G/1$ system with Exponential-Pareto mixture service times with parameters $x_0 = 1$, $\alpha = 2.1$, $\lambda = 2.2$. We compare two approaches for stationary waiting times extremal index estimation: block and the regeneration methods. In fact there are several possible estimators in R language. Package "extRemes" and block sizes approximately equal to the mean regenerative cycle lengths were chosen. We simulate systems with different input rates, $\lambda_\tau \approx 0, 29; 0.73; 1.17; 1.39$, and traffic intensities, $\rho \approx 0.2; 0.5; 0.8; 0.95$ according to the expression (9) (Figs. 4a, 4b, 4c, 4d, relatively). In the case of $M/M/1$ system which corresponds to $M/G/1$ system with Exponential-Pareto mixture service times with mixing proportion $p = 1$ the exact form of extremal index is well known, $\theta = (1 - \rho)^2$, see [10]. The simulation results shown in the Fig. 4 allow us to conclude that the regenerative method demonstrates the best proximity to the theoretical value for $p = 1$ in all considered cases of $\rho =$

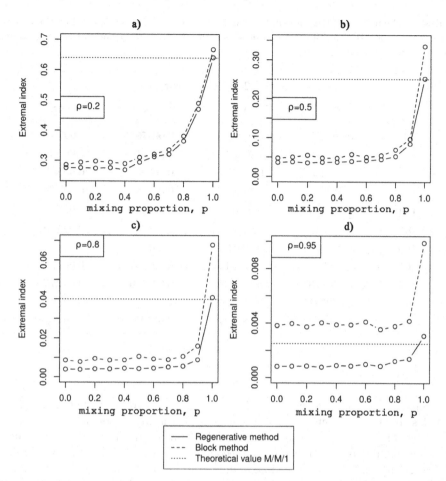

Fig. 4. Extremal index for waiting times in $M/G/1$ with Exponential-Pareto mixture service times with parameters $x_0 = 1$, $\alpha = 2.1$, $\lambda = 2.2$ and different mixing proportions $p = 0; 0.1; \ldots, 1$.

$0.2; 0.5; 0.8; 0.95$. Namely, mean absolute error MAE$= 8.8 \cdot 10^{-4}$ for $\theta_\delta(n)$ is two orders of magnitude better than MAE $= 4.6 \cdot 10^{-2}$ for $\theta_b(n)$.

Denote by $\Delta(n)$ the distance between estimates of extremal index obtained by the block and the regenerative methods,

$$\Delta(n) = |\hat{\theta}_\delta(n) - \hat{\theta}_b(n)| \tag{24}$$

Table 1 aims to compare the distance $\Delta(n)$ for different traffic intensities ρ and different mixing proportions p. The maximum distance is achieved when mixing proportion $p = 1$ (that corresponds to the case $M/M/1$) for all considered traffic intensities.

Table 1. The distance $\Delta(n)$ between estimates of extremal index obtained by the block and the regenerative methods for different traffic intensities ρ and different mixing proportions $p,\ n = 10^6$.

p	Traffic intensity, ρ			
	0.2	0.5	0.8	0.95
0	0.009980	0.011114	0.004751	0.003002
0.1	0.017966	0.012379	0.004010	0.003136
0.2	0.023353	0.019604	0.005361	0.002884
0.3	0.017713	0.010911	0.004584	0.003270
0.4	0.020520	0.010197	0.004276	0.002998
0.5	0.015944	0.017835	0.006344	0.003026
0.6	0.008095	0.008694	0.004793	0.003120
0.7	0.014793	0.009818	0.003851	0.002711
0.8	0.016276	0.016933	0.005110	0.002573
0.9	0.019797	0.012596	0.007096	0.002791
1	0.026713	0.082984	0.027015	0.006817

6   Conclusion

We have checked that Exponential-Pareto mixture distribution belongs to the subclass of subexponential distributions $\mathcal{S}^*$. For the $M/Exponential - Pareto\ mixture/1$ system we it was obtained that the limit behaviour of stationary waiting time maxima is determined by the second component of service times (Pareto distribution), and therefore the maximum under proper normalisation converges to the Fréchet law (Lemma 2). Perfect simulation technique were applied to confirm this theoretical result. Moreover, simulation results of the extremal index estimation for stationary waiting times were presented via block method and regeneration approaches. Based on the comparison of simulation results with explicit form of extremal index in $M/M/1$ we conclude that the regeneration method provides better accuracy that block method. In the considered system the greater the traffic intensity, the smaller the extremal index. With decreasing mixing proportion p, the extremal index decreases to zero, that is typical for heavy-tailed distributions. Simulation results show that approximation of waiting time maxima distribution by Frechet-type distribution works in $GI/G/2$ case as well.

References

1. Adler, R.J., Feldman, R.E., Taqqu, M.S. (eds.): A Practical Guide to Heavy Tails: Statistical Techniques and Applications. Birkhauser Boston Inc., USA (1998). https://dl.acm.org/doi/book/10.5555/292595#sec-terms

2. Asmussen, S.: Applied Probability and Queues, 2nd edn. Springer, New York (2003). https://doi.org/10.1007/b97236
3. Asmussen, S., Klüppelberg, C., Sigman, K.: Sampling at subexponential times, with queueing applications. Stochast. Process. Appl. **79**(2), 265–286 (1999). https://doi.org/10.1016/S0304-4149(98)00064-7. https://www.sciencedirect.com/science/article/pii/S0304414998000647
4. Berger, A.W., Whitt, W.: Maximum values in queueing processes. Probab. Eng. Inf. Sci. **9**, 375–409 (1995)
5. Blanchet, J.H., Pei, Y., Sigman, K.: Exact sampling for some multi-dimensional queueing models with renewal input. Adv. Appl. Probab. **51**, 1179–1208 (2019)
6. Cohen, J.W.: Extreme value distributions for the $M/G/1$ and $GI/M/1$ queueing systems. Ann. Inst. H. Poincare Sect B(4), 83–98 (1968)
7. Embrechts, P., Klüüppelberg, C., Mikosch, T.: Modelling Extremal Events for Insurance and Finance. Applications of Mathematics. Springer, Heidelberg (1997). https://doi.org/10.1007/978-3-642-33483-2
8. Foss, S., Korshunov, D.: On large delays in multi-server queues with heavy tails. Math. Oper. Res. **37**(2), 201–218 (2012). https://doi.org/10.1287/moor.1120.0539
9. de Haan, L., Ferreira, A.: Extreme Value Theory: An Introduction. Springer, Heidelberg (2006). https://doi.org/10.1007/0-387-34471-3
10. Hooghiemstra, G., Meester, L.E.: Computing the extremal index of special Markov chains and queues. Stochast. Process. Appl. **2**(65), 171–185 (1996). https://doi.org/10.1016/S0304-4149(96)00111-1
11. Iglehart, D.: Extreme values in GI/G/1 queue. Ann. Math. Statist. **43**(2), 627–635 (1972). https://doi.org/10.1214/aoms/1177692642
12. Leadbetter, M.R., Lindgren, G., Rootzén, H.: Extremes and Related Properties of Random Sequences and Processes. Springer, New York (1983). https://doi.org/10.1007/978-1-4612-5449-2
13. Markovich, N., Razumchik, R.: Cluster modeling of Lindley process with application to queuing. In: Vishnevskiy, V.M., Samouylov, K.E., Kozyrev, D.V. (eds.) DCCN 2019. LNCS, vol. 11965, pp. 330–341. Springer, Cham (2019). https://doi.org/10.1007/978-3-030-36614-8_25
14. Morozov, E.V., Pagano, M., Peshkova, I., Rumyantsev, A.S.: Sensitivity analysis and simulation of a multiserver queueing system with mixed service time distribution. Mathematics (2020)
15. Morozova, E., Panov, V.: Extreme value analysis for mixture models with heavy-tailed impurity. Mathematics **9**(18) (2021). https://doi.org/10.3390/math9182208. https://www.mdpi.com/2227-7390/9/18/2208
16. Peshkova, I., Morozov, E., Maltseva, M.: On comparison of multiserver systems with exponential-pareto mixture distribution. In: Gaj, P., Gumiński, W., Kwiecień, A. (eds.) CN 2020. CCIS, vol. 1231, pp. 141–152. Springer, Cham (2020). https://doi.org/10.1007/978-3-030-50719-0_11
17. Resnick, S.: Extreme Values, Regular Variation and Point Processes. Springer Series in Operations Research and Financial Engineering (2008). https://doi.org/10.1007/978-0-387-75953-1
18. Rodionov, I.: On threshold selection problem for extremal index estimation. In: Shiryaev, A.N., Samouylov, K.E., Kozyrev, D.V. (eds.) ICSM-5 2020. SPMS, vol. 371, pp. 3–16. Springer, Cham (2021). https://doi.org/10.1007/978-3-030-83266-7_1
19. Rootzen, H.: Maxima and exceedances of stationary Markov chains. Adv. Appl. Probab. **20**(2), 371–390 (1988). https://doi.org/10.2307/1427395

20. Smith, L., Weissman, I.: Estimating the extremal index. J. R. Stat. Soc. Ser. B: Methodol. **56**(3), 515–528 (1994). https://doi.org/10.1111/J.2517-6161.1994.TB01997.X
21. Smith, R.: The extremal index for a Markov chain. J. Appl. Probab. **29**(1), 37–45 (1992). https://doi.org/10.2307/3214789
22. Weissman, I., Novak, S.: On blocks and runs estimators of extremal index. J. Stat. Plan. Inference **66**(2), 281–288 (1998). https://doi.org/10.1016/S0378-3758(97)00095-5

Simulation of Multimedia Traffic Delivery in WLAN

Leonid I. Abrosimov$^{(\boxtimes)}$ (iD) and Margarita Orlova$^{(\boxtimes)}$ (iD)

National Research University "Moscow Power Engineering Institute", Moscow, Russia
{AbrosimovLI,OrlovaMA}@mpei.ru

Abstract. Nowadays, multimedia traffic is growing. This traffic causes overloads and affects the performance of WLAN. The WLAN performance characteristics depend on wireless channel parameters, the number of wireless stations and inbound traffic intensity. The approach of performance modeling using a combination of analytical modeling and simulation is presented in this paper.

Keywords: WLAN · performance modeling · QoS · multimedia

1 Introduction

The growth of IP traffic continues because of a new demand for multimedia applications [1]. This trend is similar for mobile users. The specificity of multimedia traffic (the most is UDP) is an occurrence of network overload. The WLAN (IEEE 802.11) is the simplest solution for network setup. The most popular wireless access method is Wi-Fi i.e. IEEE802.11. The WLAN utilizes the shared wireless medium, and therefore specific medium access control is needed. The performance characteristics of WLAN depend on medium access control parameters, the number of wireless stations and inbound traffic intensity.

The performance evaluation of WLAN has been extensively studied recently [2–6]. The works [3–6] consider unsaturated WLAN and most of medium access control parameters. But these results are not suitable for evaluating performance characteristics of WLAN which transmit multimedia traffic because of network overload which leads packet to get stuck in the queue. The authors [3] proposed modification of the model [2] and assumed that wireless station can buffer one packet. In WLAN with multimedia traffic it is important to consider the buffer size because of successful delivery time of the packet increases depending on the buffer size during network overload.

The performance modeling of WLAN which transmits multimedia traffic have not been studied enough. This work suggest the approach of performance modeling using a combination of analytical modeling and simulation. The obtained results are used for multimedia traffic delivery evaluation in the enterprise network.

V. M. Vishnevskiy et al. (Eds.): DCCN 2022, CCIS 1748, pp. 161–170, 2023.
https://doi.org/10.1007/978-3-031-30648-8_13

2   Modeling Multimedia Traffic in WLAN Using NS-3

The Network Simulator ns-3 is a very popular tool to simulate computer networks, cellular networks and WLAN. The simulator is a set of libraries written in C++, most of them are implementations of the standard network protocols. Using ns-3 it is possible to study a real system in highly controlled environment which tracks each packet of data. An experiment in the simulator is performed using command line and a developed program in C++ which is named a simulation script. Results of experiments can be obtained using text reports or processing generated PCAP packet trace files.

The WLAN testbed (Fig. 1) in ns-3 (v. 3.30.1) has K nodes: access point (AP) and $K - 1$ wireless stations (STA). Each node generates UDP packets with λ_i intensity $i = 1, ..., K$. Each node has a buffer (queue); when the wireless channel is busy, new arrived packets are placed in the buffer (queue). The size of this buffer is Q packets. The arrived packet may stay in queue during t_W time (configuration parameter - dot11EDCATableMSDULifetime). After t_W time, the packet will be dropped from the buffer (queue).

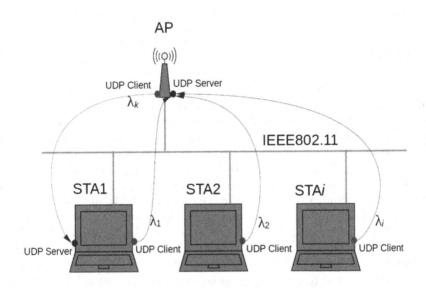

Fig. 1. The WLAN testbed in ns-3

A simulation script is built using following components (C++ classes): YansWifiChannelHelper, YansWifiPhyHelper, WifiHelper, WifiMacHelper, NetDeviceContainer, InternetStackHelper and ApplicationContainer. There isn't ns-3 component to generate traffic using specific arrival time distribution in the simulator. The following components are proposed for this task: ConfigurableUdpClientHelper and ConfigurableUdpServer. The timing diagram example of these components is shown (see Fig. 2). The STA1 uses ConfigurableUDPClient to send packets. The AP uses ConfigurableUDPServer to receive packets. The ns-3 tracks each packet. For each j−packet, we collect the following timestamps:

- t_{1j} - the time when UDP client app sends the packet,
- t_{2j} - the time when the packet is queued by MAC sublayer,
- t_{3j} - the time when the packet is sent by MAC sublayer,
- t_{4j} - the time when MAC sublayer on the UDP server-side successfully receives the packet.

Packets may be resent due to collisions. Because of this, packet IDs can be duplicated. To estimate delivery time, we fix t_{3j} and t_{4j} for each packet ID for the first time. The average service time $\bar{t}$ includes the time of packet retransmission (by wireless medium) and duplicate packet service.

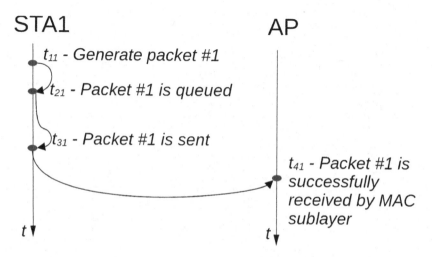

Fig. 2. Timing diagram of network packet processing by developed components (ConfigurableUdpClientHelper and ConfigurableUdpServer) and ns-3

The developed simulation script of WLAN includes the following IEEE 802.11 parameters (see Table 1) including PhyStandard (PHY specification), K - number of wireless stations, L - packet size, Λ - incoming UDP packet intensity. The stationary phase is started after parameter startSimulation which is obtained experimentally during tests, and for less than 21 nodes is equal to 20 s. The input data for the variables are shown in the Table 1.

The set of experiments produced 2920 reports which are processed using program in Python with following libraries: numpy, pandas and matplotlib. The results are shown on the Fig. 3, 4, 5 and 6. The developed simulation script of WLAN is validated using real WLAN with PNTP wired LAN and additional developed software in C++ (UDP client, UDP server, which are similar to ns-3 apps) and SHELL (to run experiments remotely).

Table 1. The WLAN simulation parameters

Variable	Value
dot11BeaconPeriod	102400 ms
dot11DTIMPeriod	102400 ms
dot11EDCATableMSDULifetime	0.5 s
dot11OperationalRateSet	ErpOfdmRate54Mbps
PhyStandard	WIFI_PHY_STANDARD_80211g
K	1 - 20
L	1500 B
Λ	var
startSimulation	20 s
Q	100 packets

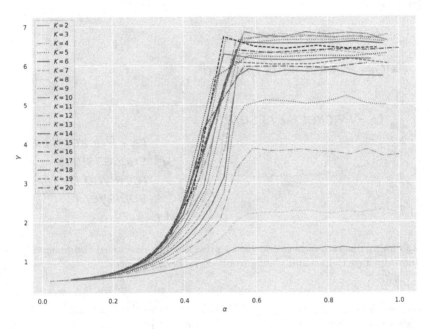

Fig. 3. The coefficient $\gamma(K, \alpha)$, $\alpha \leq 1.0$

The Fig. 3 shows the coefficient γ depend on the WLAN load α and number of wireless station. The coefficient γ is used to obtain time of multimedia packet through WLAN delivery and obtained using following characteristics: $\gamma = \frac{1+\sigma_t^2}{\bar{t}^2}$, where $\bar{t}$ is the average service time (wireless medium), σ_t^2 is the variance of service time. The coefficient α shows the average number of received packets per average packet service time and shows WLAN load: $\alpha = \Lambda \cdot \bar{t}$. As we can see (Fig. 3) if $\alpha > 0.57$ the γ becomes always constant. The Fig. 5 shows obtained coefficient

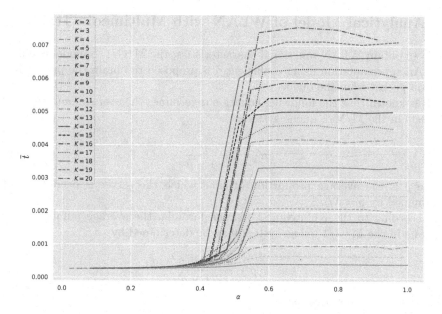

Fig. 4. The average service time (wireless medium) $\bar{t}$ (K, α), $\alpha \leq 1.0$

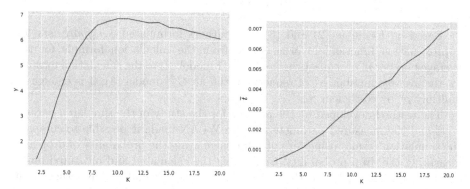

Fig. 5. The coefficient $\gamma(K)$, $\alpha > 1.0$ **Fig. 6.** The average service time (wireless medium) $\bar{t}$ (K), $\alpha > 1.0$

γ depends on K ($\alpha > 1.0$). The Fig. 4 shows the dependency of the average service time (wireless medium) $\bar{t}$ (K, α). As we can see (Fig. 6) if $\alpha > 0.57$ the $\bar{t}$, as the coefficient γ, becomes always constant. The characteristics of the WLAN are obtained for the wide range of WLAN parameters and can be used to obtain additional characteristics using mathematical modeling.

3   Analytical Model of WLAN with Multimedia Traffic

In [7] to obtain WLAN functional characteristics, the M/G/1/S model is modified. An approximation using coefficient γ is proposed to obtain the probability
of drop packet from the queue.

It is known that the following equation determines the average number n' of
requests in the M/M/1 queue system:

$$n' = \rho + \frac{\rho^2}{1 - \rho} \tag{1}$$

where ρ is the average proportion of time which the server is occupied and
determined by $\rho = \lambda/\mu$.

According to the Khintchine-Pollaczek formula, the average number n'' of
requests in the M/G/1 queue system may be determined by:

$$n'' = \rho + \frac{\rho^2 \cdot (1 + \sigma^2)}{2(1 - \rho)} \tag{2}$$

which can be represented as:

$$n'' = \rho + \frac{\rho^2 \gamma}{(1 - \rho)} \tag{3}$$

Based on formulae (2) and (3), the authors formulated a **hypothesis** for
approximation function: in order to go from the calculation formula for the
average number n' of requests in the $M/M/1$ model (2) to the calculation formula
for the average number n'' of requests in the $M/G/1$ model (3), it is enough to
multiply ρ^2 by coefficient γ.

The testbed developed by the authors for the study of the simulated parameters and the measured parameters of the WLAN made it possible to compare
the simulation results using formulae for calculating the probability of drop and
measuring the drop probability. Comparing the calculated and measured probabilities (see Fig. 7) led to the conclusion that the formulated hypothesis can
be used for practical application. Of course, work continues on the theoretical
substantiation of the proposed hypothesis.

We modified model [7] to obtain new equations which consider a wait queue
time limit (the parameter dot11EDCATableMSDULifetime - t_W). The intensity
ν of lost packets is $\nu = 1/t_W$ The WLAN model bases on M/G/1/S queue
system [7] has S + 1 states $S = K \cdot Q$, where K - number of wireless stations,
Q - size of WNIC buffer. The coefficient β used to show the average number of
lost packets per average packet service time $\beta = \nu \cdot \bar{t}$. The differential equations
for WLAN based on M/G/1/S are:

$$p_i = \frac{\alpha^i \cdot \gamma^{\lfloor i/2 \rfloor}}{\prod\limits_{j=1}^{i-1}(1 + j \cdot \beta)} \tag{4}$$

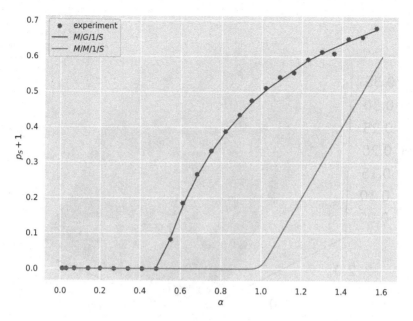

Fig. 7. The packet drop probability $P_S + 1$, K = 12, Q = 100

To obtain packet delivery time the following equations are used. The average number n of packets in the system:

$$n = \sum_{i=1}^{s+1} i \cdot p_i \qquad (5)$$

The average intensity u of packet delivery depends on inbound packet intensity which

The average intensity u of packet delivery depends on inbound packet intensity. However, the number of packets which is served by the wireless medium may decrease or increase in a different case. First, the number of packets decreases because of the drop (queue size, waiting time)-the number of packets served by the wireless medium increases because of retransmission. Retransmissions are limited. The packet with maximum retransmission attempts is removed and not retransmitted anymore. The average intensity u is determined as:

$$u = \Lambda \cdot (1 - p_{S+1}) \cdot (1 + p_c) \cdot (1 - p_d) \qquad (6)$$

where p_c is the probability of packet collision in the wireless medium, p_d is the probability of packet loss in the wireless medium, parameters p_c and p_d are depend on α and K ($\alpha \leq 1$) and obtained using ns-3 testbed.

The average time of packet successfully delivery is $T = n/u$.

The Fig. 8 shows the dependency of average packet successfully delivery time $T(K, \alpha)$ for a range of $K = 2, 20$ and $\alpha = 0.1, 2.5$.

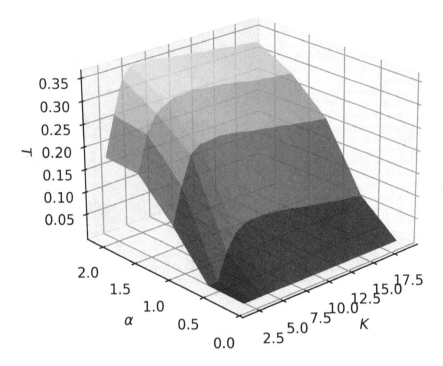

Fig. 8. The average time $T(K, \alpha)$ of packet successfully delivery

4   Conclusion

In spite of detailed performance analysis of WLAN, the multimedia traffic transmission specificity has not been studied. The main WLAN performance parameter is the packet successfully delivery time. To evaluate this time the approximation function is found. The approach of performance modeling using a combination of analytical modeling and simulation is presented in this paper. The dependency of average packet successfully delivery time is obtained for a wide range of following parameters: number of wireless station and inbound traffic intensity.

Access points have many counters (Table 2) which are accessible using the remote management protocol. However, there is no counter or parameter for evaluating multimedia traffic delivery time. We propose equations to count multimedia traffic delivery time using obtained results and access point counters.

Further, we obtain the initial data for obtained equations from the actual enterprise WLAN using counters (Table 2). The average packet size $\bar{l}$ is:

$$\bar{l} = C_{TI}/u_{IO} + C_{TO}/u_{OO} - 2H \tag{7}$$

where H is an IEEE 802.11 MAC sublayer header length, C_{TI} - number of successfully received packets:

Table 2. The access point counters

Counter	Variable
dot11FailedCount	C_R
dot11TransmittedFragmentCount	C_{TF}
dot11ACKFailureCount	C_F
ifInUcastPkts	u_{IU}
ifInMulticastPkts	u_{IM}
ifInBroadcastPkts	u_{IB}
ifInOctets	u_{IO}
ifInDiscards	u_{ID}
ifOutUcastPkts	u_{OU}
ifOutMulticastPkts	u_{OM}
ifOutBroadcastPkts	u_{OB}
ifOutOctets	u_{OO}
ifOutErrors	u_{OE}

$$C_{TI} = u_{IU} + u_{IM} + u_{IB} + u_{ID} \qquad (8)$$

C_{TO} - number of successfully sent packets:

$$C_{TO} = u_{OU} + u_{OM} + u_{OB} - (u_{OE} + u_{OD}) \qquad (9)$$

The probability of packet collision p_c in the wireless medium determined by:

$$p_c = C_F / (C_{TF} + C_F) \qquad (10)$$

The probability of packet loss p_d in the wireless medium is obtained using:

$$p_d = C_R / (C_{TF} + C_R + u_{OD}) \qquad (11)$$

Using the ns-3 result we obtain α by p_d and p_c. Using α we obtain γ and perform a calculation to obtain delivery time. The results of this paper is used to perform real-time calculation of average multimedia traffic delivery time in enterprise WLAN.

References

1. Cisco Visual Networking Index Predicts IP Traffic to Triple from 2014–2019; Growth Drivers Include Increasing Mobile Access, Demand for Video Services—The Network.   https://newsroom.cisco.com/press-release-content?type=webcontent& articleId=1644203. Accessed 04 Mar 2021
2. Bianchi, G.: Performance analysis of the IEEE 802.11 distributed coordination function. IEEE J. Sel. Areas Commun. **18**, 535–547 (2000)

3. Duffy, K., Malone, D., Leith, D.J.: Modeling the 802.11 distributed coordination function in non-saturated conditions. IEEE Commun. Lett. **9**, 715–717 (2005)
4. Zhang, X.: A new method for analyzing nonsaturated IEEE 802.11 DCF networks. IEEE Wirel. Commun. Lett. **2**, 243–246 (2013)
5. Gao, Y., Sun, X., Dai, L.: Throughput optimization of heterogeneous IEEE 802.11 DCF networks. IEEE Trans. Wirel. Commun. **12**, 398–411 (2013)
6. Stojanova, M., Begin, T., Busson, A.: Conflict graph-based Markovian model to estimate throughput in unsaturated IEEE 802.11 networks. In: 15th International Symposium on Modeling and Optimization in Mobile, Ad Hoc, and Wireless Networks (WiOpt), Paris, pp. 1–9 (2017)
7. Rudenkova, M., Khayou, H., Abrosimov, L.I.: Investigation of the guaranteed traffic rate in enterprise WLAN. In: Vishnevskiy, V.M., Samouylov, K.E., Kozyrev, D.V. (eds.) DCCN 2020. CCIS, vol. 1337, pp. 70–81. Springer, Cham (2020). https://doi.org/10.1007/978-3-030-66242-4_6

A Model for 5G Millimeter Wave Service Rate Abstraction

Vladislav Prosvirov[1]([✉]) [ID], Emil Khayrov[1] [ID], and Evgeny Mokrov[2] [ID]

[1] Higher School of Economics (HSE University),
20 Myasnitskaya St., Moscow 101000, Russian Federation
vprosvirov97@gmail.com
[2] Peoples' Friendship University of Russia (RUDN University),
6 Miklukho-Maklaya St, Moscow 117198, Russian Federation
evmokrov@sci.pfu.edu.ru

Abstract. An accurate channel abstraction is an important part of solving resource allocation problems in 5G networks. The purpose of this work is to present a channel abstraction model that takes into account changes in total cell capacity over time and does not require complex simulation. We analyze the cell capacity, taking into account the 3GPP standards. Based on the obtained values, we propose a model based on the first-order autoregression method. The numerical results show that the suggested method provides an accurate approximation of the total cell rate. The obtained results can be used in applied research in 5G wireless networks.

Keywords: Total cell capacity · First-order autoregression · Simulation

1 Introduction

Currently, wireless networks have become an important part of our lives. Every day, the number of technologies and applications that require high data rates, low latency and high reliability is increasing. An example of such technologies are ultra-high-definition (UHD) video, Internet of Things (IoT), Tactile Internet, etc. [12]. In connection with this, in order to satisfy the growing demand, a transition is being made to 5G networks using mmWave. In such networks are characterized by such functional features as multi-connectivity [9,10], network slicing [2,4], implemented using redundancy algorithms, prioritization and transmission scheduling. To develop such mechanisms for improving the reliability and quality of user service, a adequate model of the communication channel is needed. Simulation modeling often turns out to be quite a difficult solution due to the need to model many factors, such as mobility and environmental conditions. In mathematical models, it is generally necessary to assume a constant value for the communication channel capacity, however, in real conditions, this parameter may change over time due to external factors.

The research was funded by the Russian Science Foundation, project No22-29-00222 (https://rscf.ru/en/project/22-29-00222).

In this paper, we propose an accurate abstraction of the communication channel for analyzing the total cell capacity, which is a compromise between the use of simulation and the constant capacity assumption often applied in the mathematical approach.

The paper is organized as follows: in Sect. 2 we define the system model, where we describe all the main sub-models that we used, such as user mobility, the propagation model, and blockage conditions. In Sect. 3 we describe the simulation approach that we used for the analyze the total capacity estimation, also we define the first- order autoregressive model (AR(1)) that we propose as a model for the abstraction of the communication channel. In Sect. 4 we evaluate the change in the total capacity in time and compare the results with the AR(1) model. Conclusions are drawn in the last section.

2   System Model

This section of the article will introduce the sub-models that are needed for the simulation, including the features of user movement and signal propagation. After that, we present the metrics that we use to evaluate the results obtained.

2.1   Deployment Model

In our study, we consider a mature phase of mmWave NR market penetration by considering a well-provisioned mmWave BS deployment. In this deployment, we consider a circularly-shaped service area of a mmWave BS with radius r_{BS}. We assume that r_{BS} is such that no outage happens at the cell boundary. This radius is determined in Sect. 3 by utilizing the set of NR modulation and coding schemes (MCS) and propagation model defined below. The heights of the BS is constant and given by h_{BS}. The carrier frequency is f_c while the available spectrum is B Hz.

2.2   Mobility and Blockage Models

Users' Mobility: In this paper, we assume the use of Random Direction Mobility (RDM) [8] as a model of user mobility. According to this model, each i-th user $(i = 1, \ldots, N)$ chooses a random direction uniformly distributed over $[0, 2\pi)$ and moves in this direction at a fixed constant speed v during an individual moving period exponentially distributed with mean τ_{RDM}. Once this run time period expires, the procedure for the user is repeated. Note that the velocity and direction of the movement of one user are independent of the speed and direction of other users.

LoS/nLoS Conditions: Following [1] the probability of LoS and nLoS between the base station (BS) and user equipment (UE) is defined as

$$
p_{\mathrm{LoS}}(r) = \begin{cases} 1, & r \leq 18, \\ 18r^{-1} + e^{-\frac{r}{36}}(1 - 18r^{-1}), & r > 18. \end{cases} \tag{1}
$$

where r is the 2D distance between UE and BS.

The state of LoS/nLoS determined for each user according to (1) remains unchanged during an exponentially distributed time period with mean τ_{LoS}.

Human-Body Blockage Conditions: In addition to LoS/nLoS states, we also take into account human blocking states, where HB is blocked by human, nHB is not blocked by a human. We suppose that human blockers are represented by cylinders with the base radius of r_B and the height of h_B meters and the human-body blockage probability can be evaluated as [3]

$$p_{HB}(r) = 1 - e^{-2\zeta_{HB}r_B\left(\sqrt{r^2+\Delta h^2}\frac{h_B-h_{UE}}{\Delta h}+r_B\right)}, \tag{2}$$

where ζ_{HB} is the density of the blockers per square meter, r is the two-dimensional (2D) distance between BS and UE, and $\Delta h = h_{BS} - h_{UE}$ where h_{BS} and h_{UE} are the BS and UE heights respectively.

2.3   Propagation Model

To represent the mmWave propagation losses we utilize 3GPP Urban-Micro (UMi) Street-Canyon model Canyon-Street propagation model. According to [1], the path loss for the frequency band 0.5–100 GHz can be expressed in dB as

$$\text{PL}_{[dB]}(d) = 10\alpha_{LoS/nLoS}\log_{10}d + \beta_{HB/nHB} + 20\log_{10}f_c, \tag{3}$$

where d is the three-dimensional (3D) distance in meters between the NR BS and the UE, $\alpha_{LoS/nLoS}$ is a coefficient being 21 under the line-of-sight (LoS) and 31.9 under non-line-of-sight (nLoS) conditions, $\beta_{HB/nHB}$ is 32.4 when the UE is not blocked by human (nHB) and 52.4 when the UE is blocked by human, f_c is the carrier frequency measured in GHz.

The signal-to-interference-plus-noise ratio (SINR) can be written as:

$$\text{SINR}(d) = \frac{P_{BS}G_{BS}G_{UE}}{\text{PL}(d)N_0B_{PRB}M_{SF}L}, \tag{4}$$

where P_{BS} is the emitted power, G_{BS} and G_{UE} are respectively the BS and UE antenna gains, N_0 is the thermal noise power spectral density, B_{PRB} is the size of the physical resource block (PRB), $M_{SF_{[dB]}} \curvearrowright Norm(0, \sigma_{SFLoS/SFnLoS})$ is the slow fading and L represents aggregated losses given, in decibels, by

$$L_{[dB]} = M_I + F_N + L_C \tag{5}$$

with M_I being the interference margin, F_N the noise figure, and L_C the cable losses.

174   V. Prosvirov et al.

2.4   Resource Allocation Model

In our study we use the total cell capacity as a main metric of interest. To
calculate it, we assume a uniform distribution of BS resources among all UEs.
We divide the simulation time into time slots of length Δt and at the beginning
of each slot $[t_j, t_j + \Delta t)$, $j = 1, \ldots, T$, $t_1 = 0$ the distance between each UE and
BS is calculated to estimate SINR of a user i, and his spectral efficiency $\eta_{i,j}$, by
utilizing NR modulation and coding scheme (MCS) to SINR mapping, shown in
Table 1. Finally, the total BS capacity for $t \in [t_j, t_j + \Delta t)$ is approximated as

$$C(t) = C(t_j) = \frac{B}{N} \sum_{i=1}^{N} \eta_{i,j}, \tag{6}$$

where B denotes the bandwidth.

Table 1. CQI, MCS, and SINR mapping for 5G NR [11]

CQI	MCS	Spectral efficiency	SINR in dB
0	out of range		
1	QPSK, 78/1024	0.15237	−9.478
2	QPSK, 120/1024	0.2344	−6.658
3	QPSK, 193/1024	0.377	−4.098
4	QPSK, 308/1024	0.6016	−1.798
5	QPSK, 449/1024	0.877	0.399
6	QPSK, 602/1024	1.1758	2.424
7	16QAM, 378/1024	1.4766	4.489
8	16QAM, 490/1024	1.9141	6.367
9	16QAM, 616/1024	2.4063	8.456
10	64QAM, 466/1024	2.7305	10.266
11	64QAM, 567/1024	3.3223	12.218
12	64QAM, 666/1024	3.9023	14.122
13	64QAM, 772/1024	4.5234	15.849
14	64QAM, 873/1024	5.1152	17.786
15	64QAM, 948/1024	5.5547	19.809

3   The Proposed Approximation

This section first presents a method of simulation to estimate the total cell
capacity, and then we introduce a mathematical model based on a first-order
autoregressive model AR(1). Finally, we evaluate the model's accuracy by comparing its results with those from the simulation.

3.1   System Level Simulations

This subsection describes a simulation approach for estimating the total cell capacity in the process of serving mobile users by a single base station. We assume that coverage radius of the BS r_{BS} is such that no outage happens at the cell boundary with probability p_{out}. We are interested in characteristics of the stochastic process which describes the total cell capacity behavior for a different number of users. We consider a single base station deployed to serve N users during the simulation time T.

In calculating the BS coverage radius, we assume the worst scenario, such that each UE is in states nLoS and HB, the value of SINR equal to threshold value SINR$_{thre}$. The path loss in (3) can be represented in the linear scale by utilizing the model in the form $Ay^{\alpha_{nLoS}}$, where $A = 10^{2\log_{10} f_c + \beta}$. The distance d in (3) can be represented as

$$d = \sqrt{r_{BS}^2 + [h_{BS} - h_{UE}]^2} \tag{7}$$

By solving (4) with respect to r_{BS} we can determine the BS coverage radius as

$$r_{\text{BS}} = \sqrt{\left(\frac{P_{\text{BS}}G_{\text{BS}}G_{\text{UE}}}{10^{\frac{\beta_{\text{HB}}}{10}} f_c^2 N_0 B_{\text{PRB}} M_{\text{SF}}\text{SINR}_{\text{thre}}}\right)^{\frac{2}{\alpha_{\text{nLoS}}}} - (h_{BS} - h_{UE})^2}, \tag{8}$$

where M_{SF} is the slow fading margin given by

$$M_{\text{SF[dB]}} = \sqrt{2}\,\text{erfc}^{-1}(2p_{\text{out}})\sigma_{\text{SFnLoS}} \tag{9}$$

with erfc$^{-1}(\cdot)$ denoting the inverse complementary error function.

UEs are modeled as cylinders, initially distributed uniformly in the service area of the BS, after that at each discrete time t_j, $j = 1, \ldots, T$, the new coordinate of the i-th user is calculated according to the RDM model. The probabilities of HB/nHB and LoS/nLoS are determined by (2) and (1) respectively. The times spent by the UE in the LoS/nLoS and HB/nHB states are exponentially distributed with average τ_{LoS} and τ_{HB}. Also σ_{SF} is selected based on LoS/nLoS state. At each time t_j, $j = 1, \ldots, T$, for each i-s user we calculate the 2D distance from the UE to the BS, which is located in the center of the simulation area with coordinates $(x_{BS}; y_{BS}) = (0, 0)$, based on the obtained distance we compute $SINR$, spectral efficiency look up the η_i according to Table 1 and finally obtain the total capacity value. This is detailed in Algorithm 1.

3.2   Approximate Model

As a mathematical model for the total cell capacity, we propose to use the first-order autoregressive process AR(1) [7]. A process is said to be AR(1) if it obeys the following recursion

$$Y_n = \phi_0 + \phi_1 Y_{n-1} + \epsilon(n), n = 0, 1, \ldots, \tag{10}$$

Algorithm 1: Simulation algorithm

Input: Number of users N, simulation time T, BS coverage radius R (8),
     frame time t_{frame}, user speed v.
Output: Total capacity $C := [C(t_1), C(t_2), ..., C(t_T))]$

---

for $i = 1; n \leq N;$ **do**
    $(x_i(t_0), y_i(t_0)) \backsim Uniform(-R, R)$
    $(t_{move_i}, t_{los_i}, t_{bl_i}) = 0$
    $d_{frame} = v t_{frame}$

for $j = 1; j \leq T;$ **do**
    for $i = 1; n \leq N;$ **do**
        $t_{move_i} = t_{move_i} - t_{frame}$
        $t_{los_i} = t_{los_i} - t_{frame}$
        $t_{bl_i} = t_{bl_i} - t_{frame}$
        if $t_{move_i} \leq 0$ **then**
            $\alpha_i \backsim Uniform(0, 2\pi)$
            $t_{move_i} \backsim Exp(\tau_{RDM}^{-1})$
            $\Delta x_i = d_{frame} \cos(\alpha_i)$
            $\Delta y_i = d_{frame} \sin(\alpha_i)$
        if $t_{los_i} \leq 0$ **then**
            $t_{los_i} \backsim Exp(\tau_{los}^{-1})$
            p_{los_i} - compute by (1)
            sample LoS/nLoS state according to p_{los_i}
            M_{SF_i}(LoS/nLoS state) $\backsim \mathcal{N}(0; \sigma_{SF_{LoS/nLoS}})$
        if $t_{bl_i} \leq 0$ **then**
            $t_{bl_i} \backsim Exp(\tau_{bl})$
            compute p_{bl_i} by (1)
            sample HB/nHB state according to p_{bl_i}
        if $(x_i(t_{j-1}) + \Delta x_i)^2 + (y_i(t_{j-1}) + \Delta y_i)^2 \geq R^2$ **then**
            $\alpha_i = \alpha_i + \pi$
            $\Delta x_i = d_{frame} \cos(\alpha_i)$
            $\Delta y_i = d_{frame} \sin(\alpha_i)$
        $x_i(t_j), y_i(t_j) = x_i(t_{j-1}) + \Delta x_i, y_i(t_{j-1}) + \Delta y_i$
        compute $SINR_i(d_i(t_{j-1}))$ by (4)
        η_i - obtain from Table 1
        $C_i(t_j) = \eta_i * \frac{B}{N}$
    $C(t_j) = \sum\limits_{i=1}^{N} C_i(t_j)$

---

where ϕ_0 and ϕ_1 are constants, $\epsilon(n), n = 0, 1, \ldots$ are independently and identically distributed random variable having the same normal distribution with zero mean and variance $\sigma^2[\epsilon]$. If a process given by (10) is covariance stationary we have

$$E[Y] = \mu, \ \sigma^2[\epsilon] = \gamma_0, \ Cov(Y_0, Y_i) = \gamma_i, \tag{11}$$

where μ, γ_i, $i = 0, 1, \ldots$, are some constants. The mean, variance and covariance of AR(1) are given by

$$\mu = \frac{\phi_0}{1 - \phi_1}, \ \sigma^2[X] = \frac{\sigma^2[\epsilon]}{1 - \phi_1^2}, \ \gamma_i = \phi_1^i \gamma_0. \tag{12}$$

The parameters of AR(1) models can be found as follows

$$\begin{cases} \phi_1 = K_Y(1), \\ \phi_0 = \mu_Y(1 - \phi_1), \\ \sigma^2[\epsilon] = \sigma^2[Y](1 - \phi_1^2). \end{cases} \tag{13}$$

where $K_Y(1)$, μ_Y and $\sigma^2[Y]$ are the lag-1 autocorrelation coefficient, mean and variance of cell rate observations, respectively. Using formulas above we can get a time series for the total capacity based on the data obtained from the simulation.

4   Numerical Results

In this section, we present our results. We start by presenting the time series for the total cell capacity and their distributions, then we present the results of comparing the simulation results with the AR(1) model. The system parameters for cell capacity characterization are provided in Table 2. We analyze the results for 30 and 60 UEs as these are the most commonly used values in technical reports [5].

4.1   Statistical Characteristics of the Cell Rate Process

We start with a time series analysis of the total cell capacity obtained from the simulation. Figure 1 reports the change in the total cell rate over time. Figure 1(a) shows the change in speed over the entire duration of the simulation time and Fig. 1(b) demonstrates in detail the evolution of the total cell rate over a period of 200 s. The hopping character of the total cell rate is caused by the movement of each UE and the change in LoS/nLoS and HB/nHB states during the simulation time. Also, Fig. 1 shows that the average total cell rate remains unchanged at $\bar{C} \approx 2000$ Mbps.

Figure 2 reports the probability density functions (pdf) and cumulative distribution functions (CDF) of the total cell capacity for 30 and 60 UEs. By analyzing the presented data, one may observe that the distribution is close to Log-Normal according to the central limit theorem. In Fig. 2(a), due to the increase in the number of UEs from 30 to 60, the distribution of their coordinates over the service area becomes "denser" and most of the UEs have cell capacity close to each other, therefore, the value of the standard deviation of the cell capacity decreases. The "tails" on the left side of the pdf plot are due to the fact that a large number of UEs may be in LoS and HB at the same time period, which leads to a decrease in cell

Table 2. Capacity simulation parameters

Symbol	Value	Description
B	400 MHz	Bandwidth
B_{PRB}	1.44 MHz	PRB size
f_c	28 GHz	Operating frequency
F_N	7 dB	Noise figure
G_{BS}	11.57 dBm	BS antenna gain
G_{UE}	5.57 dBm	UE antenna gain
h_B	1.7 m	Blocker height
h_{BS}	10 m	BS height
h_{UE}	1.5 m	UE height
L_C	2 dB	Cable losses
M_I	3 dB	Interference margin
N_0	–174 dBm/Hz	Noise power spectral density
P_{BS}	24 dBm	Transmitting BS power
p_{out}	0.05	Outage probability
$SINR_{thre}$	–9.478 dB	SINR threshold
T	10000	Simulation time
Δt	1 s	Time slot length
v	1.5 m/s	User speed
σ_{SFnLoS}	8.2 dB	nLoS shadow fading STD
σ_{SFLoS}	4 dB	LoS shadow fading STD
τ_{RDM}	30 s	Mean run time in the RDM model
τ_{LoS}	30 s	Mean LoS/nLoS state time
τ_{HB}	3 s	Mean HB/nHB state time

capacity. Figure 2(b) shows the CDFs of the total cell capacity. The figure shows a high probability that the total cell capacity is not lower than approximately 1900 Mbps. The differences between the lines for 30 and 60 UEs are due to a decrease in the standard deviation with an increase in the number of UEs.

Figure 3 presents the autocorrelation function for 30 and 60 UEs. As it can be seen, the total cell capacity decreases exponentially and the process is a short-term memory process. As a result, the total cell capacity can be approximated by a Markov process and we can make a tentative conclusion that the use of the AR(1) model will be adequate. The next section presents the results for the AR(1).

4.2   Approximation Accuracy

This section presents the comparison of the simulation approach and AR(1) model. Figure 4 shows a comparison of the pdf of the total cell capacity obtained

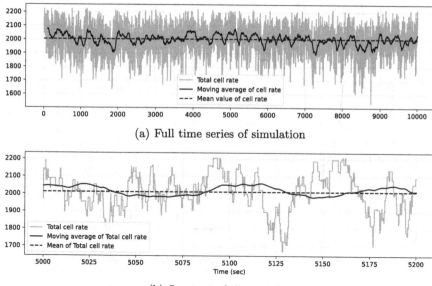

(a) Full time series of simulation

(b) Segment of times series

Fig. 1. Times series of the cell capacity

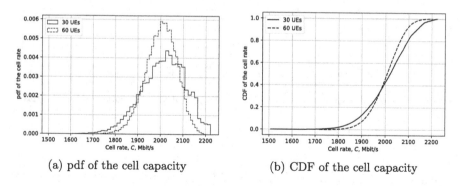

(a) pdf of the cell capacity

(b) CDF of the cell capacity

Fig. 2. Characteristics of the cell capacity process.

from the simulation and calculated using the AP(1) model for 30 and 60 UEs. Observe that the time series for the simulation and AR(1) have discrepancies due to the random component $\epsilon(n)$. However, the values for the AR(1) model are very close to the values obtained in the simulation. It is important that AR(1) takes into account the "tails" of the pdf, which is especially noticeable in Fig. 4(a). In Fig. 4(b), the approximation is more accurate, because the total cell capacity for 60 users has a lower standard deviation value.

The Fig. 5 shows that the values of ACF, as mentioned above, decrease exponentially, and the process also is a short-term memory process. This behavior is

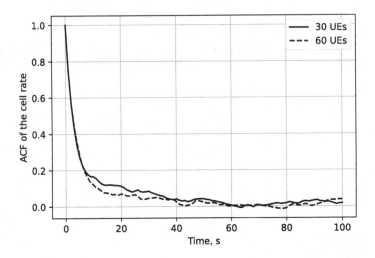

Fig. 3. Autocorrelation function of total cell capacity

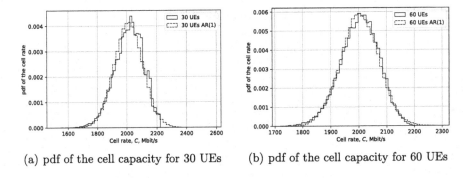

(a) pdf of the cell capacity for 30 UEs   (b) pdf of the cell capacity for 60 UEs

Fig. 4. pdf of total capacity for simulation and AR(1)

typical for the Markov process, of which AR(1) is a special case, so we observe a good approximation.

It can also be concluded that with a larger number of UEs the approximation will be accurate due to a "denser" arrangement of UEs and a smaller value of the standard deviation. This can also be seen from Table 3, which shows the values of the average speed and standard deviation for a different number of users. It can be observed that the standard deviation decreases as the number of users increases from 10 to 60. Furthermore, the approximation can be made even more accurate using Markov Modualated Process or Gamma-Beta Autoregressive Process (GBAR) [6].

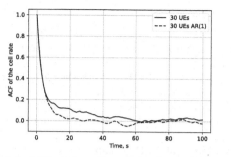

(a) ACF of the cell capacity for 30 UEs

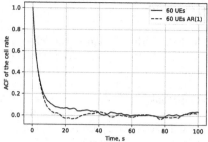

(b) ACF of the cell capacity for 60 UEs

Fig. 5. ACF of total capacity for simulation and AR(1)

Table 3. Comparison of simulation and AR(1) results

Cell capacity parameters obtain from simulation						
UEs number	10	20	30	40	50	60
Average cell rate	2011.5	2008.6	2010.5	2008.4	2010.1	2000.4
Standart deviation of cell rate	159.9	115.4	94.6	85.1	74.2	66.5
Autocorrelation function(Lag=1)	0.726	0.729	0.74	0.749	0.738	0.720
Cell capacity parameters obtain from AR(1)						
UEs number	10	20	30	40	50	60
Average cell rate	2011.2	2010	2008.4	2004.4	2012.1	2000.9
Standart deviation of cell rate	158.9	115.6	94.4	87.4	75	65.4
Autocorrelation function(Lag=1)	0.732	0.725	0.742	0.763	0.741	0.714

5   Conclusions

A method has been proposed for carefully abstracting the communication channel in 5G mmWave networks. The results obtained in the article showed that the total speed can change greatly in time and our method by us allows to well approximate the capacity changes. It can also be seen from the results that the accuracy of the approximation increases with an increase in the number of users. The baseline results of the present article demonstrate that the proposed method is a good accommodation between mathematical and simulation approaches and allows precise evaluation and analysis without repeated simulation modeling. The task of further research is to develop a method for obtaining parameters for AR(1) without using a simulation approach.

References

1. 3GPP: Study on channel model for frequencies from 0.5 to 100 GHz (Release 14). 3GPP TR 38.901 V14.1.1, July 2017
2. Foukas, X., Patounas, G., Elmokashfi, A., Marina, M.K.: Network slicing in 5G: survey and challenges. IEEE Commun. Mag. **55**(5), 94–100 (2017)
3. Gapeyenko, M., et al.: Analysis of human-body blockage in urban millimeter-wave cellular communications. In: Communications (ICC), 2016 IEEE International Conference on, pp. 1–7. IEEE (2016)
4. Gaydamaka, A., Yarkina, N., Khalina, V., Moltchanov, D.: Comparison of machine learning algorithms for priority-based network slicing in 5G systems. In: 2021 13th International Congress on Ultra Modern Telecommunications and Control Systems and Workshops (ICUMT), pp. 72–77 (2021). https://doi.org/10.1109/ICUMT54235.2021.9631633
5. ITU-R Rec. Y.2410: Minimum requirements related to technical performance for IMT-2020 radio interface(s) (2020)
6. Koucheryavy, Y., Moltchanov, D., Harju, J.: A novel two-step mpeg traffic modeling algorithm based on a gbar process. In: International Conference on Network Control and Engineering for QoS, Security and Mobility, vol. 5, pp. 293–304 (2002). https://doi.org/10.1007/978-0-387-35620-426
7. Moltchanov, D.: State description of wireless channels using change-point statistical tests. In: Braun, T., Carle, G., Fahmy, S., Koucheryavy, Y. (eds.) WWIC 2006. LNCS, vol. 3970, pp. 275–286. Springer, Heidelberg (2006). https://doi.org/10.1007/11750390_24
8. Nain, D., Towsley, B.L., Liu, Z.: Properties of random direction models **3**, 1897–1907 (2005)
9. Petrov, V., et al.: Dynamic multi-connectivity performance in ultra-dense urban mmWave deployments. IEEE J. Sel. Areas Commun. **35**(9), 2038–2055 (2017). https://doi.org/10.1109/JSAC.2017.2720482
10. Ravanshid, A., et al.: Multi-connectivity functional architectures in 5G. In: 2016 IEEE International Conference on Communications Workshops (ICC), pp. 187–192 (2016). https://doi.org/10.1109/ICCW.2016.7503786
11. Samuylov, A., et al.: Characterizing resource allocation trade-offs in 5G NR serving multicast and unicast traffic. IEEE Trans. Wirel. Commun. **19**(5), 3421–3434 (2020). https://doi.org/10.1109/TWC.2020.2973375
12. Yastrebova, A., Kirichek, R., Koucheryavy, Y., Borodin, A., Koucheryavy, A.: Future networks 2030: architecture amp; requirements, pp. 1–8 (2018). https://doi.org/10.1109/ICUMT.2018.8631208

Cybersecurity System with State Observer and K-Means Clustering Machine Learning Model

Artur Sagdatullin[(✉)]

Intelligent Oil and Gas Technologies, Kazan, Russian Federation
saturn-s5@mail.ru

Abstract. Machine learning is an automated mechanism to help human intelligence detect potential risks. These techniques could be implemented in the following ways: spam detection based on text analysis, anomalies detection, and data prediction. Such systems give capabilities to analyze large scale data-sets of logs or texts to make sense of similar parameters. The system's goal is to iterate unlabeled data and find similarities or patterns hidden in its structure as similar components may be drawn from log files with the data entries as machine messages. Determining their type could facilitate the machine learning system and help automatically classify raw data. Timely analysis of log files and text messages is essential as many systems' problems are disclosed retroactively. Therefore, modern techniques are proposed with forehanded streaming observing of the system. A text analysis algorithm based on the k-means clustering is modeled for detecting classes of messages. It implemented a discrete-time observer algorithm for the cybersecurity of industrial networks.

Keywords: Machine learning · Cybersecurity · Discrete-time observer · K-Means Clustering · Industrial network

1 Introduction

In the last decades, there has been significant interest in cybersecurity as today's world of interconnected systems is potentially a target for emerging threats. Industrial applications and networks are automated by internet technologies and internet of things devices. In this environment, it's hard to assess hidden dangers. Also, the last investments of researchers from all over the world in computer science spawn the vast area of artificial intelligence systems such as machine learning and learning algorithms. Applications based on machine learning are used in different areas of the human being. One of the prospective fields of implementing such technologies is cybersecurity. Researchers could detect potential threats and malware using specially intended cybersecurity machine learning models. There are many definitions for cybersecurity, but this paper comprises information about the internal network devices' security.

Machine learning is a proven reasoning and automated mechanism to help human intelligence detect potential risks. The techniques could be implemented

V. M. Vishnevskiy et al. (Eds.): DCCN 2022, CCIS 1748, pp. 183–195, 2023.
https://doi.org/10.1007/978-3-031-30648-8_15

in the following ways: spam detection based on text analysis, anomalies detection, and data prediction. This paper will discuss the text analysis algorithm based on the clustering machine learning technique for detecting potential areas of threatening messages in the modeled industrial system.

2   Related Work

Applications of machine learning such as clustering algorithms, convolution neural networks, and language models have been extensively used in data analysis: anomaly detection in industrial control systems [1], support vector machine k-means clustering combination for intrusion detection in supervisory control and data acquisition systems [2,3], fuzzy clustering classification of spam messages in email spamming [4], cyber attacks segmentation models [5], cyberattack patterns by analyzing social networks [6], internet hosts profiling and clustering [7], KNN approach for network traffic classification [8], fuzzy intuition clustering of network [9], anomalous network flows online detection [10], data set classification by fuzzy c-means clustering [11], entropy clustering for forecasting of DDoS attacks [12], malware detection based on cluster ensemble classifiers [13].

The clustering approach helps to solve grouping problems on big data sets with similar entities. K-means clustering is an algorithm related to unsupervised machine learning, where sets of data are divided by a number of groups with the shortest distance of data points from the centroids of each cluster. The similarity of entities could be measured by different metrics, representing the distance between data points: Euclidian, Manhattan, Chebyshev, Minkowsky. The complexity of the k-means clustering algorithm is NP-Hard, which means for every problem L of size n in NP (nondeterministic polynomial) number of iterations to solve this problem is a polynomial time to G. The time complexity of the k-means algorithm reduces to $O(kmdt)$, where k is the number of clusters, m is some entries of d-dimensional data with the number of iterations t. It means that with the growth of a number of entries, dimensionality of the data, and the number of iterations algorithm, approximately $t \sim n$ reduces to $O(m^2)$. But limiting the number of iterations to a constant value reduces the algorithm time complexity to $O(m)$ [14]. Close to the k-means algorithm is Lloyd's (Voronoi tessellation) algorithm for finding centroids of evenly spaced points in Euclidean subsets. It repeatedly reiterates data for new entries to find the nearest centroid over the cluster space region. If the number of clusters corresponds to $O(\log n)$ the algorithm reduces to quasi-polynomial time [15].

Cybersecurity clustering algorithms are often used to analyze data and detect and operate in a given situation. The main difference between clustering and classification is that there are no determined classes beforehand. The system's main task is to go through unlabeled data and find similarities or patterns hidden in its structure. As human capabilities to analyze large scale data-sets of logs or machine messages will likely not be enough to make sense of similar parameters. Similar components may be drawn from log files with the data entries as machine messages. Determining their type could potentially facilitate the classification system and allow to discover and classify raw data analysis automatically. Log files and messages contain essential information about events in a system for the last time range. Many problems in log analysis are disclosed as backdating. Therefore, modern online web services management techniques are proposed with timely streaming for real-world log files data-sets. Near real-time systems are necessary for online monitoring complex industrial systems and predicting the essential parameters [16,17,19]. Distributed file systems generate millions of logs and texts daily, and watching such parameters may help prevent specific industrial systems' faults.

3 State Observer Architecture for Cybersecurity of Industrial Network

3.1 Discrete-Time System Mathematical Definition

The state feedback control loop allows for managing indicators of the open-loop system and adjusting the poles of the closed-loop system through planning state feedback blocks. The significant coupling of objects and components of the system can be a considerable difficulty, both in software development and in the implementation of automated systems. For a continuous linear time-invariant system, equations are following:

$$\dot{x} = Ax + Bu$$
$$y = Cx \tag{1}$$

Unknown parameter $X(t)$ as a random variable with its own expectation $E(x)$ and distribution could be defined as:

$$x[k]; x \geq 0$$
$$x[k] = x[k-1] + z[k] \tag{2}$$

where $z[k]$ is a noise process, supposed to be $\sim N(0, \sigma_{nn}^2)$. Consider a LTI system with discrete time:

$$x(k) = A(k-1)x(k-1) + f(k)$$
$$z(k) = C(k)x(k) + w(k) \tag{3}$$

where $x(k)$ represents a state variables vector, $A(k-1)$ is the transition matrix, $f(k)$ is a noise vector inherited to state space model, $z(k)$ is the observations vector, $C(k)$ is a state matrix related to $x(k)$.

For sampling interval of T linear discrete-time system equations are following:

$$\dot{x}(k-1) = Lx(k) + Hu(k)$$
$$y(k) = Cx(k)$$
$$Lx = e^{AT} \tag{4}$$
$$H = e^{AT} \int_0^T e^{-A\tau} d\tau$$

where u is the control vector, x is the state vector, L is the unobservable per-turbations, H is the function of disturbing influences on the control object.

The state observer model may be used to implement controllable incoming data for the LTI system. It allows predicting the system's state based on the $(k-1)$ time-step measurement.

3.2   State System Observer Control Scheme

The representation of state vectors describes the characteristics of the control object for closed-loop systems with feedback, the tasks of which are to find the optimal location of the roots of the system equation by adjusting its parameters. Digital representation is the direct form of implementing state observers as software modules. Thus, a discrete description will be implemented for digital representation. Therefore, to describe such systems, it is required to represent the discrete state of the system (see Fig. 1).

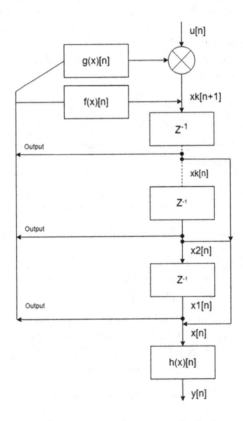

Fig. 1. State system observer control scheme

The State system observer control scheme solves the problem of monitoring parameters (A, B, C, D) of the industrial system at any given time. The system's state describes an estimate of the measurement of input and output. This expression shows that under known initial conditions, the state estimate is approximately equal to the actual value of the output signal. This fact gives us an observability definition concerning the designed machine learning model to process input text messages as parameter values.

4   Methodology of Machine Learning Algorithm Design

Machine learning algorithms comprise linear and logistic regression, polynomial regression, decision trees, support vector machines, random forests, etc. All these concepts are the tasks of supervised and unsupervised learning. Classification and clustering are the methods at the heart of the proposed system. The Clustering aims to distinguish n clusters in $[k - 1]$ observations and distribute the rest of estimates value points in the nearest mean cluster.

K-means clustering is related to unsupervised machine learning algorithms, which means no labeling data are provided. It is needed to solve an optimization problem to form clusters and determine the distance from centroids to data entries of related points. It is often called an expectation maximization problem, where a set of d-dimensional data representing observations are given $x \in X : Y_s$:

$$X(k) \in (x_1, x_2, x_3, ..., x_k),$$
$$Y_s(k) \in (y_1, y_2, y_3, ..., y_s),$$

$$J = \sum_{i=1}^{k} \sum_{j=1}^{y_s} \varsigma_{i,j} \parallel x_i - c_j \parallel^2, \tag{5}$$

$$for:$$

$$x \mid \forall y : f(x) \leq f(y).$$

where X is the observations, Y_s is the output subset of X, while $k \leq s$, representing the number of cluster centers. Data points x_k are related to the nearest cluster center by computing the sum of squared distances:

$$c_j = \frac{\sum_{i=1}^{k} \varsigma_{i,j} x_k}{\sum_{i=1}^{k} \varsigma_{i,j}} \tag{6}$$

$$\varsigma_{i,j} = \begin{cases} 1 & \text{if } x_i \in \text{nearest cluster } k, \\ 0 & \text{otherwise.} \end{cases} \tag{7}$$

For clustering, there are different techniques with their intrinsic advantages and weaknesses. In the first step, it is arbitrarily assigned centroids for the given number of clusters. An algorithm generally means delimiting n-dimensional data into separated subsets (see Fig. 2). In the second stage, each sample of data was assigned to the closest centroid, determined randomly at the previous step. Using the Euclidean norm, the distance between points is calculated.

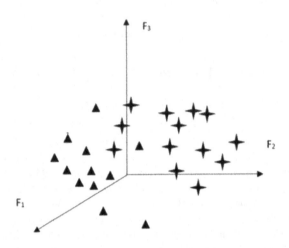

Fig. 2. Vector space model representation with samples of the data by features F_1, F_2, F_3

Where distance is the square root of squared differences of the sum for every dimension, then, for every instance, calculate the distance value from it to a new data point after choosing the k number of neighbors. For every k data set found in the data set, assign a new center to a point with a minimum distance. After, update the cluster centroid by calculating the mean of entries in the cluster.

$$\frac{\sum_{i=1}^{k} x_1}{n}, \frac{\sum_{i=1}^{k} y_1}{n}, \frac{\sum_{i=1}^{k} z_1}{n}, \tag{8}$$

The general Euclidean distance computed for k-means clustering algorithm as follows:

$$d_k = \sqrt{(x_1 - y_2)^2 + (x_2 - y_2)^2 \ldots} \tag{9}$$

The algorithm of the clustering problem states that by a given vector of parameters, it could be expected centroid distance to individual data points [20–23]. Then, the mean of the vector will be recomputed for a new centroid finding. For two classes problem (see Fig. 2), where presented stars and triangle classes, the triangle point is located at the center after calculation, and the distances for k nearest points will be considered as star points class. Moving further to the stars class center, the confidence of selecting triangles points to the stars class drops, plotting the boundary between the two categories. Choosing the proper

Algorithm 1. An algorithm according to the methodology

Require: $n \geq 0$ ▷ This is the number of clusters
Ensure: $k \geq 0$ ▷ Centroid data points
  while $i \neq 0$ **do**
    if X_i **then**
      $L(x_1, x_2) \leftarrow \sum_{i}^{[x_1] \in \mathcal{R}^N} (X_1[i], X_2[i])^2$ ▷ Categorize each point to its closest
centroid
        $k \leftarrow L(x_1, x_2)$
    else
        $k \leftarrow k$ ▷ Compute the new centroid
        $i \leftarrow i - 1$
    end if
  end while

---

parameter of k determines the sharpness of the outline border. Furthermore, essential features for new data points are reflected in weight parameters. Vector space models help to identify the nearest meaning of different types of words and sentences (Fig. 3).

An algorithm according to the described methodology is presented in the algorithm section:

The exemplary graphical algorithm could be presented in the block diagram, shown in Fig. 4. The data set is divided into k clusters such that every instance related to a particular data cluster allows structuring the frame of data into categories.

5   Experiment

For the experiment it was taking the training data about potentially threatening and spam messages inside the industrial network, represented by a mixed dataset of open data and log files collected from the system:

According to the presented algorithm, experiments were carried out to test it. The data representation problem depends on an unknown cluster number at once. Therefore, it was studied the following paths: a) Two cluster segmentation, where all data belongs to two classes (spam or not spam text); b) Three cluster segmentation, where all data belongs to three classes (spam, not spam, or suspicious). All data results presented in figures (see Fig. 5, 6).

In the Fig. 5 the clustering algorithm performance are shown. Training takes about 0.17 s for two cluster analysis and about 0.23 s for three cluster centroid problems with samples of about 5000. The next part is text processing, as the data is combined from various sources. It is essential to process data and transform it into an appropriate form for machine learning algorithms. All processing of the data may be represented by the following transformations Fig. 7.

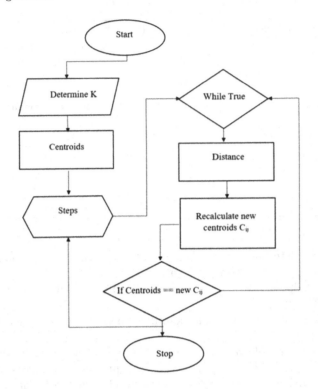

Fig. 3. Graphical representation of the k-means algorithm

	text
0	Thought praps you meant another one. Goodo! I'...
1	Good Luck! Draw takes place 28th Feb 06. Good ...
2	Eerie Nokia tones 4u, rply TONE TITLE to 8007 ...
3	URGENT! We are trying to contact you. Last wee...
4	I've sent L my part..
...	...

Fig. 4. Example of first rows of presented data-set

This transformation comprises data lowering, removing stop-words and special characters, numbers, white spaces, and other symbols.

After obtained data serve as inputs to the k-means clustering algorithm. Randomly chosen classes parameter k at the beginning helps to print result shown in figure below Fig. 8. The algorithm calculates the closest data points and separates them by the shortest distance calculation.

Fig. 5. Two cluster analysis problem

In Fig. 9 it is shown results of Elbow method for determining the optimal number of cluster. According to the graph, it is shown that the optimum k number is between 2 and 3. Therefore, the initial assumption about the number of clusters was correct.

The experimental results show that the proposed k-means clustering algorithm with the architecture of discrete-time observer supports industrial cybersecurity systems. Many disturbance parameters affect such systems' performance, including data set consistency and clearness. It implements a k-means clustering algorithm for data analysis with input text preprocessing transformations. The algorithm was tested for the initial assumption of class numbers, which shows the result conformed to the algorithm's results by the presented data-set. The Elbow model shows that an optimal number of clusters is from 2 to 3 for the data.

Fig. 6. Three cluster analysis problem.

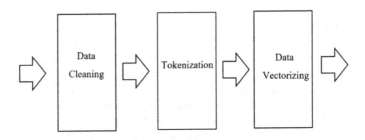

Fig. 7. Three cluster analysis problem.

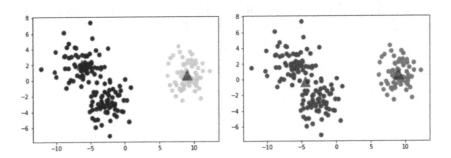

Fig. 8. Experiment results of the k-means clustering algorithm.

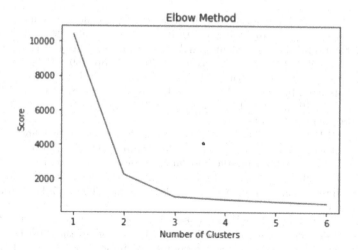

Fig. 9. Elbow method for determining the optimal number of cluster

6 Conclusion

Machine learning is an important technology for future cybersecurity and information and communication networks. This paper discusses the algorithm based on the k-means clustering technique for text analysis in the industrial system. The industrial network model implements a discrete-time observer algorithm with k-means clustering. It is an essential for data analysis and machine learning. Presented algorithms show good results in working with high dimensional and text data sets.

References

1. Kiss, I., Genge, B., Haller, P., Sebestyen, G.: Data clustering-based anomaly detection in industrial control systems. In: Proceedings of the 2014 IEEE 10th International Conference on Intelligent Computer Communication and Processing, ICCP 2014, 6937009, pp. 275–281 (2014)
2. Maglaras, L.A., Jiang, J.: OCSVM model combined with K-means recursive clustering for intrusion detection in SCADA systems. In: Proceedings of the 2014 10th International Conference on Heterogeneous Networking for Quality, Reliability, Security and Robustness, QSHINE 2014, 6928673, pp. 133–134 (2014)
3. Sagdatullin, A.: Functioning and development of a real-time information system for the oil treatment technological process control. In: Proceedings of the 2020 2nd International Conference on Control Systems, Mathematical Modeling, Automation and Energy Efficiency, SUMMA 2020, 9280577, pp. 847–852 (2020)
4. Wijayanto, A.W., Takdir: Fighting cyber crime in email spamming: an evaluation of fuzzy clustering approach to classify spam messages. In: Proceedings of the 2014 International Conference on Information Technology Systems and Innovation, ICITSI 2014, 7048231, pp. 19–24 (2014)

5. Strapp, S., Yang, S.J.: Segmenting large-scale cyber attacks for online behavior model generation. In: Proceedings of 7th International Conference Social Computing, Behavioral-Cultural Modeling, and Prediction, SBP 2014 Washington, DC, USA, 1–4 April, ICITSI 2014 - Proceedings 7048231, pp. 169–177 (2014)
6. Du, H., Yang, S.J.: Discovering collaborative cyber attack patterns using social network analysis. In: Salerno, J., Yang, S.J., Nau, D., Chai, S.-K. (eds.) SBP 2011. LNCS, vol. 6589, pp. 129–136. Springer, Heidelberg (2011). https://doi.org/10.1007/978-3-642-19656-0_20
7. Wei, S., Mirkovic, J., Kissel, E.: Profiling and clustering internet hosts. In: Proceedings of International Conference on Data Mining (DMIN) (2006)
8. Wu, D., et al.: On addressing the imbalance problem: a correlated KNN approach for network traffic classification. In: Proceedings of the Network and System Security 8th International Conference, NSS 2014 Xi'an, China, 15–17 October, pp. 138–151 (2014)
9. Panwar, A.: A kernel based Atanassov's intuitionistic fuzzy clustering for network forensics and intrusion detection. In: Proceedings of the 2015 IEEE/ACIS 14th International Conference on Computer and Information Science, ICIS 2015, 7166578, pp. 107–112 (2015)
10. Zolotukhin, M., Hamalainen, T., Kokkonen, T., Siltanen, J.: Online detection of anomalous network flows with soft clustering. In: 2015 7th International Conference on New Technologies, Mobility and Security - Proceedings of NTMS 2015 Conference and Workshops, 7266510 (2015)
11. Liu, L., et al.: Robust dataset classification approach based on neighbor searching and kernel fuzzy c-means. IEEE/CAA J. Automatica Sinica $2(3)$, 7152657, 235–247 (2015)
12. Olabelurin, A., Veluru, S., Healing, A., Rajarajan, M.: Entropy clustering approach for improving forecasting in DDoS attacks. In: 2015 IEEE 12th International Conference on Networking, Sensing and Control, ICNSC 2015, 7116055, pp. 315–320 (2015)
13. Hou, S., Chen, L., Tas, E., Demihovskiy, I., Ye, Y.: Cluster-oriented ensemble classifiers for intelligent malware detection. In: Proceedings of the 2015 IEEE 9th International Conference on Semantic Computing, IEEE ICSC 2015, 7050805, pp. 189–196 (2015)
14. Pakhira, M.K.: A linear time-complexity k-means algorithm using cluster shifting. In: 2014 International Conference on Computational Intelligence and Communication Networks, pp. 1047–1051 (2014). https://doi.org/10.1109/CICN.2014.220
15. Mahajan, M., et al.: The planar K-means problem is NP-hard. Theor. Comput. Sci. **442**, 13–21 (2012)
16. He, P., Zhu, J., Zheng, Z., Lyu, M.R.: Drain: an online log parsing approach with fixed depth tree. In: Proceedings of the International Conference on Web Services (ICWS), pp. 33–40. IEEE (2017)
17. Sagdatullin, A., Degtyarev, G.: Development of a cyber-physical system for neuro-fuzzy prediction of the concentration of the contained prime during transportation of oil wells emulsion. Stud. Syst. Decis. Control **417**, 169–180 (2022)
18. Katare, D., El-Sharkawy, M.: Embedded system enabled vehicle collision detection: an ANN classifier. In: 2019 IEEE 9th Annual Computing and Communication Workshop and Conference (CCWC), pp. 0284–0289 (2019)
19. Sagdatullin, A.: Application of fuzzy logic and neural networks methods for industry automation of technological processes in oil and gas engineering. In: Proceedings of the 2021 3rd International Conference on Control Systems, Mathematical Modeling, Automation and Energy Efficiency, SUMMA 2021, pp. 715–718 (2021)

20. Chen, Y., Khandaker, M., Wang, Z.: Pinpointing vulnerabilities. In: Proceedings of the 2017 ACM on Asia Conference on Computer and Communications Security, ASIA CCS 2017, pp. 334–345. ACM, New York (2017)
21. Ishida, C., Arakawa, Y., Sasase, I., Takemori, K.: Forecast techniques for predicting increase or decrease of attacks using Bayesian inference. In: Proceedings of PACRIM 2005 IEEE Pacific Rim Conference on Communications, Computers and Signal Processing, Victoria, 24–26 August, pp. 450–453. IEEE (2005)
22. Li, Z., et al.: VulDeePecker: a deep learning-based system for vulnerability detection. In: 25th Annual Network and Distributed System Security Symposium, NDSS 2018, San Diego, California, USA, 18–21 February (2018)
23. Thennakoon, A., et al.: Real-time credit card fraud detection using machine learning. In: 2019 9th International Conference on Cloud Computing, Data Science & Engineering (Confluence). IEEE (2019)

Collision Provenance Using Decentralized Ledger

Q. Zirak and D. V. Shashev$^{(\boxtimes)}$

National Research Tomsk State University, 36 Lenin Ave., Tomsk, Russian Federation
dshashev@mail.ru

Abstract. The coverage and communication lag has been worked on in swarms of drones with different formation patterns, however; decentralization of the participants in the swarm is little in consideration and research. A flying ad-hoc network might be a good solution if implemented on peer-to-peer basis, but the physical layer of the swarm by itself is not sufficient to ensure immutability, integrity, and validity of data within the swarm. Due to the lack of central leadership in decentralized networks, a consensus is required for all the participants to agree on an intent or an action to be executed. A shared data structure or distributed ledger can be introduced in the application layer with some known consensus mechanisms such as proof-of-work or proof-of-stake to overcome the leadership problem. A state change can only occur by the consensus of all participants and mirroring the change. The data emitted by the participants will be stored in smart contracts where the storage is practically immutable outside of consensus. Thus, the collision data carries a layer of trust within itself if stored on a greatly distributed ledger and becomes a good approach for collision provenance in shared airspace. In this paper, we will integrate distributed ledger such as Ethereum and Hedera Hashgraph with a swarm starting with a brief comparison between both.

Keywords: Decentralization · blockchain · hashgraph · consensus · collision · Ledger

1 Introduction

Flying ad-hoc network can be said as the first step towards decentralization after laying out the physical connections. Data packets are routed and forwarded from source to destination, hop by hop, via the transit drones. In typical centralized model, data packets are routed or forwarded with client-server architecture. These data packets are not sent randomly but follow well known reactive and proactive routing protocols. Some well known such protocols are Destination Dynamic Source Routing (DSR) and Sequenced Distance Vector (DSDV) and respectively. These routing protocols get the best pathway based on some metrics for packet delivery [1]. Comparatively however, our motivation[1], inspiration,

---

[1] Supported by Tomsk State University Development Program (Priority-2030).

and contribution goes beyond. We have based our work in the application layer of the network in the form of trust. Integrating a shared data structure or a distributed ledger, such as Ethereum blockchain or Hedera hashgraph, is a good solution to avoid decentralization problems. Having same state across all participants ensures immutability, validity and integrity of data if supported by consensus mechanism and governance.

Distributed Ledgers (DLTs) such as Bitcoin [2], Ethereum [3], Hedera, Solana [4], Avalanche, Cosmos, Polkadot and more, have become the leaders of progress in decentralization. Each DLT has some consensus mechanism working to maintain a shared state. Proof-of-Work, Proof-of-Stake, or Proof-of-History are some examples of consensus mechanisms. Consensus mechanism lets the participants decide who is going to update the shared database if data is emitted by any participant. The shared data state is stored by each participant with mandatory requirement of having a private and public key to interact with it for a state change. External set of nodes maintain the DLT and the drones themselves do not participate in the consensus. The participants also do not store copy of the ledger but act as remote clients to utilize a 'greatly distributed' DLT. To become greatly distributed, the nodes are required not to be pooled together too much. A not so greatly distributed ledger is prone to 51% attack on the network [5]. All the previously mentioned DLTs can be considered greatly distributed and can be used for provenance. The only job of the drones in a swarm is to periodically send and publish GPS information along with additional required information to topics created in the DLT. Once published, the information becomes immutable and a good source for provenance of collision in a shared airspace.

1.1 Blockchain

A singly linked list can be considered a very basic blockchain. Both are sequential list of objects each pointing to the next object. Blockchain has sequential list of blocks of information or chain of blocks [6]. The information within each block is stored as transaction which is an input that changes the global state of the network. Instead of a pointer, as in linked list, blockchain uses hash of all transaction in a block as a pointer. When a block is formed, it stores this hash pointer of the previous block in its block header. Patricia Merkle Trie is used for hashing and retrieving stored data. Leaf nodes are the transactions in a block, and the root hash of these transactions is kept by the next block. Since hash value changes due to any single change in data, it prevents data maliciously being changed. As a result, a change will require consensus by all the participants. Not only that, blocks after the exploited block, will require recalculation of their hashses. This creates immutability of information as shown in Fig. 1.

It is to be noted that every technology has its limitations and blockchain is no different. One of the most commonly pointed out limitation is the fact that it can be forked thus resulting into two paths of blocks formation. This happened with Ethereum (ETH) when Ethereum classic (ETC) separated as a result of hack happened in 2016. Community was split whether to roll back the hack through governance or proceed with it. Thus, ETH separated from ETC as a result of rollback. Another limitation of blockchain is the fact that even if there is no activity going on, the blocks formation is still continued. Blockchains also do not respect the order of transactions, rather the transaction mining depends upon how much gas fee is supplied to the miners. Thus, the miners order transactions to boost their rewards which introduces unfair ordering. Another limitation in some blockchains is finality takes a lot of time because of higher block formation time. Higher block formation time also makes blockchain difficult to be hacked.

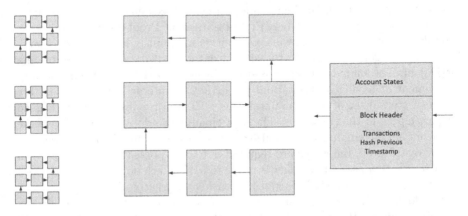

Fig. 1. Blockchain with its origin block at bottom right and current block at top left. Every participant stores the same copy of blockchain and requires consensus for update.

1.2   Hashgraph

Hashgraph is another form of DLT that does not utilize conventional blockchain but it uses Directed Acyclic Graph (DAG) and it can be considered as the next generation of blockchain. Due to the blazing fast speed it offers for transactions and the introduction of virtual voting [7], it is becoming adopted as well. Just like blockchain, it is also stored by all the participants of the network. But in contrast, with time each participant shares global state with random participants. The receiving participants create events which include two hashes. One of its own previous event and one is the hash of event it received from other participant. This is continued throughout the network using the Gossip protocol. Hashgraph consensus [8] also Gossips about Gossips that are spread throughout the network, which results in exponential spread throughout. Compared to blockchains, there is no need for artificial block time or delays as shown in Fig. 2. Voting is generally used for governance but since Gossip protocol is being used, each participant already knows about choices of every other participant, thus virtual voting occurs. This removes the need of transactions for voting.

The main limitation of hashgraph is the adoption of the technology. Hedera, being the parent company, and Fantom are the only companies working with DAG based distributed ledgers. On the otherhand blockchain is well known and well adopted technology by industries. However, hashgraph overcomes a lot of limitations of blockchain. Hashgraph does not continue block formation if there is no activity going on in the network as mentioned previously. It is also forkless in nature because of literally no block formation time (live events formation). The order of transactions submitted is respected and can not be changed. Lastly, since there is no block formation time, the finality time is almost instant.

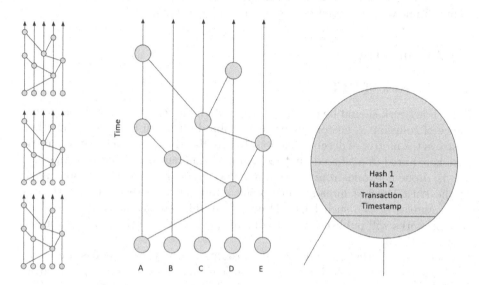

Fig. 2. Hashgraph forming events with the passage of time. Each event requires two hashes. Every participant stores same copy of hashgraph. Hashgraph is forkless.

2   Related Works

Ruba Alkaldi et al. in [9], provided a very good up-to-date review of blockchain-based UAV networks. Literature on state-of-the-art cross blockchain frameworks has been surveyed. Blockchain interoperability has been discussed followed by discussion on paradigms of safety, security and reliability. Open issues have also been discussed to the end of the paper. Scenarios of actual implementation have been discussed theoretically. The actual implementation has not been provided.

Jian Wang et al. in [10], have proposed a new routing algorithm for swarms based on blockchain technology. The algorithm enhances security of 5G NR cellular networking by adopting pheromone model to estimate the traffic status. This essentially replaces the already existing consensus algorithms with a new Proof-of-Traffic consensus. Synchronization happens passively under energy constraints which resulted in energy efficient routing.

Rahman et al. in [11], proposed a UAV network using smart contracts to have a collision-free environment. Routes are planned in advance to avoid collision with private properties. The drones positioning details are logged to blockchain network. However, the number of drones at same altitude are minimized to lower same altitude collision risk. There is a punishment system where when a drone violates a rule, points are subtracted from that drone's reputation.

Allouch et al. in [12], proposed a solution for securely planning a path and data sharing among participants based on permission-ed blockchain. However, the computation was mainly done off-chain in a cloud server. The implementation included only the ground control stations to store copy of ledger. Hyper Ledger Fabric framework was used to create custom blockchain.

3   Architecture

3.1   Network of DLT

A DLT network should be greatly distributed and spread to have a good provenance of collision. A custom internal specialized DLT can be built up but that destroys the motive of decentralization [13]. A DLT that is not proprietary or private is required in a 'shared airspace' between different manufacturers. In other words, drone networks from different manufacturers and vendors take advantage of the trust layer of already functioning public DLTs. Ethereum and Hedera are examples of two such DLTs. Drones from different manufacturers in their own networks will publish GPS coordinates periodically to the DLT network as shown in Fig. 3.

Note that in Hedera network infrastructure, two types of nodes are being used. Mainnet nodes which take part in actual consensus and the copy of ledger is subject to state change via transactions. And the mirror nodes are read only nodes which keep transactions history. These mirror nodes provide a cost effective query of transaction history which in our case can be the data subscribed to topics created in Hedera Consensus Service (HCS).

Fast-paced dynamic movement is one of the characteristic of a swarm. This requires a careful consideration of co-ordinates publishing interval [14]. There is a cost for every global state change. This cost is paid in native token of the DLT and calculated using a metric called gas. Gas (EIP-1559) is the measurement unit that denotes the cost necessary to pay for computational resources used by a node to process transactions [15]. For example in Ethereum, the update of Storage costs around 20,000 gas. Costs can be further reduced such as only publishing GPS data if there is another object in proximity.

3.2   Smart Contracts

Pieces of code that are deployed onto a distributed ledger that are only reactive based on the transactions received. These are called Smart contracts. Their access control is based on how well the code is written. Private and public keys are required to interact with them. They are reactive because external agents such as externally owned blockchain accounts are required to send transactions and trigger their functionalities. In our case of swarms, a small back-end program can trigger the functions of a smart contract. As mentioned previously,

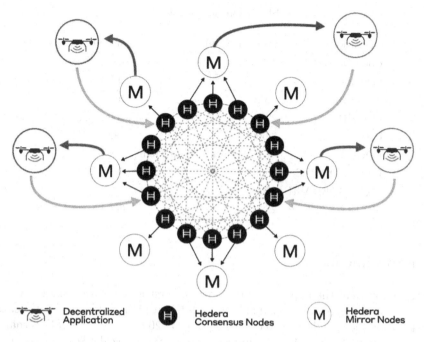

Fig. 3. Hedera Network with Consensus and Mirror nodes. Drones in same colored circles are of same network and manufacturer publishing GPS data.

each line of a smart contract that changes the state of the DLT has a gas cost. The smart contracts are usually written in Solidity language, but the Ethereum Virtual Machine (EVM) can be directly programmed for gas optimization using the intermediate assembly language Yul.

The cost of each line really matters when we consider Ethereum blockchain. Considering the gas cost of Ethereum to be around 40 gwei at the time of writing this, the cost of SLOAD operation would be 0.000084$. Since Ethereum DLT has been adopted in large numbers, which made other DLTs to include compatibility of EVM. It has become a common practise for modern DLTs to be EVM compatible. Hedera is one of those which supports EVM. In other words, Hedera Network takes care of smart contracts written for EVM. They can be deployed and interacted with on Hedera Hashgraph. The gas cost of contract deployment and interaction on Hedera is quite cheap and transaction speed, because of adopting DAG, is quite fast. Gas cost of some of the opcodes can be seen in Table 1.

Table 1. Opcodes gas fees.

Operation	London (Gas)	Current (Gas)
Code Deposit	200*bytes	Max
SLOAD	2,100	2,100
SSTORE (new)	22,100	Max
SSTORE	2900	2900
CALL	2,600	2,600
LOG0-4	375*2*topics + data Mem	Max
EXTCODESIZE	2600	2600
EXTCODEHASH	2600	2600
SELFDESTRUCT	2600	2600

4   Experiment

We can implement the apparatus in two ways. First is to create a single shared smart contract among all participants from different manufacturers. That means that the data for each drone is recorded against its 20 bytes address of Externally Owned Account (EOA). This is a simpler solution but non scalable as there might be some variables that are specific to some drones. Then there are variables that must be present for all drones individually such as GPS coordinates. For these we can make an base interface to be inherited by all participants.

Secondly, we can create a Factory Contract just like well-known Uniswap's factory contract that can create and deploy child GPS Storage contracts using the Create2 Opcode of EVM. Create2 derives a deterministic address of a smart contract to be deployed prior to its actual deployment. But the problem is, at the time of this paper Hedera did not yet support Create2 opcode of EVM. Therefore, a good approach was to create a Factory Contract that uses Create opcode instead of Create2.

Next to keep in mind is to follow a proxy pattern for the contracts as shown in Fig. 4. Deploying a contract without a proxy would be again a non-scalable approach. The proxy pattern such as EIP1967 or EIP2535 (Diamond Standard) give the ability to separate logic contract from its storage. The proxy contract acts as a storage that does delegate calls to implementation contract. Thus, the functions of implementation contract are run with the context of proxy storage. We will be using Openzeppelin's well known proxy libraries. It is to be noted that the proxy admin should be a multi-signature address or an address governed through votes. Governance module is out of the scope of this paper.

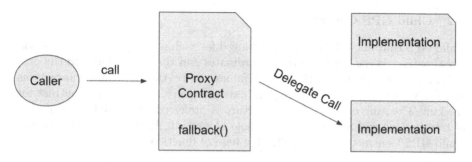

Fig. 4. Call with 'calldata' having the selector of a function in implementation contract. It triggers fallback function of the proxy, which does a delegate call to implementation.

The equipment consists of:

1. SDK of Hedera
2. Account for Hedera Network
3. A Parent Factory Contract
4. Child GPS Contracts
5. An optional Router Contract
6. Back-end Services for automation

Since Hedera is EVM compatible, same contracts can be deployed on Ethereum with addition of the following:

1. Account for Ethereum Network
2. Node Provider

SDK of Hedera will be used to deploy the smart contracts:

1. Publish ByteCode to Hedera Network using *FileCreateTransaction()*
2. Instantiate the Smart Contract byte code using *ContractCreateTransaction()*
3. Interact with the deployed contract using *ContractExecuteTransaction()*

4.1   Factory Contract

Factory contract is a singleton parent contract that creates and deploys child contracts. Each child contract's address will be stored by factory contract. To have decentralization, there will be no owners of the factory contract. The following functions must be implemented in the Factory Contract:

1. Function *createTracking(droneID)* - Deploys/Overrides a new Child GPS Contract for the 'droneID'. 'droneID' is public address which the drone will be using to send transactions. The 'msg.sender' or caller of the function must be the 'droneID'.
2. Event *TrackingCreated(droneID indexed)* - An indexed event emitted when tracking is created.

4.2 Child GPS Contract

Child GPS contract is individually deployed for each participant of the network. This contract keeps track of GPS co-ordinates and the code related to the process of these co-ordinates. History of values can be tracked using transactions history. We can also create additional data structure within smart contract for tracking. Creating our own data structure is preferred since relying on history of transactions will require parsing whole blockchain. Each drone can have one Child GPS Contract with the following base of functions:

1. Function *publish(coordinates)* - Updates state of DLT with GPS coordinates supplied as a struct parameter to the function. Coordinates struct can have required variables such as Altitude, Latitude, Longitude, Speed, Direction and more. The caller of the function must be the owner of the Child GPS Contract.
2. Function *alertProximity()* - This function is called when proximity of another object is detected by the back-end service. This in turn calls *publish(coordinates)* and emits an event of alert. The reason of creating this function is to easily track collision history by searching for *Proximity(coordinates)* event. Caller must be the owner of the contract. Details of back-end service is out of the scope of this paper.
3. Event *Published(coordinates indexed)* - An event emitted when coordinates are published.
4. Event *Proximity(coordinates indexed)* - An event emitted when a proximity is alerted.

4.3 Router Contract

A router contract is supposed to be the entry point of all calls for all drones or participants. This is an optional contract but we think it makes the logic of calls very easy. It can be also thought of the Router contract used by Uniswap. The router contract does not store any core logic and it does not keep track of any variables related to drones. It only forwards calls to Factory contract to get addresses of Child GPS contract and updates the positioning. The router contract has only one instance and should have the following functions:

1. Function *rPublish(coordinates)* - Updates state of DLT with GPS coordinates supplied as a struct parameter to the function. The caller of the function is a drone. This in turn checks if Factory Contract returns any address for Child GPS Contract for the callee. If returns address is zero address, it calls *createTracking(droneID)* from factory contract. Then it calls *publish(coordinates)* function from newly created or already existing Child GPS Contract for the callee.
2. Function *rAlertProximity()* - This function in turn checks if GPS Child contract for the caller already exists. If not, it creates GPS child contract and then calls the *alertProximity()* function of the newly created contract.

5   Conclusion

Combining traditional ad-hoc network in swarm of drones with a trustless application layer in the form of a distributed ledger creates a powerful history of tracking for swarms in a shared airspace. The participants and manufacturers behind those participants need not to rely on any third party company or proprietary database, but rather the whole notion of trust is being removed. The only thing they need to trust is the code of smart contract.

We did calculate the costs of functionalities with our implementation. Considering that the technology is new to adoption, the results were quite tolerable. The data recorded is averaged in Table 2.

Table 2. Gas costs of functions.

Finality (Avg)	publish (Gas)	alertProximity(Gas)	createTracking (Gas)
4.5 seconds	104,075	106,273	1,228,565

Keeping in mind Hedera charges 0.0000000852 dollars per unit of gas and Ethereum charges 0.000084\$ as in Sect. 3.2, the actual costs are in Table 3.

Table 3. Gas costs of functions in dollars.

Ledger	publish	alertProximity	createTracking
Hedera	0.00886719 \$	0.0090544596 \$	0.104673738 \$
Ethereum	8.7423 \$	8.926932 \$	103.19946 \$

As it can be seen that if GPS coordinates are published every 2 s, total cost of a day would be 400\$ approximately. This is quite a lot for a 2 s interval but this can be reduced by increasing interval. The price can be greatly reduced if we introduce proximity sensor to our drones, then only calling 'publish' function when proximity of another object or more specifically drone is detected. Hence, if a collision happens between two drones from different manufacturers, it is not possible to alter collision data by these manufacturers as shown in Fig. 5. The reason is we are using public DLT not proprietary. And the data stored is immutable and requires consensus for state change. Thus, a reliable source for provenance of collision.

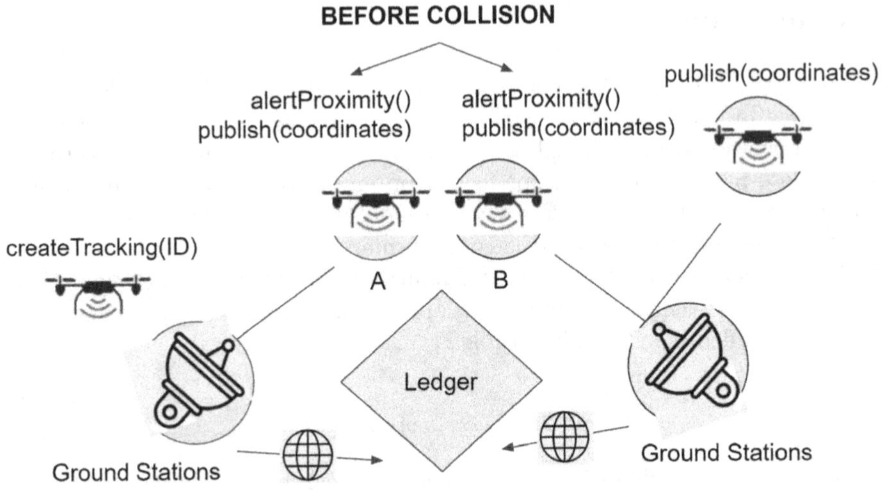

BEFORE COLLISION

Fig. 5. Drones from two manufacturers before collision alert proximity and publish coordinates to Ledger via ground stations which in turn have access to the internet. Thus, the Child GPS contracts for colliding drones are updated with GPS coordinates.

The area of decentralization is in its very early form of research. DLTs are still not very mature yet and require a constant push to research and development. Gas costs and finality of ledgers are still an active field of research and improvement. This makes their usage very expensive and costly in Internet of Things. But for swarm of drones, it is the beginning of a true decentralised model. With very fewer works regarding integration of DLTs with swarm, this paper shows a basic practical application of DLTs in swarm of drones.

References

1. Lundell, D., Hedberg, A., Nyberg, C., Fitzgerald, E.: A routing protocol for LoRA mesh networks. In: 2018 IEEE 19th International Symposium on A World of Wireless Mobile and Multimedia Networks (WoWMoM), pp. 14–19 (2018)
2. Satoshi, N.: Bitcoin: a peer-to-peer electronic cash system (2008)
3. Ethereum white paper (2014). https://ethereum.org/en/whitepaper/
4. Yakovenko, A.: Solana: a new architecture for a high performance blockchain v0.8.13 (2017)
5. Sayeed, S., Marco-Gisbert, H.: Assessing blockchain consensus and security mechanisms against the 51% attack. Appl. Sci. **9**, 1788 (2019). https://doi.org/10.3390/app9091788
6. Jensen, I.J., Selvaraj, D.F., Ranganathan, P.: Blockchain technology for networked swarms of unmanned aerial vehicles (UAVs). In: 2019 IEEE 20th International Symposium on A World of Wireless, Mobile and Multimedia Networks (WoWMoM), pp. 1–7 (2019). https://doi.org/10.1109/WoWMoM.2019.8793027
7. Baird, L., Harmon, M., Madsen, P.: Hedera: A Public Hashgraph Network & Governing Council (2020)

8. Diving deep into Hashgraph consensus. https://hedera.com/learning/what-is-hashgraph-consensus

9. Alkadi, R., Alnuaimi, N., Yeun, C.Y., Shoufan, A.: Blockchain interoperability in unmanned aerial vehicles networks: state-of-the-art and open issues (2022). https://doi.org/10.1109/ACCESS.2022.3145199

10. Wang, J., Liu, Y., Niu, S., Song, H.: Lighweight blockchain assisted secure routing of swarm UAS networking (2020)

11. Rahman, M.S., Khalil, I., Atiquzzaman, M.: Blockchain-powered policy enforcement for ensuring flight compliance in drone-based service systems. IEEE Netw. **35**(1), 116–123 (2021)

12. Allouch, A., Cheikhrouhou, O., Koubâa, A., Toumi, K., Khalgui, M., Gia, T.N.: UTM-chain: blockchain-based secure unmanned traffic management for internet of drones. Sensors **21**(9), 3049 (2021)

13. Alsamhi, S., et al.: Blockchain for decentralized multi-drone to combat COVID-19 and future pandemics: framework and proposed solutions. Trans. Emerg. Telecommun. Technol. **32**, e4255 (2021). https://doi.org/10.1002/ett.4255

14. Wang, T., Ai, S., Cao, J.: A blockchain-based distributed computational resource trading strategy for industrial internet of things considering multiple preferences (2022)

15. Liu, Y., Lu, Y., et al.: Empirical analysis of EIP-1559: transaction fees, waiting time, and consensus security (2022)

A First-Priority Set of Telepresence Services and a Model Network for Research and Education

A. E. Koucheryavy[1], M. A. Makolkina[1], A. I. Paramonov[1],
A. I. Vybornova[1], A. S. A. Muthanna[1]([✉]), R. A. Dunaytsev[1],
S. S. Vladimirov[1], V. S. Elagin[1], O. A. Markelov[2], O. I. Vorozheykina[1],
A. V. Marochkina[1], L. S. Gorbacheva[1], B. O. Pankov[1],
and B. N. Anvarzhonov[1]

[1] Department of Communication Networks and Data Transmission,
The Bonch-Bruevich Saint Petersburg State University of Telecommunications,
22 Prospect Bolshevikov, Saint Petersburg 193232, Russia
ammarexpress@gmail.com, {roman.dunaytsev,v.elagin}@spbgut.ru
[2] Radio Systems Department, Saint Petersburg Electrotechnical University "LETI",
5 Professora Popova Str., Saint Petersburg 197376, Russia
oamarkelov@etu.ru

Abstract. A first-priority set of telepresence services is proposed, and the delay requirements and fault probabilities for these services are defined. The end-to-end latency and reliability requirements are derived from analysis of ITU-T, 3GPP, ETSI standards and recommendations. The characteristics of a next-generation model network for research and education in the field of telepresence services are discussed. The model network is based on a DWDM core, a variety of server equipment, holographic fans, 3D cameras and projectors, avatar robots and multifunctional robots, and augmented reality terminal devices. The results of the first tests on the model network are presented.

Keywords: Telepresence · Model network · Holographic communication · Augmented reality · Avatar robots · Multifunctional robots

1 Introduction

The following services are proposed to be included in the first-priority set of telepresence services for deployment on communication networks:

1. teleconferences;
2. holographic images using holographic fans and hybrid augmented reality and holographic glasses;
3. augmented reality for medicine, industry and education;

V. M. Vishnevskiy et al. (Eds.): DCCN 2022, CCIS 1748, pp. 208–219, 2023.
https://doi.org/10.1007/978-3-031-30648-8_17

4. glove-based avatar robots and robotic arms to perform economical activities under adverse environmental conditions and in remote care of people and animals in need;
5. robot manipulators and their swarms for remote control when performing various tasks in economical activities.

The deployment of such services on communication networks requires different levels of development of public communication networks, determined mainly by the required values of end-to-end latency for the provision of certain telepresence services.

2 End-to-End Latency and Reliability Requirements

Currently, one of the most promising areas of development of communication systems and networks is the development of communication networks with ultra-low latency [1,2]. Initially, research on the development of such networks was based on the concept of the Tactile Internet [3,4]. However, later the composition of applications for communication networks with ultra-low latency has been significantly expanded. The criterion for classifying network services as ultra-low latency network services was the value of end-to-end latency in the range from 1 to 10 ms [1].

It is fundamentally important for our study that this range includes requirements for augmented reality services with a latency of 5 ms [5]. It is expected that augmented reality services will gain widespread adoption, at least for medicine, industry and education. In addition, these services also create conceptually new traffic. For example, it is required to consider the angular speed of the user, as well as the possibility of using augmented reality services with a 360° circular view [5,6].

Even more stringent end-to-end latency constraints are imposed for remote control of robots. Since in such systems, as a rule, robots perform either movements repeating the movements of human arms or legs – robot-avatars [7,8], or work under control of a human or another robot through a network – robot-manipulators, in both cases the concept of the Tactile Internet manifests itself in one form or another. According to the latest version of the 3GPP (Third Generation Partnership Project) standard TS 22.261 V18.6.1 [9] Sect. 7.11 "KPIs for tactile and multi-modal communication service" for remote control of robots with high mobility up to 50 km/h, end-to-end latency should be from 1 to 20 ms. It is important to note that these values must be met when remotely controlling an area of no more than 1 km². Note that the 1 ms requirement for remote robot control has been experimentally validated on a model network deployed in the Department of Communication Networks and Data Transmission at Saint Petersburg State University of Telecommunications (SPbSUT) for significantly larger distances with a delay value of 1 ms.

As for the 3GPP standard itself, it seems to be extremely useful as a system specification on the requirements for modern services of the fifth and subsequent generations of communication networks (6G and Network 2030). In particular, in the same section of [9], the maximum allowed end-to-end latency for virtual and augmented reality services between immersive multi-modal device and application server is limited to 5 ms.

The issue of end-to-end latency for holographic telepresence applications is more distinctive. In [10], it is noted that for holographic-type communication (HTC) services only those guidelines that are set for audio and video conferencing can be used. This is a 320 ms delay for streaming video with a reliability requirement of 99.999% (probability of failure-free operation). In [11], the end-to-end latency requirements for audio and video conferencing are somewhat extended, namely, up to 320 ms for streaming video and up to 280 ms for streaming audio. In addition, a certain synchronization between audio and video streams must be ensured. Thus, the audio stream can be no more than 40 ms ahead of the video stream, and follow the video stream no more than 60 ms behind.

It should be noted that the deployment of 6G and 2030 communication networks is expected to provide a significantly wider range of telepresence services than is currently possible. Moreover, the developers of 6G networks largely define them as networks that can provide holographic services [10,12]. Data rates ranging from 2 Gbit/s to 2 Tbit/s are considered for such services [10], and according to [13] up to 4.32 Tbit/s.

It is important to note that, according to [10], these services will largely be the result of the convergence of the Tactile Internet and HTC, and they·will already be subject to very different end-to-end latency and reliability requirements. Moreover, it is also noted there that such services will be among the main ones in the implementation of digital medicine and Industry 4.0.

In this case, we will be talking about end-to-end latencies in units of milliseconds, and perhaps in 6G and 2030 networks even less [8,12]. As an example, here is a comparison from [10] when providing telepresence services using HTC and VR/AR. A user controls a remote device through haptic interactions. The whole difference in providing a telepresence service using VR, AR, and holographic technologies is how the user perceives the remote location of the controlled device. This is accomplished by:

1. VR, where a 360° circular view of a remote location is projected onto a virtual reality head mounted display (VR-HMD);
2. AR, where 3D video of remote objects is projected onto an augmented reality head mounted display (AR-HMD);
3. a remote object is projected in the real world as a hologram.

Holograms give the user more flexibility in inspecting and controlling a remote object. However, this comes at the cost of very low latency and much higher throughput requirements compared to VR and AR applications.

Nevertheless, as latency requirements for the holographic image delivery service using holographic fans and hybrid AR and holographic glasses, the end-to-end latency requirements are proposed with both the above aspects and the experience gained from testing on the model network deployed in the Department of Communication Networks and Data Transmission at SPbSUT. The same applies to telepresence services based on the use of avatar robots and manipulator robots. The table below summarizes the end-to-end latency and reliability requirements for the first-priority set of telepresence services. Note that in the literature, the probability of failure-free operation is quite often associated with the packet loss probability (Table 1).

Table 1. Latency and reliability requirements for the first-priority set of services.

#	Telepresence service	End-to-end latency, ms	Reliability, %	Reference
1	Teleconferences	320 (video), 280 (audio)	99.999	[11]
2	Holographic fans and glasses	5	99.9999	[9]
3	Augmented reality	5	99.9999	[14]
4	Glove-based avatar robots	1	99.9999	[9]
5	Robot manipulators	1	99.9999	[9]

The analysis of the table shows that the telepresence services of the first-priority set can be divided into two groups. The first group is teleconferencing. In order to provide such a service with the required parameters of quality of service (QoS) and quality of experience (QoE), the current level of development of traditional communication networks is quite sufficient, because end-to-end latencies are normalized within the range from 280 ms for audio to 320 ms for video. With such delays within the territory of the Russian Federation there is no need to take special measures to limit the distance from the source of information to the client.

In [14], the values of delay that occur at the physical layer when using fiber-optic transmission systems are defined. This value is 5 μs per kilometer, which limits the distances over which ultra-low latency communication services can be provided. As mentioned above, the definition of ultra-low latency communication services implies that the latency should be from 1 to 10 ms. Among the first-priority set of telepresence services, services with latency values of 1 ms and 5 ms prevail. Since these latencies are end-to-end, the distances are limited by 50 km and 250 km radius, respectively. In addition, since the backbone network is often built on the basis of fiber-optic communication lines, and given the varied nature of the terrain in the Russian Federation, as a rule, fiber-optic communication lines are laid along highways or railroads, it is necessary to consider the reduction of the radius of service due to the irregularity of these structures. Such a methodology exists and is described in [15].

In view of the above, it is necessary to study the characteristics of density of population, distances between settlements, gross regional product to assess the possibility of widespread implementation of the first-priority set of telepresence services, either in its entirety or in a limited form.

It is important to note that the problems of implementing ultra-low latency communication services, which mainly include the first-priority set of telepresence services, have now served as the basis for introducing new technologies to provide such services on the fifth and subsequent generations of communication networks. The most important among these new technologies are edge and fog computing [16,17], microservices and their migration [18], drones [19], device-to-device (D2D) interactions [20,21], which makes it possible to provide ultra-low latency communication services, and thus the first-priority set of telepresence services, where previously it seemed to be impossible. Therefore, a comprehensive analysis of these technologies and the above-mentioned characteristics of the territories of the Russian Federation should make it possible to determine the standard rational solutions for the widespread implementation of the first-priority set of telepresence services.

3   Model Network for Research and Education in the Field of Telepresence Services

Figure 1 shows the structure of the model network that is composed of seven segments described below.

1. The core of the model network is based on Russian DWDM equipment manufactured by T8 [22]. The core is the most important part of the model network as it largely depends on it whether telepresence services could be provided at all, since the transmission of 3D holographic images with the appropriate quality may require Tbit/s bandwidth [13].
2. Robotic arms and robot avatars segment includes three robotic manipulators "Dobot Magician" with network control, as well as a set of 5 DoF (degrees of freedom) robotic arms which, with addition of AR and holographic devices, represents a swarm of avatar robots.
3. Holographic images and augmented reality segment consists of holographic fans (including self-engineered one), and AR devices, such as AR glasses and smartphones with AR support. Some of the AR glasses support 3D or spatial augmented reality, e.g. Microsoft HoloLens. This allows to create a system for 3D video online transmission and display.
4. SDN segment consists of OpenFlow switches and controllers. This segment allows to conduct research at the network level.
5. Access network is presented by Gigabit Ethernet and Wi-Fi 6 technologies and could be supplemented with Bluetooth, LoRa, etc.
6. Control, test, and measurements segment consists of workstations that allow to configure model network equipment, monitor it, and make measurements required for the research.
7. Application and data processing segment allows to create 3D video images for holographic and AR applications, as well as use the capabilities of machine learning to create applications for avatar robots and robotic manipulators.

4   Results of the First Tests on the Model Network

To conduct experimental studies in order to determine the effect of command transmission delay on the control process of a multifunction robot, a network

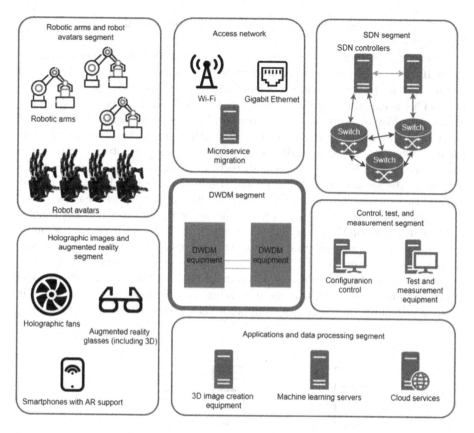

Fig. 1. Structure of the model network deployed in the Department of Communication Networks and Data Transmission at SPbSUT.

core based on the Volga multiservice DWDM platform and D-Link DES-3526 switches was used [22,23]. It should be noted that the terminal nodes of the DWDM system actually do not contribute to the end-to-end latency because of the principles of their construction, and therefore it makes sense to consider only the fixed delay due to signal propagation in fiber in accordance with [14]. Thus, as the main sources of delay can be considered the switches and the terminal of the actuator (multifunctional robot). As protocols with erase and retry are used in this case, a variable attenuator FHA2S01 is installed between the DWDM terminal nodes to obtain different delay values, which can be determined by using the terminal software. This variable attenuator allows attenuating up to 80 dB in 0.05 dB steps. By varying the attenuation of the attenuator, the probability of error in the system can be changed within a wide range, up to the "loss of signal" state. It should be noted here that the built-in controls and monitoring systems allow to get all information about the state of the system, including the probability of error. Changing the error probability makes it possible to vary the delay in the delivery of packets containing control commands to the multifunction robot.

During the preliminary experiment, it was determined that the reduction up to 20 dB in signal strength introduced by the attenuator (which is equivalent to a tract length of about 100 km) does not significantly increase the error probability. This is explained by the fact that the introduced attenuation is partially compensated by the automatic gain control (AGC) system, while the main contribution to maintaining a low error probability is made by the FEC coding, present today in almost any DWDM system. It should be noted here that the attenuator does not introduce linear distortion, so in practice (i.e., using a line of roughly the same length) the expected result will be somewhat worse due to the additional reduction in noise immunity caused mainly by phase distortion.

Figure 2 shows the structure of the model network used to study the robotic manipulator "Dobot Magician". At the transmitting side, the personal computer generates UDP datagrams that encapsulate the control commands and sends them to the receiver. At the receiver side, the personal computer connected to the robotic manipulator opens one port of the UDP socket to listen to the incoming datagrams. Each command is indexed by a sequence number, so it is possible to record the number of commands sent and received on the two sides.

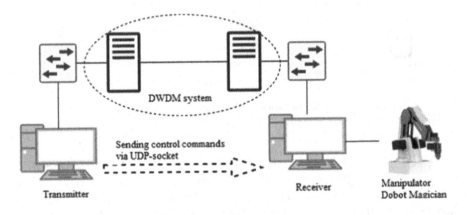

Fig. 2. Model network used to study the robotic manipulator performance.

The experiment scenario was as follows:

- The transmitted command is encapsulated as a payload of a UDP datagram. The command is compiled in JSON format due to easy programming (e.g., dictCmd = "number": 1, "X": 187.02, "Y": 10.32, "Z": 10).
- Commands are transmitted sequentially every 100 ms.
- At each attenuation value the experiment is conducted 3 times, the duration of each experiment is 2 min.
- As a result of each experiment the end-to-end latency, packet delivery ratio, and bit error probability are calculated.

During the experiments, two applications of the robot were considered: carrying an object and drawing a circle. If packets or datagrams are lost, the control commands are not delivered to the receiver, so the robot misses points when drawing a circle or commands when moving an object. When moving an object from point A to point B, the manipulator travels a certain distance. If the attenuation is greater than 22.7 dB the robot begins to miss commands and a delay becomes noticeable. In this case, of course, the transfer of the object is not correct and accurate.

In order to draw circles with the robotic arm, the circles were divided into points. Figures 3, 4 and 5 show three examples of circles that were drawn by the manipulator at different attenuation values. It is easy to see that for attenuation greater than 23 dB, the manipulator was unable to draw complete circles.

Figures 6, 7 and 8 show the end-to-end latency, packet delivery ratio, and bit error probability as a function of distance. The distance is determined in accordance with the fact that 0.2 dB corresponds approximately to a distance of 1 km.

According to the provided graphs, a significant increase in end-to-end latency starts at a distance of about 115 km. For the packet delivery ratio a sudden decrease occurs also in the distance range of 115 km, and the bit error probability increases more than twice. These experimental results suggest that the ITU-T recommended value of 5 μs per kilometer should be refined (Figs. 4 and 7).

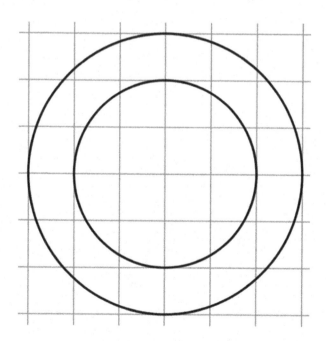

Fig. 3. Attenuation set to 23.0 dB.

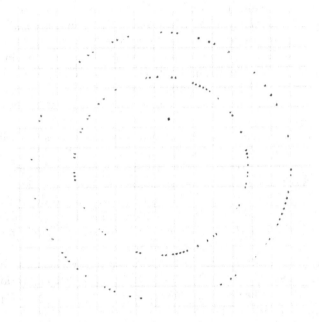

Fig. 4. Attenuation set to 23.3 dB.

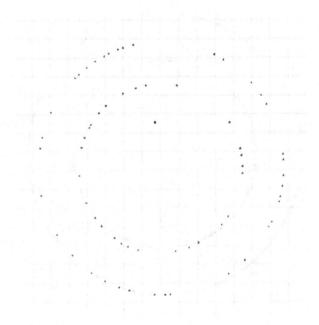

Fig. 5. Attenuation set to 23.4 dB.

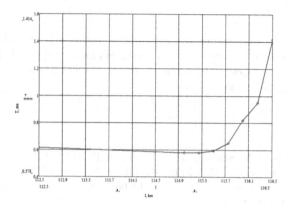

Fig. 6. End-to-end latency as a function of distance.

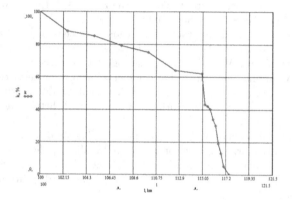

Fig. 7. Packet delivery ratio as a function of distance.

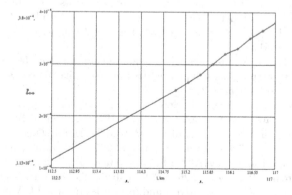

Fig. 8. Bit error probability as a function of distance.

5   Conclusion

A first-priority set of telepresence services is proposed, and the end-to-end latency and reliability requirements for these services are defined. The results of the first tests on the model network are presented, which show the need to refine the ITU-T recommendations.

References

1. Li, Z., Uusitalo, M.A., Shariatmadari, H., Singh, B.: 5G URLLC: design challenges and system concepts. In: 15th International Symposium on Wireless Communication Systems (ISWCS), pp. 1–6. IEEE, Lisbon, Portugal (2018). https://doi.org/10.1109/ISWCS.2018.8491078
2. Popovski, P., Stefanovic, C., et al.: Wireless access in ultra-reliable low-latency communication (URLLC). IEEE Trans. Commun. **67**(8), 5783–5801 (2019). https://doi.org/10.1109/TCOMM.2019.2914652
3. ITU-T Technology Watch Report. The Tactile Internet. https://www.itu.int/oth/T2301000023/en. Accessed 01 Sept 2022
4. Fettweis, G.P.: The tactile internet: applications and challenges. IEEE Veh. Technol. Mag. **9**(1), 64–70 (2014). https://doi.org/10.1109/MVT.2013.2295069
5. Makolkina, M., Pham, V.D., Kirichek, R., Gogol, A., Koucheryavy, A.: Interaction of AR and IoT applications on the basis of hierarchical cloud services. In: Galinina, O., Andreev, S., Balandin, S., Koucheryavy, Y. (eds.) NEW2AN/ruSMART -2018. LNCS, vol. 11118, pp. 547–559. Springer, Cham (2018). https://doi.org/10.1007/978-3-030-01168-0_49
6. Makolkina, M., Koucheryavy, A., Paramonov, A.: The models of moving users and IoT devices density investigation for augmented reality applications. In: Galinina, O., Andreev, S., Balandin, S., Koucheryavy, Y. (eds.) NEW2AN/ruSMART/NsCC -2017. LNCS, vol. 10531, pp. 671–682. Springer, Cham (2017). https://doi.org/10.1007/978-3-319-67380-6_64
7. Maier, M.: Toward 6G: a new era of convergence. In: Optical Fiber Communications Conference and Exhibition (OFC), pp. 1–3. IEEE, San Francisco, USA (2021)
8. Yastrebova, A., Kirichek, R., Koucheryavy, Y., Borodin, A., Koucheryavy, A.: Future networks 2030: architecture and requirements. In: 10th International Congress on Ultra Modern Telecommunications and Control Systems and Workshops (ICUMT), pp. 1–8. IEEE, Moscow, Russia (2018). https://doi.org/10.1109/ICUMT.2018.8631208
9. 3GPP TS 22.261 V18.6.1. Service requirements for the 5G system; Stage 1 (Release 18). https://www.3gpp.org/ftp/Specs/archive/22_series/22.261/22261-i61.zip. Accessed 01 Sept 2022
10. Taleb, T., Nadir, Z., Flinck, H., Song, J.: Extremely interactive and low-latency services in 5G and beyond mobile systems. IEEE Commun. Stand. Mag. **5**(2), 114–119 (2021). https://doi.org/10.1109/MCOMSTD.001.2000053
11. 3GPP TS 22.263 V17.4.0. Service requirements for video, imaging and audio for professional applications (VIAPA); Stage 1 (Release 17). https://www.3gpp.org/ftp/Specs/archive/22_series/22.263/22263-h40.zip. Accessed 01 Sept 2022
12. Li, R.: Network 2030: market drivers and prospects. In: 1st ITU Workshop on Network 2030. New York, USA (2018). https://www.itu.int/en/ITU-T/Workshops-and-Seminars/201810/Documents/Richard_Li_Presentation.pdf

13. Clemm, A., Vega, M.T., Ravuri, H.K., Wauters, T., Turck, F.D.: Toward truly immersive holographic-type communication: challenges and solutions. IEEE Commun. Mag. **58**(1), 93–99 (2020). https://doi.org/10.1109/MCOM.001.1900272
14. Recommendation ITU-T Y.1541. Network performance objectives for IP-based services. https://www.itu.int/rec/T-REC-Y.1541. Accessed 01 Sept 2022
15. Paramonov, A., Chistova, N., Makolkina, M., Koucheryavy, A.: The method of forming the digital clusters for fifth and beyond telecommunication networks structure based on the quality of service. In: NEW2AN/ruSMART -2020. LNCS, vol. 12525, pp. 59–70. Springer, Cham (2020). https://doi.org/10.1007/978-3-030-65726-0_6
16. ETSI Gs MEC 003 V2.1.1. Multi-access Edge Computing (MEC); Framework and Reference Architecture. https://www.etsi.org/deliver/etsi_gs/mec/001_099/003/02.01.01_60/gs_mec003v020101p.pdf. Accessed 01 Sept 2022
17. Byers, C.C.: Architectural imperatives for fog computing: use cases, requirements, and architectural techniques for fog-enabled IoT networks. IEEE Commun. Mag. **55**(8), 14–20 (2017). https://doi.org/10.1109/MCOM.2017.1600885
18. Artem, V., Ateya, A.A., Muthanna, A., Koucheryavy, A.: Novel AI-based scheme for traffic detection and recognition in 5G based networks. In: Galinina, O., Andreev, S., Balandin, S., Koucheryavy, Y. (eds.) NEW2AN/ruSMART -2019. LNCS, vol. 11660, pp. 243–255. Springer, Cham (2019). https://doi.org/10.1007/978-3-030-30859-9_21
19. Ateya, A.A., Muthanna, A., Kirichek, R., Hammoudeh, M., Koucheryavy, A.: Energy- and latency-aware hybrid offloading algorithm for UAVs. IEEE Access **7**, 37587–37600 (2019). https://doi.org/10.1109/ACCESS.2019.2905249
20. Paramonov, A., Hussain, O., Samouylov, K., Koucheryavy, A., Kirichek, R., Koucheryavy, Y.: Clustering optimization for out-of-band D2D communications. Wirel. Commun. Mob. Comput. **2017**, 1–11 (2017). https://doi.org/10.1155/2017/6747052
21. Muthanna, A., et al.: Analytical evaluation of D2D connectivity potential in 5G wireless systems. In: Galinina, O., Balandin, S., Koucheryavy, Y. (eds.) NEW2AN/ruSMART -2016. LNCS, vol. 9870, pp. 395–403. Springer, Cham (2016). https://doi.org/10.1007/978-3-319-46301-8_33
22. Koucheryavy, A.E., Makolkina, M.A., et al.: Model network for research, training, and testing in the area of telepresence services. Electrosvyaz Mag. **1**, 14–20 (2022). (In Russian). https://doi.org/10.34832/ELSV.2022.26.1.001
23. Gorbacheva, L.S., Fam, V.D., Matyuhin, A.Y., Koucheryavy, A.E.: Investigation of the influence of network characteristics on the functioning of a multifunctional robotic arm. Electrosvyaz Mag. **2**, 37–41 (2022). (In Russian). https://doi.org/10.34832/ELSV.2022.27.2.005

Analytical Modeling of Distributed Systems

Computational Analysis of a Service System with Non-replenish Queue

Alexander M. Andronov[1] , Iakov M. Dalinger[2] ,
and Nadezda Spiridovska[1(✉)]

[1] Transport and Telecommunication Institute,
Lomonosov Str., 1, Riga LV-1019, Latvia
spiridovska.n@tsi.lv
[2] Russian University of Transport, Obrazcov Str., 9, Moscow 127994, Russia
http://www.tsi.lv

Abstract. The paper proposes a probabilistic method for an analysis of service system without input flow, but having initial queue of claims. Such situation takes place in an airport, when an express train delivers passengers from the city. Screening of the arriving passengers is carried out at security checkpoints where a queue of passengers is formed. The queue is not reinforced to the last checked passenger arriving by train.

Considered queueing system has many servers and Erlang distribution of the service time. Formulas are derived for calculating the distribution function of a time until all services ending and the average waiting time. Calculations are supported by the developed MathCAD programs.

Keywords: Service · Sojourn time · Normal approximation

1 Introduction

A service system without input flow of claims is considered, that is all claims are in a queue initially. It is the main difference from classical queueing models [5–8,10]. The number of such claims equals $nmax$. The number of servers equals k. It is supposed that $nmax >> k$. A time of one service has Erlang distribution with parameters $\mu > 0, m = 1, 2, ...$, and the mean value $\tau = m/\mu$. The density $f(t)$ and the cumulative distribution function $F(t)$ are the following:

$$f(t) = \mu(\mu t)^{m-1}\frac{1}{(m-1)!}exp(-\mu t), t \geq 0, \tag{1}$$

$$F(t) = 1 - \sum_{\eta=0}^{m-1}(\mu t)^{\eta}\frac{1}{(\eta)!}exp(-\mu t), t \geq 0. \tag{2}$$

The primary aim of this paper is to derive methods for calculating the distribution function of the time until the end of all services and the average waiting time of one claim.

V. M. Vishnevskiy et al. (Eds.): DCCN 2022, CCIS 1748, pp. 223–233, 2023.
https://doi.org/10.1007/978-3-031-30648-8_18

A motivation of such model's consideration is caused by application to problem of entrance screening of passengers arriving at the airport by express trains from the city. Screening is carried out by aviation security services at security checkpoints [2,4,12] where a queue of arriving passengers is formed. The queue is not reinforced to the last checked passenger arriving by train. The same situation takes place when a cruise ship or ferry with passengers arrives at the seaport.

Let us remark also that the considered model can be interpreted in terms of the inventory theory and the reliability theory [10].

Further exposition is organized as follows. Section 2 is dedicated to service process by one server. Many servers are considered in Sect. 3. Analysis of the final transient period is investigated in Sect. 4. Linearization for a big data is considered in Sect. 5. Section 6 contains a numerical example. Section 7 contains concluding remarks. Calculations are supported by the developed MathCAD programs.

2   Service Process by One Server

We begin with a consideration of one server using the renewal models [1,3,11]. "Renewal" means that one service is completed. Now $f(t)$ is the density of time between renewals, and $F(t)$ is the corresponding distribution function. We denote the n-time convolution of the density $f(t)$ by $f^{(n)}(t)$. Then, the number $N(t)$ of completed services at the time t has the following distribution:

$$P(N(t) = n) = \left[\begin{array}{l} 1 - F(t), n = 0, \\ \int_0^t f^{(n)}(z)(1 - F(t-z))dz, n = 1, 2, ... \end{array} \right. \tag{3}$$

The convolution $f^{(n)}(t)$ is the density of time of n services. It follows the Erlang distribution with parameters $\mu > 0$ and mn:

$$f^{(n)}(t) = \mu(\mu t)^{mn-1} \frac{1}{(mn-1)!} exp(-\mu t), t \geq 0.$$

Therefore, for $t > 0$

$$P(N(t) = 0) = \sum_{\eta=0}^{m-1} (\mu t)^\eta \frac{1}{\eta!} exp(-\mu t), \tag{4}$$

$$P(N(t) = n) = \int_0^t f^{(n)}(z)(1 - F(t - z)dz) =$$

$$\int_0^t \mu(\mu z)^{mn-1}\frac{1}{(mn-1)!}exp(-\mu z)\sum_{\eta=0}^{m-1}(\mu(t-z))^\eta\frac{1}{\eta!}exp(-\mu(t-z))dz = \quad (5)$$

$$\sum_{\eta=mn}^{mn+m-1}(\mu t)^\eta\frac{1}{\eta!}exp(-\mu t), n = 1, 2, ..., nmax - 1.$$

The corresponding distribution function is calculated as:

$$P(N(t) \le n) = \sum_{\eta=0}^{n}(P(N(t) = \eta), n = 0, 1, ..., nmax - 1.$$

Also, formulas (4) and (5) describe the distribution of the number of services during time t by one server. Now we calculate the expectation and the variance of this distribution.

Let us bear in mind the total claims $nmax$ at the initial point; therefore, $nmax$ is the maximal value of services. These services are performed by many servers. Their number k equals 6 for our presented below example. Considered values $nmax$ equal 70 or 100. The event $\{N(t) < nmax\}$ has a probability which is very close to 1. It allows calculating the expectation and the variance of $N(t)$ as follows:

$$E(N(t)) = \sum_{n=1}^{nmax} n \sum_{\eta=mn}^{mn+m-1}(\mu t)^\eta\frac{1}{\eta!}exp(-\mu t)$$

$$= \sum_{n=1}^{nmax} n \sum_{\xi=0}^{m-1}(\mu t)^{\xi+mn}\frac{1}{(\xi+mn)!}exp(-\mu t), \quad (6)$$

$$Var(N(t)) = \sum_{n=1}^{nmax} n^2 \sum_{\eta=mn}^{mn+m-1}(\mu t)^\eta\frac{1}{\eta!}exp(-\mu t) - E(N(t))^2. \quad (7)$$

For the usual renewal process $N(t)$ the Central Limit Theorem (CLT) holds

$$\lim_{t\to\infty} P\left\{\frac{N(t) - E(N(t))}{\sqrt{Var(N(t))}} \le x\right\} = \Phi(x),$$

where $\Phi(x)$ is the distribution function of the standard normal distribution.

Due to quick convergence in the CLT for renewal processes we could suppose that for some limited random time interval $[t*, t^*]$, where $t* \approx 10\,m/\mu$ and $t^* = inf\{t : N(t) = nmax\}$, that the CLT also fulfilled for considered process $N(t)$

and its distribution can be approximated in this interval by the normal distribution:

$$P\{N(t) \leq n\} = \Phi\left(\frac{n - E(N(t))}{\sqrt{Var(N(t))}}\right). \tag{8}$$

The example below illustrates such approximation.

3   Service Process for Many Servers

It is supposed that we have k servers; therefore during random time before $nmax - k + 1$ services end the k bounded renewal processes occur. We will denote the number of services for the i-th server by $N_i(t)$ and the total number of services by $N^*(t)$:

$$N^*(t) = min\left\{\sum_{i=1}^{k} N_i(t), \quad nmax\right\}, t \geq 0, \tag{9}$$

where $N_i(t), i = 1, ..., k$, has the distribution presented above.

In the case $nmax = \infty$, the processes $\{N_i(t)\}$ are independent. For any finite $nmax$ it is not true. For example it is not true for the heavy tailed service distributions. However, for big value $nmax$, small k and Erlang service time distribution it is possible to consider them as independent (or weak dependent). Thus because we consider a case, when $nmax$ is a very big number, namely $nmax = 70$ or $nmax = 100$, but a number of servers is small, namely $k = 6$, we can suppose that the process $S(t) = \sum_{i=1}^{k} N_i(t)$ is a sum of k independent processes.

We use the following expressions further:

$$E(S(t)) = \sum_{i=1}^{k} E(N_i(t)) = kE(N(t)), t \geq 0, \tag{10}$$

$$Var(S(t)) = \sum_{i=1}^{k} Var(N_i(t)) = kVar(N(t)), t \geq 0. \tag{11}$$

Furthermore, based on the above argumentation we will use the normal approximation for the number of total services during an appropriate random time interval by referring to the formula (8) where $S(t)$ is used instead of $N(t)$:

$$P\{N^*(t) \leq n\} = \begin{bmatrix} \Phi\left(\dfrac{n - E(S(t))}{\sqrt{Var(S(t))}}\right), n < nmax, \\ 1, n \geq nmax, \end{bmatrix} \tag{12}$$

This expression allows calculating the expectation of $N^*(t)$:

$$E(N^*(t)) = \sum_{n=0}^{nmax-1} P\{N^*(t) > n\} = \sum_{n=0}^{nmax-1} \left(1 - \Phi\left(\frac{n - E(S(t))}{\sqrt{Var(S(t))}}\right)\right), t \geq 0.$$

(13)

Let us calculate the distribution of time $T(n)$ until the completion of n services. For $t \geq 0$ we have

$$P\{T(n) \leq t\} = P\{N^*(t) \geq n\} = \begin{bmatrix} 1 - \Phi\left(\dfrac{n - 1 - E(N^*(t))}{\sqrt{Var(N^*(t))}}\right), k \leq n \leq nmax, \\ 0, n > nmax, \end{bmatrix}$$

(14)

The mean time of the service of n claims, $k \leq n \leq nmax$, is calculated as usual:

$$E(T(n), U) = \int_0^U (1 - P\{T(n) \leq t\})dt = \int_0^U \Phi\left(\frac{n - 1 - E(N^*(t))}{\sqrt{Var(N^*(t))}}\right)dt, \quad (15)$$

where U is an assigned determinate upper limit of $T(n)$, which replaces the infinity ∞.

Let $\hat{t}(n, \gamma)$ be the time when the number of completed services is no less than n with the probability γ. The corresponding values are calculated as follows:

$$\hat{t}(n, \gamma) = \inf_t\{P\{T(n) \leq t\} = \gamma\}, 0 < \gamma < 1. \tag{16}$$

The calculations in this formula are performed in the following way. We fix small $\Delta > 0$ and calculate values of (14) sequentially for $t = 0, \Delta, 2\Delta, \ldots$ until the obtained result is less than γ. The last value t is an unknown quantity.

The following question becomes of interest: how can we calculate the average sojourn time avr of one customer until the completion of its service, if initially $nmax$ customers take place? The integral

$$\int_0^t \left(nmax - E(N^*(z))\right)dz,$$

gives the average total sojourn time of all customers until the time t.

Therefore, the average sojourn time avr of one customer

$$avr(nmax, t) = t - \frac{1}{nmax}\int_0^t E(N^*(z))dz. \tag{17}$$

We will write $avr(nmax, \gamma, \Delta)$ instead of $avr(nmax, t)$ if $t = \hat{t}(nmax, \gamma)$, where $\Delta > 0$ is defined under the formula (16).

4   Analysis of the Final Transient Period

Now we consider the final period when the number of working servers decreases. If $nmax$ is the total number of claims, k is the number of servers, then the period begins when the number of working servers decreases from $nmax - k + 1$ to $nmax - k$.

The previous period is a steady-state one, because the service intensity is not dependent on an instant t of this period. In our case this means that the steady-state distribution of a number of a current phase for a separate server is uniform on the set $\{1, ..., m\}$, the corresponding probability being $p_i = \frac{1}{m}, i = 1, ..., m$.

Let us consider a server which begins to serve a last claim. Its distribution of service time $F(t)$ is given by formula (2). Other $k - 1$ servers continue to provide services. The steady-state distribution function of the service residual time is the following:

$$
\begin{aligned}
G(t) &= \frac{1}{m} \sum_{i=1}^{m} \left(1 - \sum_{j=0}^{i-1} (\mu t)^j \frac{1}{j!} exp(-\mu t) \right) \\
&= 1 - \frac{1}{m} \sum_{j=0}^{m-1} \sum_{i=j+1}^{m} (\mu t)^j \frac{1}{j!} exp(-\mu t) \\
&= 1 - \frac{1}{m} \sum_{j=0}^{m-1} (m - j)(\mu t)^j \frac{1}{j!} exp(-\mu t), t \geq 0.
\end{aligned}
\tag{18}
$$

Also, the distribution function of the service time of all k last claims is calculated as follows:

$$
GG(t) = F(t)G(t)^{k-1}, t \geq 0.
\tag{19}
$$

The corresponding mean value is calculated as usual, it equals $\int_0^\infty (1 - GG(t))dt$.

5   Linearization for a Big Data

If a number of considered completed services n or the considered time t is high, this can cause potential processing difficulties. In this case, it is possible to use the linearity of considered average indices with respect to n or t. Let $EY(x)$ be the calculated value of the index Y for a fixed x, $\widetilde{EY}(x)$ be the value, which is obtained by means of linearization. Then, for $x_1 < x_2 < x_3$

$$
\widetilde{EY}(x_3) = EY(x_2) + \frac{EY(x_2) - EY(x_1)}{x_2 - x_1}(x_3 - x_2).
\tag{20}
$$

In our case, three periods take place for a considered index $EY(x)$. The first one is the transient period $(0, x_1)$, since, initially, all services have the first phase.

The second one is the steady-state period. The third one is again the transient period, since the number of working servers decreases.

We can consider these periods in the following way. Let $(\underline{x}, \bar{x})$ be the steady-state interval, for which the index $EY(x)$ could be found relatively easy. The bounds $\underline{x}$ and $\bar{x}$ can be evaluated empirically.

Let $x^* > \bar{x}$ be the maximal value of x, for which the index $EY(x)$ must be calculated, and w be the whole part of the quotient $(x^* - \underline{x})/(\bar{x} - \underline{x})$. Then

$$EY(x^*) = EY(\underline{x}) + w\big(EY(\bar{x}) - EY(\underline{x})\big) + EY\big(x^* - \underline{x} - w(\bar{x} - \underline{x})\big). \qquad (21)$$

Note that the last addend in this formula can be calculated in various ways. However usually it does not significantly change the final result.

6 Numerical Example

We consider the following initial data. The number of servers $k = 6$, the number of the service phases $m = 3$, the service intensity during one phase $\mu = 12, 1/\text{min}$. The mean τ and the standard deviation σ of the service time are the following: $\tau = m/\mu = 0.25$ min, $\sigma = \frac{\sqrt{m}}{\mu} = \frac{\sqrt{3}}{12}$ min. All computations are performed using MathCAD program.

The mean number $E(N(t))$ of services performed by one server, as a function of time t, is presented in Fig. 1 for $nmax = 20$. It can be observed that a linear dependence occurs in this case.

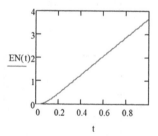

Fig. 1. Graph $EN(t) = E(N(t))$

The distribution $p(n, t) = P(N(t) = n)$ of number $N(t)$ of completed services during time $t = 3$ for $nmax = 30$ and its normal approximation $\phi(n, 3)$ are presented in Fig. 2. One can observe that the approximation is highly effective in this case; therefore, it will be employed further. The following Fig. 3 corresponds to the distribution function $p(n, 3) = P(N(t) \le n)$ and its normal approximation (8).

Now, we present the results of the calculations for six servers. The probabilities (14) for six servers ($k = 6$) are given in Fig. 4 and Fig. 5, where $Pr(t, n) = P\{T(n) \le t\}$.

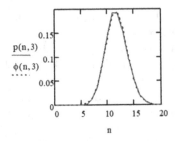

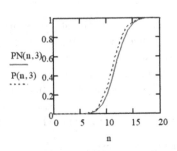

Fig. 2. Graphs $p(n,t) = P(N(t) = n)$ for $t = 3$

Fig. 3. Graphs of distribution functions for $t = 3$

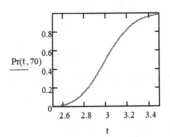

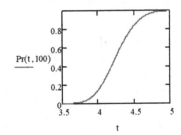

Fig. 4. Graph of $PT(70) \leq t$

Fig. 5. Graph of $PT(100) \leq t$

The mean time (15) of the service of n claims as the function of n is presented in Fig. 6, where $ETM(n, U) = E(T(n), U)$. Again, the linear dependence can be observed.

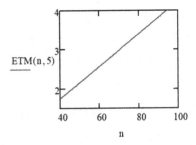

Fig. 6. The mean time of the service of n claims

Calculations performed by using the formula (16) for $\Delta = 1/60$ show that $\tilde{t}(20, 0.99) = 1.233$, $\tilde{t}(100, 0.99) = 4.867$.

The dependencies (10) and (13) as the function of time t are presented in Fig. 7, and Fig. 8 where $ENM(t) = E(S(t))$ and $ERNM(t) = E(N^*(t))$ for $nmax = 70$.

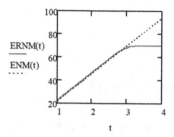

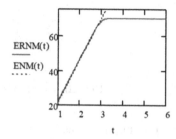

Fig. 7. Graph of $E(S(t))$ and $E(N^*(t))$ **Fig. 8.** Graph of $E(S(t))$ and $E(N^*(t))$

The graph of the average sojourn time $avr(n, \gamma, \Delta)$ of one customer till the completion of its service (see formula (17)), if initially n customers take place, is presented in Fig. 9.

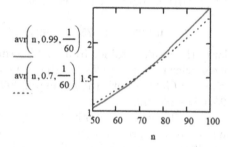

Fig. 9. Graph of the expectation of a sojourn time

Now we consider the final period when the number of non-served claims equals to k or is less than k. This period begins when the new service commences and $k - 1$ services continue. The distribution function (19) of the time till completion of all services k is presented in Fig. 10. The corresponding expectation is as follows: $\int_0^\infty (1 - GG(t))dt = 0.341$ min. It is evident that the final period has little effect on the final result.

We finalize our numerical example by the case where the big data cause processing difficulties as well as software bugs. This is likely to happen when the considered number of claims is bigger than 100. The method of linearization described above should be used in this case. Let us consider one example for the mean time $E(T(n))$ until the completion of n services, where $n > 100$. The

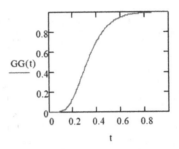

Fig. 10. Graph of the distribution function (19)

considered function is as follows: $Y(x) = E(T(n))$ where $x = n$. Let $\underline{x} = 40$, $\bar{x} = 100$, $x^* = 300$. Then,

$$\frac{(x^* - \underline{x})}{(\bar{x} - \underline{x})} = \frac{300 - 40}{100 - 40} = \frac{260}{60} = \frac{13}{3}.$$

The whole part of $13/3$ is $w = 4$. Now with respect to formula (21):

$$E(T(300)) = E(T(40)) + 4(E(T(100)) - E(T(40))) + E(T(300 - 40 - 4(100 - 40))).$$

The calculations produce the following results: $E(T(40)) = 1.575$, $E(T(100)) = 4.257$, $E(T(20)) = 0.924$. We have thus definitively confirmed that:

$$E(T(300)) = 1.575 + 4(4.257 - 1.575) + 0.924 = 12.68 \text{ min.}$$

This result is similar to the result obtained from the deterministic calculation. In fact, the mean of one service time $\tau = 0.25$ min, the number of servers $k = 6$. Therefore, the mean time $E(T(300))$ until the ending of 300 services equals to $300 \times \tau/k = 300 \times 0.25/6 = 12.5$ min.

The deterministic approach gives the result 12.5 min that is less than the result 12.68 min of the stochastic approach. It is clear, because the deterministic approach supposes that all six servers work until the completion of the services. Let us calculate the corresponding mean time.

An instant is fixed when a number of non-served claims equals to $k = 6$, see Sect. 4. The mean number of the non-served phases equals $3 + 5 \times \frac{3}{2} = 10.5$. It averages $10.5 \times \frac{1}{12} = \frac{10.5}{12}$ min. This time is divided among the servers uniformly, so the mean time of services completion equals $\frac{1}{6} \times \frac{10.5}{12} = \frac{10.5}{72} = 0.146$ min.

We acquired the result of 0.341 min using the stochastic approach. The difference contains $0.341 - 0.146 = 0.195$ min, which appears to be similar to the previously obtained odds $12.68 - 12.5 = 0.18$ min.

It should be emphasized, that our stochastic approach allows for the calculations of various probabilistic indices, not only the average ones.

7   Conclusion

The difference between the considered model in the paper and the classical queue models lie in the fact that there is no incoming claims flow: all claims are initially queued. The suggested approach is based on the renewal theory and the normal approximation of probabilistic distributions. The obtained results allow the calculations of various efficiency indices: the average sojourn time of one customer till the completion of its service, the distribution function of the time till all services completion, etc.

A motivation of the performed investigation is caused by an application to problem of entrance screening of passengers arriving at the airport by express trains from the city or similar problems.

The elaborated queueing model can be generalised in various ways: for instance, the Erlang distribution of service time can be replaced by the distribution of phase type [9]. The authors intend to focus their future research on other types of generalisations of the queueing model, such as, for example, the existing of incoming claims flow.

Acknowledgment. Authors would like to thank reviewers for taking the time and effort necessary to review the paper. Authors sincerely appreciate all valuable remarks and suggestions, which helped to improve the quality of the paper.

References

1. Cox, D.R.: Renewal Theory. Methuen and CO Ltd - John Wiley and Sons INC, London - New York (1961)
2. De Barros, A.G., Tomber, D.D.: Quantitative analysis of passenger and baggage security screening at airports. J. Adv. Transp. **41**(2), 171–193 (2007)
3. Feller, W.: An Introduction to Probability Theory and its Applications, vol. II. Second Edition, Wiley, New York (1971)
4. Gilliam, R.R.: An application of queueing theory to airport passenger security screening. INFORMS J. Appl. Analytics **9**(4), 117–123 (1979)
5. Kijima, M.: Markov Processes for Stochastic Modelling. Chapman & Hall, London (1997)
6. Kim, C.S., Dudin, A., Klimenok, V., Khramova, V.: Erlang loss queueing system with batch arrivals operating in a random environment. Comput. Oper. Res. **36**(3), 674–697 (2009)
7. Krishnamoorthy, A., Joshua, A.N., Vishnevsky, V.: Analysis of a k-stage service queueing system with accessible batches of service. Mathematics **9**(55), 1–16 (2021)
8. Lucantonti, D.M.: New results on the single server queue with a batch Markovian arrival process. Commun. Stat. Stochast. Models **7**, 1–46 (1991)
9. Neuts, M.F.: Probability distribution of phase type. In: Liber Amicorum Professor Emerlitus H. Florin. Department of Mathematics, University of Louvain, pp. 173–206 (1975)
10. Prabhu, N.: Queues and Inventories. Wiley, New York (1971)
11. Smith, W.I.: Renewal theory and its ramifications. J. Roy. Stat. Soc. **20**, 243–284 (1958)
12. Stotz, T., Bearth, A., Ghelfi, S.M., Siegrist, M.: The perceived costs and benefits that drive the acceptability of risk-based security screenings at airports. J. Air Transp. Manage. **100**, 102183 (2022)

Neuro-Fuzzy Model Based on Multidimensional Membership Function

Bui Truong An[1(✉)], F. F. Pashchenko[2(✉)], Tran Duc Hieu[1],
Nguyen Van Trong[1(✉)], and Pham Thi Nguyen[3(✉)]

[1] Moscow Institute of Physics and Technology (National Research University),
9 Institutskiy per., Dolgoprudny, Moscow Region 141701, Russia
info@mipt.ru
[2] V. A. Trapeznikov Institute of Control Sciences of Russian Academy of Sciences,
65 Profsoyuznaya Street, Moscow 117997, Russia
pif-70@yandex.ru
[3] Ho Chi Minh University of Natural Resources and Environment,
Ho Chi Minh, Vietnam
info@hcmunre.edu.vn

Abstract. Currently, systems based on multidimensional membership functions are being actively researched and developed. Most algorithms for determining the parameters of fuzzy membership functions are developed on the basis of one-dimensional membership functions. The fuzzy rules generated by these algorithms often overlap and cannot act as independent rules. The overlap of fuzzy rules in fuzzy systems does not allow one to evaluate the reliability of individual fuzzy rules and at the same time creates limitations in extracting knowledge from fuzzy systems. In this article, a neuro-fuzzy neural system will be built based on a multidimensional Gaussian membership function with the ability to describe the relationship of interaction between input variables, and at the same time, the generated fuzzy rules are capable of independent operation.

Keywords: fuzzy set · multidimensional membership functions · Gaussian functions · fuzzy inference system · neuro-fuzzy model · Kaczmarz algorithm · recurrent least squares

1 Introduction

Fuzzy logic is often used in conjunction with neural networks to create systems that have the benefits of fuzzy logic clarity and neural network learning capabilities. Such neuro-fuzzy systems are very diverse and are increasingly being improved in accordance with the development of neural network learning algorithms.

Most of the above works are built on the basis of one-dimensional membership functions, such as Gaussian, Bell, Triangular, RBF. The limitation of this

ⓒ The Author(s), under exclusive license to Springer Nature Switzerland AG 2023
V. M. Vishnevskiy et al. (Eds.): DCCN 2022, CCIS 1748, pp. 234–245, 2023.
https://doi.org/10.1007/978-3-031-30648-8_19

approach is the complexity of the model in terms of the number of rules, which increases exponentially with the number of inputs (spatial curse).

As an effective solution to the above problem, the use of multidimensional membership functions in a fuzzy inference system is proposed. The most commonly used multivariate membership function is the multivariate Gaussian function [1–3]. Gaussian functions require fewer parameters to describe a fuzzy rule than linear membership functions [4].

The above models of multidimensional fuzzy neural networks show the relationship between input variables. However, the generated fuzzy rules often overlap. This affects the independent operation of each fuzzy rule, which leads to inefficient data extraction from the fuzzy model.

2   Neuro-fuzzy Model Based on Multidimensional Membership Function

The multidimensional Gaussian membership function (Fig. 1) can be described as [3]:

$$H(x_0) = e^{-(x_0-C)S^{-1}(x_0-C)^T},   \tag{1}$$

where C is the center of the membership function, S is the matrix of expansion coefficients. The matrix S is initialized and folded as a positive-definite matrix, which ensures that the value of $H(x_0)$ is always within the half-interval (0,1].

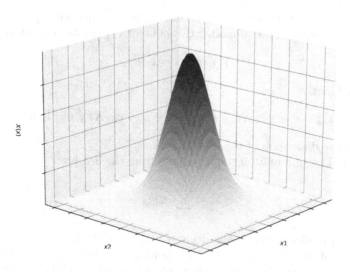

Fig. 1. Gaussian membership function with two variables

The proposed architecture of the model consists of 4 layers shown in Fig. 2.

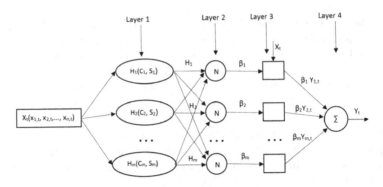

Fig. 2. Architecture of the neuro-fuzzy model based on the MMF

Layer 1. Calculation of the degree of membership of the input vector for each fuzzy rule

$$H_i = e^{-(X_t - C_i)S_i^{-1}(X_t - C_i)^T}, i = \overline{1, m},\tag{2}$$

where m is the number of fuzzy rules.

Layer 2. Finding the normalized rule truth value

$$\beta_i = \frac{H_i}{\sum\limits_{k=1}^{m} H_k}, i = \overline{1, m},\tag{3}$$

Layer 3. Calculation of the parameters of the consequences $\beta_i Y_{i,t}$
Depending on the type of model, the value of $Y_{i,t}$ can be a constant

$$Y_{i,t} = c_i, i = \overline{1, m},\tag{4}$$

or a linear function

$$Y_{i,t} = c_{i,0} + c_{i,1} X_{t,1} + c_{i,2} X_{t,2} + \ldots + c_{i,n} X_{t,n}, i = \overline{1, m},\tag{5}$$

where n is the dimension of the input space.

Layer 4. The sum of all incoming values from the 3rd layer

$$Y_t = \sum_{i=1}^{m} \beta_i Y_{i,t},\tag{6}$$

Fuzzy neural model synthesis algorithm is presented in Fig. 3.

The idea of the fuzzy neural model synthesis algorithm is to divide the input space with each direction divided into n equal segments, and then consider pairs of fuzzy rules with distances less than a given value dmin.

If two fuzzy rules are spaced less than dmin and do not give a rapid increase in the error function, then they are pooled together. The algorithm is stopped when the model achieves a sufficiently small number of fuzzy rules or the model error exceeds the threshold.

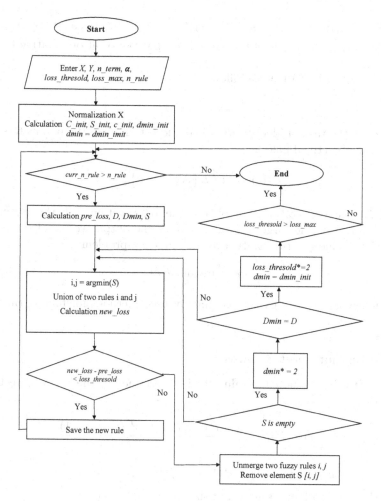

Fig. 3. Algorithm for synthesizing a neuro-fuzzy model

3 Approach to the Application of Algorithms for Identifying the Coefficients of Linear Equations

The output value of the neuro-fuzzy model can be described as follows:

$$\hat{y}(t) = \sum_{i=1}^{m} \beta_i(t) Y_i(t) = \sum_{i=1}^{m} \sum_{j=0}^{n} c_{i,j} \cdot \beta_i(t) \cdot X_j(t), \tag{7}$$

where n is the dimension of the input space, $c_{i,j}$ —is the j-th free coefficient of the i-th fuzzy rule, $X_j(t)$ is the value of the j-th input at the t-th observation ($X_0(t) = 1$ for any observation t).

Let $= (c_{1,0}, ..., c_{m,0}, c_{1,1}, ..., c_{m,1}, ..., c_{1,n}, ..., c_{m,n})^T$ - vector of adjustable coefficients.

$$\tilde{x}(t) \quad = \quad \left(\beta_1(t), ..., \beta_m(t), \beta_1(t) \cdot X_1(t), ..., \beta_m(t) \cdot X_1(t), ..., \beta_1(t) \cdot X_n(t), ..., \right.$$
$$\left.\beta_m(t) \cdot X_n(t)\right)^T \text{ - the value of the extended input vector at observation } t.$$

Then (1) can be written as follows:

$$\hat{y}(t) = c\tilde{x}(t), \tag{8}$$

In this case, the task of identification is reduced to finding the parameters minimizing the objective functional:

$$J(c) = M\left\{L\left(\tilde{X}, Y, \hat{Y}, c\right)\right\}, \tag{9}$$

where M is the expectation symbol, L is the loss function, $\hat{X}$ is the augmented input vector, Y is the target vector, and $\hat{Y}$ is the output vector of the model. The desired value c is the solution to the following problem:

$$c* = \arg\min_{c} J(c)., \tag{10}$$

In the following, we will explore commonly used methods for finding the value of $c*$.

3.1   Recurrent Least Squares

According to this method, the value of c is determined by the following recursive formula [5]:

$$P(t) = P(t-1) - \frac{P(t-1)\,\tilde{x}(t)\,\tilde{x}^T(t)\,P(t-1)}{1 + \tilde{x}^T(t)\,P(t-1)\,\tilde{x}(t)} \tag{11}$$

$$c(t) = c(t-1) + P(t-1)\left(y(t) - c^T(t-1)\,\tilde{x}(t)\right) \tag{12}$$

where $P(t)$ is a correction matrix of size $m(n+1) \times m(n+1)$,

$$P = \begin{bmatrix} \alpha & 0 & ... & 0 \\ 0 & \alpha & ... & 0 \\ ... & ... & ... & ... \\ 0 & 0 & ... & \alpha \end{bmatrix} \tag{13}$$

α is a large enough number.

The value to be determined is $c* = c(N)$, where N is the number of observations.

3.2   Kaczmarz Algorithm

According to the Kaczmarz algorithm, c is determined by the formula [5]:

$$c(t) = c(t-1) + \frac{y(t) - \hat{y}(t)}{\|\tilde{x}(t)\|^2}\tilde{x}(t). \tag{14}$$

3.3   Generalized Adaptive Identification Algorithms

The general form of this algorithm can be represented by the formula [5]:

$$c(t) = \alpha(t) c(t-1) + \beta(t) \tilde{x}(t) \tag{15}$$

where $\alpha(t)$ and $\beta(t)$ is the optimal adaptation parameters, in this case, are determined by the following expressions

$$
\begin{aligned}
\alpha(t) &= \frac{\|c(t-1)\|^2 \|\tilde{x}(t)\|^2 - y(t) \left(c^T(t-1)\tilde{x}(t)\right)}{\|c(t-1)\|^2 \|\tilde{x}(t)\|^2 - \left(c^T(t-1)\tilde{x}(t)\right)^2} \\
\beta(t) &= \frac{\|c(t-1)\|^2 \left(y(t) - c^T(t-1)\tilde{x}(t)\right)}{\|c(t-1)\|^2 \|\tilde{x}(t)\|^2 - \left(c^T(t-1),\tilde{x}(t)\right)^2}
\end{aligned}
\tag{16}
$$

or by the expressions [6]

$$
\begin{aligned}
\alpha(t) &= \frac{\|c(t-1)\|^2 \|\bar{x}(t)\|^2 - y(t) \left(c^T(t-1)\bar{x}(t)\right)}{\|c(t-1)\|^2 \|\bar{x}(t)\|^2 \sin^2\left(c^T(t-1),\bar{x}(t)\right)} \\
\beta(t) &= \frac{y(t) - c^T(t-1)\bar{x}(t)}{\|\bar{x}(t)\|^2 \sin\left(c^T(t-1),\bar{x}(t)\right)}
\end{aligned}
\tag{17}
$$

Then algorithm (15) can be rewritten as:

$$c(t) = c(t-1) + \frac{y(t) - c^T(t-1)\bar{x}(t)}{\|\bar{x}(t)\|^2 \sin^2\left(c^T(t-1)\bar{x}(t)\right)} \left[I - \frac{c(t-1) c^T(t-1)}{\|c(t-1)\|^2}\right] \bar{x}(t) \tag{18}$$

where I is the identity matrix.

To compare the rate of convergence and performance of the algorithms, we use two examples of identifying the coefficients of systems with the following parameters:

$X = [x_0, x_1, x_2, x_3]$- vector of input data
$x_0(i) = 1$
$x_1(i) = 1 + 0.4\sin(0.1i\pi + 0.2)$
$x_2(i) = 1.5 + \sin(0.05i\pi + 0.4)$
$x_3(i) = 1 + \sin(0.07i\pi + 1.6)$ where $i = 1..300$.
$K = [k_1, k_2, k_3, k_4]$ - vector of object parameters.
$k_1 = 2.5$ for a stationary system and $k_1 = 2 + 0.5\sin(0.001i\pi + 0.2)$ for a non-stationary system
$k_2 = -1.7; k_3 = 1.3; k_4 = 0.25$

The simulation results of the above methods for identifying coefficients for a stationary model are shown in Fig. 4.

The running time of the algorithms, respectively:

– Recurrent least squares: 0.26 s

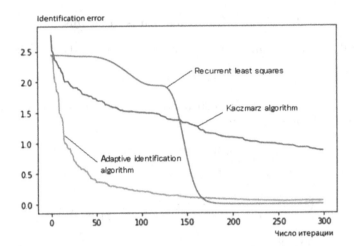

Fig. 4. Uncertainties in Estimating the Parameters of a Stationary System

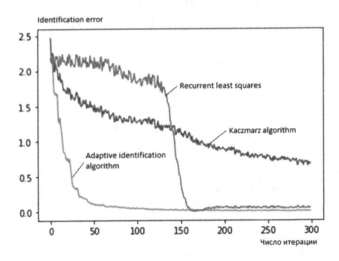

Fig. 5. Uncertainties in Estimating the Parameters of a Nonstationary System

– Generalized adaptive identification algorithms: 0.19
– Kaczmarz algorithm: 0.13

On Fig. 5 shows the results of modeling identification methods for a non-stationary system.

The running time of the algorithms, respectively:

– Recurrent least squares: 0.28 s
– Generalized adaptive identification algorithms: 0.21
– Kaczmarz algorithm: 0.13.

The simulation results show that the Kaczmarz algorithm has the shortest execution time. However, the disadvantage of this method is the rate of convergence. The least squares method provides a high rate of convergence, but has the disadvantage of computational complexity, which leads to a long implementation time. The generalized adaptive identification algorithm has a shorter execution time than the least squares algorithm and has a high convergence rate.

Based on the results of the analysis, an approach to the use of algorithms for identifying the coefficients of linear equations is proposed:

For an available data set, a neuro-fuzzy model does not require a high learning rate. In this case, using the recurrent least squares method allows you to achieve optimal accuracy.

For data collected sequentially, the model needs to be updated in a timely manner in accordance with new changes in the input data. The coefficient homogenization algorithm must simultaneously meet the requirements of processing speed and accuracy. In this case, a generalized adaptive identification algorithm should be used.

4   Application of a Fuzzy System in the Problem of Modeling a Linear Function

To evaluate the performance of the algorithm, we use a model to simulate a first order linear function. With an input of 25 data pairs generated from a first order linear equation $y = x_1 + x_2$.

The training data is a set of 25 values taken at regular intervals in the interval $[0, 4] \times [0, 4]$. The test data is a set of 10,000 pairs of values taken at equal intervals in the interval $[0, 4] \times [0, 4]$.

The model is initialized with the following parameters:

$n_{term} = 10$

$loss_{thresold} = 10e - 4$

$loss_{max} = 10e - 3$

The error function used is MSE:

$$LSE = \frac{1}{n} \sum_{i=1}^{n} (Y_i - \widehat{Y_i})^2 \qquad (19)$$

where: MSE is the model error, Y is the target result, $\widehat{Y}$ and is the model output. After the fuzzy rule grid is created, the fuzzy rules that do not contain training data are removed. The number of fuzzy rules before the union is 31, and after the union it is 9. The algorithm stops when the error $loss_{thresold}$ exceeds $loss_{max}$ (Table 1).

- The model training error is 0.0096.
- The model testing error is 0.0117.

Table 1. Fuzzy rules obtained after simulating the linear function.

Fuzzy rule	Fuzzy rule center	Extended Matrix	Coefficient
R_1	$C_1 = [0.05\ 0.05]$	$S_1 = \begin{bmatrix} 0.0004 & 0 \\ 0 & 0.0004 \end{bmatrix}$	$c_1 = 0$
R_2	$C_2 = [0.15\ 0.15]$	$S_2 = \begin{bmatrix} 0.003 & -0.003 \\ -0.003 & 0.003 \end{bmatrix}$	$c_2 = 1$
R_3	$C_3 = [0.25\ 0.25]$	$S_1 = \begin{bmatrix} 0.011 & -0.010 \\ -0.010 & 0.011 \end{bmatrix}$	$c_3 = 2$
R_4	$C_4 = [0.39\ 0.38]$	$S_4 = \begin{bmatrix} 0.032 & -0.030 \\ -0.030 & 0.030 \end{bmatrix}$	$c_4 = 3$
R_5	$C_5 = [0.445\ 0.534]$	$S_5 = \begin{bmatrix} 0.039 & -0.039 \\ -0.039 & 0.040 \end{bmatrix}$	$c_5 = 4$
R_6	$C_6 = [0.621\ 0.606]$	$S_6 = \begin{bmatrix} 0.030 & -0.030 \\ -0.030 & 0.032 \end{bmatrix}$	$c_6 = 5$
R_7	$C_7 = [0.725\ 0.740]$	$S_7 = \begin{bmatrix} 0.015 & -0.013 \\ -0.013 & 0.014 \end{bmatrix}$	$c_7 = 6$
R_8	$C_8 = [0.85\ 0.85]$	$S_8 = \begin{bmatrix} 0.003 & -0.003 \\ -0.003 & 0.003 \end{bmatrix}$	$c_8 = 7$
R_9	$C_9 = [0.95\ 0.95]$	$S_9 = \begin{bmatrix} 0.0004 & 0 \\ 0. & 0.0004 \end{bmatrix}$	$c_1 = 0$

The matrix of expansion coefficients of the multidimensional membership function is considered equivalent to the covariance matrix of input variables X_1, X_2:

$$\Sigma_{X_1,X_2} = \begin{bmatrix} cov[X_1, X_1] & cov[X_1, X_2] \\ cov[X_2, X_1] & cov[X_2, X_2] \end{bmatrix} \tag{20}$$

The values $cov[X_1, X_2]$ and $cov[X_2, X_1]$ represent a linear relationship between the two input variables. The equality $cov[X_1, X_2] = 0$ shows that X_1 and X_2 are independent (examples for this case are R_1 and R_9). Then the fuzzy rule can be represented as:

$$R_1: \text{if } x\,(x_1, x_2) \text{ is } H_1 \text{ then } y = c_0^1$$

or can be represented as one-dimensional membership functions:

$$R_1: \text{if } x_1 \text{ is } H_1^1 \text{ and } x_2 \text{ is } H_2^1 \text{ then } y = c_0^1$$

The case $cov[X_1, X_2] \neq 0$ shows that X_1 and X_2 are dependent. For example, for fuzzy rules $R_2,...,R_8$, the linear dependence of X_1 and X_2 is represented by the following equation:

$$x_2 = m_{x_2} \pm \sqrt{\frac{cov[X_2, X_2]}{cov[X_1, X_1]}} (x_1 - m_{x_1})$$

where: m_{x_1}, m_{x_2} - center coordinates C_2.

The sign $< + >$ appears when $cov[X_1, X_2] > 0$, that is, X_1 and X_2 are linearly dependent. The sign $< - >$ appears when $cov[X_1, X_2] < 0$, that is, X_1 and X_2 are inversely linearly dependent.

Then the premise of the fuzzy rule considers the membership degree of a pair of points (x_1, x_2) lying on some "fuzzy" segment $Ax + B$. For example:

R_3: if (x_1, x_2) lies on the line $x_2 = -x_1 + 0.5$ then $y = 2$.

This description provides more flexibility to the algorithm in splitting the input space. Another advantage of the algorithm is the creation of a fuzzy set of fuzzy rules that can operate independently. For a given (usually small) cut threshold α, the fuzzy rules do not overlap. Therefore, in its "coverage area" each fuzzy rule does not depend on other fuzzy rules. This property allows you to calculate the confidence of a rule separately. This is especially important for decision support systems.

5 Building a Fuzzy Neural Model to Predict the Exchange Rate

In this section, we apply a fuzzy neural model based on a multivariate membership function to solve the exchange rate forecasting problem.

To build a model, it is necessary to collect data on the factors that affect the exchange rate. Then use the information variable selection algorithm to select the best factors for the model.

The following factors influence the exchange rate:

- oil price
- gold price
- import
- export

The input data is the value of the exchange rate and 5 factors that affect it for 62 consecutive months from January 2014 to February 2019 [7].

For complex models, using the maximum correlation coefficient instead of building the model at each step will greatly reduce the computational cost of attribute selection. The search algorithm for information variables will look like this [8]:

- Select the original model M_0.
- Calculate the maximum correlation coefficient between input variable X and output variable Y.
- Add the variable most correlated with the output variable to the M_0 model to get the M_1 model.
- We build model M_1 with the selected input variable.
- Based on the maximum correlation coefficient between the current model error M_1 and the rest of the input variables, select the next variable.
- Add the selected variable to M_1 to get M_2.

- Let the model Mk be built. We select the next input variable based on the maximum correlation coefficient between the remaining input variables and the model error M_k. The selected variable is added to M_{k+1}.
- Repeat until all required input variables are included in the model.
- The algorithm stops if it finds a model M_k that satisfies the given stopping condition. Otherwise, the loop continues until all required input variables in the model have been combined.
- We select from the models $\{M_1, ..., M_k, ..., M_n\}$ the model that provides the best accuracy for the selected quality metric.

The information variable selection algorithm stops at two inputs: the price of gold and the price of oil. The model is initialized with the following parameters:

$$n_{term} = 30$$
$$loss_{thresold} = 0.001$$
$$loss_{max} = 0.5$$

The error function used is LSE:

$$LSE = \frac{1}{n} \sum_{i=1}^{n} (Y_i - \widehat{Y_i})^2 \tag{21}$$

The right side of the model is a linear equation

$$Y_{i,t} = c_{i,0} + c_{i,1}X_{t,1} + c_{i,2}X_{t,2} + ... + c_{i,n}X_{t,n}, i = \overline{1, m}, \tag{22}$$

- The number of fuzzy rules after training is 15.
- The model training error is 0.601.
- The model testing error is 0.842.
- Model error after additional learning: 0.81.

The model has been built in Python programming language on the Google Colab Free.

6 Conclusion

This paper presents a method for constructing fuzzy neural models based on multidimensional membership functions. A method for selecting input variables and an approach to the application of algorithms for identifying the coefficients of linear equations have been developed. Algorithms for solving the problem of building a model for forecasting the exchange rate are applied. The resulting model is able to describe the interaction between input variables, and the generated fuzzy rules are able to work independently. This is an important prerequisite for extracting knowledge from digital data.

References

1. Kang, D., Yoo, W., Won, S.: Multivariable TS fuzzy model identification based on mixture of Gaussians. In: 2007 International Conference on Control, Automation and Systems, IEEE, Seoul, Korea (2007). https://doi.org/10.1109/ICCAS.2007.4407036

2. Lemos, A., Caminhas, W., Gomide, F.: Multivariable Gaussian evolving fuzzy modeling system. IEEE Trans. Fuzzy Syst. **19**, 91–104 (2010). https://doi.org/10.1109/TFUZZ.2010.2087381

3. Pratama, M., Anavatti, S.G., Angelov, P.P., Lughofer, E.: PANFIS: a novel incremental learning machine. IEEE Trans. Neural Netw. Learn. Syst. **25**, 55–68 (2013). https://doi.org/10.1109/TNNLS.2013.2271933

4. Abonyi, R., Babuska, R., Szeifert, F.: Fuzzy modeling with multivariate membership functions: gray-box identification and control design. IEEE Trans. Syst. Man Cybern. Part B (Cybernetics), **31**, 755–767 (2001). https://doi.org/10.1109/3477.956037

5. Pikina, G., Moscow, A.F., Avetisyan, A., Filippov, G., Pashchenko, F.F.: Thermal-hydraulic models of power station equipment. FIZMATLIT (2013)

6. Hieu, T.D., An, B.T., Pashchenko, F.F., Pashchenko, A.F., Trong, N.V.: : Intelligent technologies in decision-making support systems. In: 2020 International Conference Engineering and Telecommunication (En&T), pp. 1–4. IEEE, Moscow (2020). https://doi.org/10.1109/EnT50437.2020.9431248

7. Bui, Tr.A., Pashchenko, F.F., Pashchenko, A.F., Kudinov, Y.I.: Using neural networks to predict exchange rates. In: 13th International Conference "Management of large-scale system development" (MLSD), pp. 1–5. IEEE, Moscow (2020). https://doi.org/10.1109/MLSD49919.2020.9247748

8. Kamenev, A.V., Pashchenko, A.F., Pashchenko, F.F.: Neuro-fuzzy simulation system with the selection of informative variables. In: SENSORS & SYSTEMS, vol. 7–8, pp. 8–14. Moscow

Controlled Markov Queueing Systems Under Uncertainty

V. Laptin[1] and A. Mandel[2](✉)

[1] M.V. Lomonosov Moscow State University, Lenin's Mountings 1, Moscow, Russia
straqker@bk.ru
[2] V.A. Trapeznikov Institute of Control Sciences RAS, Profsoyuznaya 65,
Moscow, Russia
almandel@yandex.ru

Abstract. We investigate a model of a multilinear queueing system (QS) with channel switching under uncertainty, when statistical characteristics of the homogeneous Markov chain, which describes the transition probabilities of the environment from state to state, are unknown. Several reinforcement learning algorithms have been proposed to control such a system.

Keywords: Multi-Channel Queueing Systems · Switching of Service Channels · Case of Uncertainty · Reinforcement Learning

1 Introduction

This paper deals with a controllable queuing system in which the number of switching service channels monitor and modify at control time points spaced apart by a fixed time step. At transition from step to step, the intensity of the simplest incoming ow changes in accordance with a Markov's chain. The main topic of the paper is the investigation of such multilinear Markovian systems under conditions when transition probabilities in the finite Markov chain, which describes the change of states of the environment, are unknown a priori. Several learning algorithms are compared.

2 Related Works

QS have wide applications in the field of analysis of the activity of socio-economic systems, such as systems for selling various kinds of tickets, hotel reservations, trade and production, transport, telecommunications systems and many other systems. In this case, the number of service devices in a given system, which is chosen as a function of the time-varying random intensity of the incoming demand flow, acts as an optimized and dynamically controlled parameter.

Over the past decades, the rapid growth of information technology has led to increasingly complex networked systems, which poses challenges in gaining a

V. M. Vishnevskiy et al. (Eds.): DCCN 2022, CCIS 1748, pp. 246–256, 2023.
https://doi.org/10.1007/978-3-031-30648-8_20

solid knowledge of system dynamics. For example, due to security or economic considerations, a number of network systems are built according to the "black box" model, where information about the model simply cannot be obtained.

To overcome the above problems, it is desirable to apply schemes with self-learning algorithms. The natural approach is reinforcement learning that learns and reinforces the correct decision policy through repeated interaction with the environment. Reinforcement learning methods provide a qualitative basis on which to develop a learning policy for QS. There are two main areas of work on reinforcement learning methods: model-less reinforcement learning (e.g. value iteration [1], Q-learning [1]) and model-based reinforcement learning (e.g. PSRL [2]).

Finding a good QS control policy is essential to achieve the objectives of the model: network performance (for example, high throughput or low average task delay) or economic (such as reducing system maintenance costs or maximizing system profits).

For example, in [3], the possibility of using model-based reinforcement learning is considered to obtain an optimal queuing network control policy, where the average job delay (or, equivalently, the average queue backlog) is minimized. A QS represents several queues independent of each other with different intensities of the incoming flow and service rates, which are processed on one (common) service device.

In another paper [4], a reinforcement learning algorithm for policy derivation based on neural networks (DDPG) was used to minimize the total queue length and limit the maximum total service time of one requirement in a queuing system with a sequential arrangement of service channels (where the requirements sequentially pass through several stages).

However, for the analysis of the economic component, most studies are based on the methods of queuing theory, which are often limited to simple and unrealistic scenarios due to the mathematical unsolvability of the problem. The general approach used in this class of studies is to minimize the cost function consisting of various terms such as client service cost and server operating cost [5,6].

In this paper, we propose to use reinforcement learning methods to obtain an optimal QS control policy with switched service channels to minimize the total system costs, allowing us to move away from mathematically complex and cumbersome models from queuing theory.

3   Multichannel QS with Switching of Channels

As stated in [7], in the studied QS the number of working service channels can be changed at the moments of control, which are separated from each other for a fixed time (the control step). In this case, it is considered that the QS receives the simplest incoming flow, the intensity of which $\lambda(t)$ is constant throughout the step, and at the moments of control it undergoes jump-like Markov changes, taking a finite number k of values λ_i from the discrete set $\Lambda = \{\lambda_i, i \in \overline{1,k}\}$. The task is to form a strategy for switching working channels (disabling some of the working channels or putting into operation reserve channels), which minimizes the average costs of the QS for a given N-step planning period. In this case, as in [8], it is assumed that the matrix of transition probabilities of the corresponding homogeneous Markov chain $P = \|p_{ij}\|$ is given, where p_{ij} is the transition probability (at the time of control) from the intensity λ_i, $i \in \overline{1,k}$, at the previous step to the intensity λ_j, $j \in \overline{1,k}$, at the next step.

In [8] was shown that solving the problem of choosing the optimal channel switching strategy is reduced to the following system of dynamic programming equations:

$$C_1^*(\lambda_i, m) = \min_{u \geq \underline{u_i}} C^{(1)}(\lambda_i, m, u), \tag{1}$$

$$C_n^*(\lambda_i, m) = \min_{u \geq \underline{u_i}} (C^{(1)}(\lambda_i, m, u) + \alpha \sum_{j=1}^{l} p_{ij} C_{n-1}^*(\lambda_j, u)), n \in \overline{2, N}. \tag{2}$$

where $C_n^*(\lambda_i, m)$ is the minimum possible value of the total average costs at the last n steps of the control process, when the mathematical expectation is taken along the trajectory of the incoming flow intensity, which makes Markov jumps. The variable u in Eqs. (1)–(2) is the current (n steps before the end of the planning period) value of the control decision on the number of switched working channels. It has been proved that the optimal channel switching strategy, as in the equivalent inventory control problem with returns [9], is a four-level strategy.

4   Simulation of a Markovian Process for Channel Switching Decisions

The Markovian decision-making process is a four tuple $< S, A, P, R >$, in which:

- S is a set of states, called the state space,
- A is a set of all available actions, called the action space,
- $P_\alpha(s, s') = Pr(s_{t+1} = s' | s_t = s, a_t = a)$ is the probability that an action in state s at time t will result in a transition to state s',
- $R_a(s, s')$ is immediate reward (or expected immediate reward) received after the transition from state s to state s' as a result of action a.

In a given QS a set of states S is a set of pairs $\{(\lambda_i, m), \lambda_i \in \Lambda\}$, where m is the starting number of service channels. The set of actions A is a set of all acceptable values for the number of the service channels in the system. The action space A obeys the constraint of the system being stationary, i.e. for each intensity of the incoming flow λ_i there is an acceptable number of channels $u \in [u_{crit}(\lambda_i), m_{max}]$, where $u_{crit}(\lambda_i)$ is the minimal acceptable number of service channels necessary to obtain the stationary mode of QS functioning with the incoming flow intensity λ_i ($u_{crit}(\lambda_i) = \left[\frac{\lambda_i}{\mu}\right] + 1$) [10], while m_{max} is the maximal acceptable number of channels in QS. Transition probabilities $P_a(s, s')$ are taken from the respective homogeneous Markovian chain $P = \|p_{ij}\|$. The reward $R_a(s, s')$ is equal to the one-step cost $C^1(\lambda_i, m, u)$.

When the transition probability matrix of the corresponding homogeneous Markov chain $P = \|p_{ij}\|$ is known, that is, the transition probabilities (at the moment of control) between the states of the system are known, the optimal control policy can be derived using the value iteration method [10].

For this purpose the following functions are introduced: action value function $Q_i(s,a)$ and value function $V_i(s)$. The action value function $Q_i(s,a)$ shows whether a particular action a performed in state s is optimal in terms of decision making policy π.

The problem of keeping production machines up and running is typical of any branch of production. The task of managing spare parts inventories in a facility is complicated by their high cost and expensive storage, as well as by market volatility (which translates into fluctuations in spare part prices, availability and delivery times). Added to all of the above is the permanent desire of owners to save money, which in turn reduces the manageability of the system in the event of breakdowns.

$$Q_i(s,a) = \sum_{s'} P\left(s' \mid s, a\right) * \left[r\left(s, a, s'\right) + \alpha V_i\left(s'\right)\right]. \tag{3}$$

The value function $V_i(s)$ allows us to characterize each particular state according to the actions available in that state. Since this statement of the problem requires minimizing multi-step costs, the value function will take the form:

$$V_{i+1}(s) = \min_a \sum_{s'} P\left(s' \mid s, a\right) * \left[r\left(s, a, s'\right) + \alpha V_i\left(s'\right)\right] = \min_a Q_i(s,a), \tag{4}$$

where i is the iteration number of the algorithm.

The optimal decision making policy π^* can be found using (4):

$$\pi^*(s) = argmin_a Q_i(s,a) \tag{5}$$

Let us do numerical simulations and compare the total costs obtained using the optimal value iteration approach and by calculating the math expectation of costs (1)–(2). For simplicity, we consider a system with two possible states (the values of the QS parameters are given in Table 1). The number of steps in the planning period is 10.

In calculating the mathematical expectation of costs (1)–(2), the discount factor was chosen to be 1, and 0.99 was assumed to train the value iteration algorithm. To learn the value iteration algorithm more accurately, sample averages were estimated (each set consisted of 500 ten-steps processes, repeated 100 times). Using Table 2, we can compare the cost values obtained by value iteration and by calculating the mathematical expectation (1)–(2). Table 2 shows the numerical values of total costs for the 10-step process when the initial intensity and number of channels at step 1 are given. As you can see from the Table 2, the value iteration method can provide an optimal policy for managing the number of service channels.

Table 1. Values of the simulated QS parameters.

Notation	Value	Implication
c_1	1	The cost of operation per channel.
c_2	0.2	The cost of disabling one service channel.
μ	6	Service intensity per one channel.
$A_1 = A_2$	1	The cost of a decision to enable/disable.
Λ	$\lambda_1 = 20, \lambda_2 = 40$	Intensity of incoming flow for each state of the environment.
P	$\begin{matrix} \lambda_1 \\ \lambda_2 \end{matrix}\begin{pmatrix} 0.8\ 0.2 \\ 0.4\ 0.6 \end{pmatrix}$	Transient probability matrix of a homogeneous Markov chain.
d	1	The cost of maintaining the queue.
m_{max}	10	Maximum acceptable number of service channels

Table 2. Comparison of results obtained by the iteration method and by calculating the math expectation.

Intensity λ_i	Number of channels (m)	Value iteration, mean	Value iteration, std	Math expectation
20	4	72.52	0.41	72.54
20	5	71.58	0.42	71.54
20	6	72.76	0.44	72.74
20	7	72.95	0.41	72.94
20	8	73.23	0.43	73.14
20	9	73.41	0.37	73.34
20	10	73.58	0.46	73.54
40	4	80.49	0.36	80.45
40	5	80.42	0.40	80.45
40	6	80.42	0.40	80.45
40	7	80.41	0.43	80.45
40	8	80.43	0.37	80.45
40	9	79.48	0.37	79.45
40	10	80.64	0.46	80.65

5 Channel Switching Decisions Under Uncertainty: Q-learning Method

Now consider a case where there is no information about the environment model, which means that the transition probability matrix for the corresponding Markov chain $P = \|p_{ij}\|$ is not known. We can use the Q-learning method here [1]. For any finite Markovian decision-making process, Q-learning leads to an optimal policy (to maximize/minimize the expected value of the total reward at all consecutive steps without exception, starting from the current state).

We compared 3 varieties of Q-learning: Q-learning with ϵ-greedy policy, "expected value SARSA" with ϵ-greedy policy, and "Softmax"/"Softmin" policy. Here, by analogy with the iteration method, the action value function $Q(s, a)$ is used, and the decision making policy is implemented depending on the chosen strategy: for the ϵ-greedy policy, an action that maximizes (minimizes) the value of the function $Q(s, a)$ is chosen with probability $p = 1 - \epsilon$, while with probability $p = \epsilon$ a random action is chosen; for the "Softmax" policy, each action is chosen with its corresponding probability derived from the "Softmax" function. The action value function $Q(s, a)$ is calculated by the following formula:

$$Q^*(s, a) = r(s, a) + \gamma * max_{\alpha'} Q(s', a') \tag{6}$$

while a rule for changing the action value function is obtained as follows:

$$Q(s, a) = \alpha * Q^*(s, a) + (1 - \alpha)\gamma * Q(s, a) \tag{7}$$

where α is the learning rate (from 0 to 1), and γ is the discount factor.

In case of "expectedvalueSARSA" method with ϵ-greedy policy:

$$Q^*(s, a) = r(s, a) + \gamma * E_{a_i \leftarrow \pi(a|s')} Q(s', a_i), \tag{8}$$

where $E_{a_i \leftarrow \pi(a|s')}$ is the math expectation with regards to the policy $\pi(a \mid s')$. Whereas for the case of the "Softmin" policy:

$$\pi(a_i \mid s) = \text{softmin}\left(\left\{-\frac{Q(s, a_j)}{\tau}\right\}_{j=1}^n\right)_i = \frac{\exp\left(\frac{-Q(s, a_i)}{\tau}\right)}{\sum_j \exp\left(\frac{-Q(s, a_j)}{\tau}\right)}. \tag{9}$$

A Markovian decision-making process was also used to simulate the system, but this time the model did not receive any information on the environment. Learning occurred in real time, which means that the models made decisions independently in each learning iteration. The system parameters remained the same during simulation (Table 1). Figure 1 below shows plots of the convergence of all 3 policies as a function of the number of iterations, and Fig. 2 shows the results of controlling the system using the trained models. The parameters chosen for training were as follows: $\alpha = 0.4$ (learning rate), $\gamma = 0.95$ (discount factor), $\epsilon = 0.35$. It is worth mentioning, that all 3 approaches are characterized by low stability of convergence. Table 3 (next page) shows the results of sample averages based on 30 training cycles (1000 iterations each, of which 500 iterations were used to estimate the average). The selected approaches showed a strong dependence on the learning rate, with values of $\alpha < 0.2$, even at 2000 iterations the models showed not very impressive results (the total cost was about 20 percent higher than for carefully chosen parameters).

A closer look at the nature of convergence (Fig. 2), reveals that models require a significant number of iterations (approximately 700–800), which can lead to issues when implementing these algorithms to solve practical problems. The learning speed also depends on the number of steps in each iteration, and if the dimensionality of the matrix is maintained, an increase in the number of steps leads to an increase in the learning speed (since the model is making more attempts). However, hundreds (or even thousands if the dimensionality of the system is raised) of iterations may be unacceptable for the real life systems, since this kind of learning process would last for months.

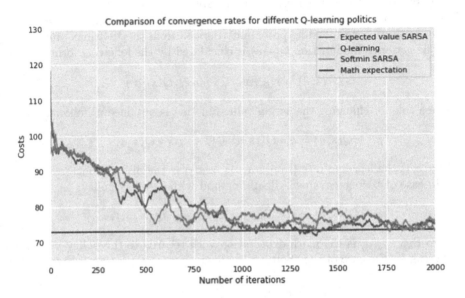

Fig. 1. Comparison of convergence rates for different Q-learning politics.

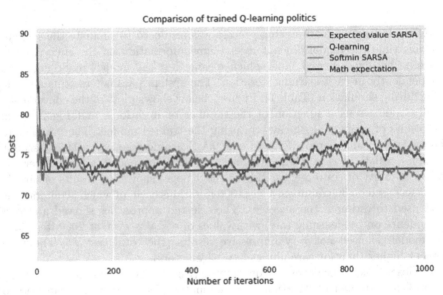

Fig. 2. Comparison of trained Q-learning politics.

To accelerate learning, one can try saving results of observations and use them for re-learning (hereinafter, "Experience replay") [11,12]. In each iteration, in addition to learning from new events, learning from several already observed events, chosen randomly from memory, is done. However, here one starts feeling the difficulty of selecting the right buffer size of the observed processes, and the question arises on how many instances from memory to perform additional learning with each iteration. Since, despite an increasing convergence rate, the

Table 3. Comparison of Q-learning algorithm results.

Intensity	Number of channels (m)	Math expectation	Q-learning	Exp. value SARSA	Softmin SARSA
20	4	72.54	75.09 ± 2.73	74.28 ± 2.31	76.51 ± 3.86
20	5	71.54	74.7 ± 5.59	74.12 ± 3.38	74.87 ± 3.98
20	6	72.74	73.86 ± 1.38	74.31 ± 2.79	75.46 ± 3.55
20	7	72.94	75.92 ± 2.26	75.32 ± 3.61	75.55 ± 3.29
20	8	73.14	75.49 ± 2.46	75.16 ± 2.37	76.55 ± 3.88
20	9	73.34	75.4 ± 5.12	75.41 ± 3.35	76.49 ± 4.88
20	10	73.54	75.78 ± 3.03	75.16 ± 1.76	75.06 ± 2.46
40	4	80.45	82.05 ± 2.16	81.93 ± 2.81	82.06 ± 2.4
40	5	80.45	81.97 ± 2.44	81.48 ± 2.47	82.0 ± 2.21
40	6	80.45	82.79 ± 2.74	81.76 ± 1.84	83.66 ± 4.29
40	7	80.45	83.52 ± 4.27	81.74 ± 1.72	82.73 ± 2.88
40	8	80.45	82.28 ± 3.17	81.78 ± 1.79	84.37 ± 4.23
40	9	79.45	80.93 ± 1.9	80.87 ± 3.04	80.75 ± 2.35
40	10	80.65	82.2 ± 2.02	82.56 ± 3.62	83.47 ± 3.79

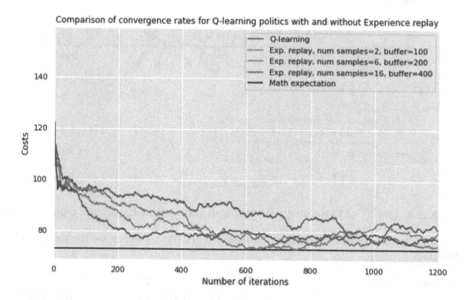

Fig. 3. Comparison of convergence rates for Q-learning politics with and without Experience replay.

very quality of the trained agent may deteriorate (so far no qualitative statistical estimates for the buffer size and number of instances have been derived). Figure 3 compares the convergence rates as a function of the buffer size and the number of instances to be trained for the Q-learning algorithm with ϵ-greedy policy, and Fig. 4 illustrates the comparative quality of the learning processes.

6   Statistical Evaluation of the Markov Chain Matrix

Due to the difficulties in controlling the speed and quality of convergence of Q-learning algorithms, one can try, as an alternative, the statistical estimation of the transition probability matrix $P = \|p_{ij}\|$. And after obtaining an estimate of such a matrix, apply the value iteration algorithm that has already proven its quality.

As an estimation method, it is proposed to use the count of the number of occurrences of each event when simulating the Markov decision-making process. The model parameters remain the same (see Table 1). Table 4 below shows the values of the estimated transition probability matrix depending on the number

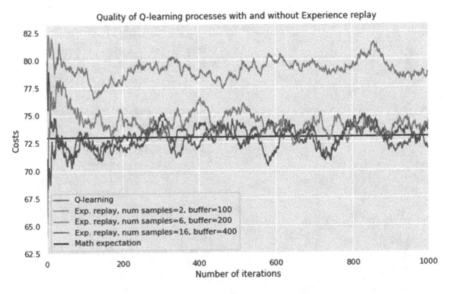

Fig. 4. Quality of Q-learning processes with and without Experience replay.

Table 4. Estimates of the transition probability matrix depending on the number of iterations.

Number of iterations	10	20	50	100
Probability Matrix Estimation	$\begin{pmatrix} 0.76 & 0.24 \\ 0.46 & 0.54 \end{pmatrix}$	$\begin{pmatrix} 0.79 & 0.21 \\ 0.39 & 0.61 \end{pmatrix}$	$\begin{pmatrix} 0.8 & 0.2 \\ 0.41 & 0.59 \end{pmatrix}$	$\begin{pmatrix} 0.79 & 0.21 \\ 0.41 & 0.59 \end{pmatrix}$

of iterations, and it can be seen (see Table 5) that already at 10–20 iterations the matrix is estimated well enough so that the results given by the value iteration method are similar to the results with a known matrix of transition probabilities $P = \|p_{ij}\|$. Which in turn is significantly superior to Q-learning methods, both in speed and quality.

7   Systems of Higher Dimension

It is also need to mention that with higher number of states (the dimension of the matrix of transition probabilities), the methods value iteration, as well as Q-learning, also show satisfactory results. In the case of a matrix with dimensions of 5 (see Fig. 5, the rest of the parameters during the simulation remained the same, see Table 1), for Q-learning without the use of "Experience replay" it takes

Table 5. Comparison of the results of the iteration algorithm for the values of the estimated matrices depending on the number of iterations.

Intensity,	Number of channels (m)	Math expectation	10 iterations	20 iterations	50 iterations
20	4	72.54	72,21	72,75	71,98
20	5	71.54	71,85	71,46	71,93
20	6	72.74	73,36	72,85	73,11
20	7	72.94	73,54	73,08	72,78
20	8	73.14	73,22	72,81	73,34
20	9	73.34	73,3	74,17	73,5
20	10	73.54	73,45	73,39	72,91
40	4	80.45	80,48	80,3	80,32
40	5	80.45	80,74	80,21	80,31
40	6	80.45	80,42	80,51	80,53
40	7	80.45	80,76	80,76	80,33
40	8	80.45	80,47	80,75	80,3
40	9	79.45	79,58	79,22	79,56
40	10	80.65	80,95	80,32	80,44

	15	25	35	45	50
15	0.5	0.2	0.1	0.1	0.1
25	0.4	0.1	0.3	0.1	0.1
35	0.1	0.2	0.2	0.1	0.4
45	0.1	0.1	0.1	0.6	0.1
50	0.1	0.4	0.3	0.1	0.1

Fig. 5. Transition probability matrix of higher dimension (5 states). Values along the axis are the incoming flow intensities

about 2.5–3 thousand iterations, but the statistical evaluation method is already at 50–100 iterations shows excellent results.

8 Summary

A solution example for a channel switching optimization problem in a queueing system with incomplete information is given. The efficiency (in terms of convergence rate and estimation accuracy) of several reward-based learning algorithms is compared.

References

1. Sutton, R.S., Barto, A.G.: Reinforcement Learning: An Introduction, 2nd edn. MIT Press, Cambridge (2018)
2. Osband, I., Russo, D., Van Roy, B.: (More) efficient reinforcement learning via posterior sampling. In: Advances in Neural Information Processing Systems, pp. 3003–3011 (2013)
3. Liu, B., Xie, Q., Modiano, E.: Reinforcement Learning for Optimal Control of Queueing Systems, pp. 663–670 (2019). https://doi.org/10.1109/ALLERTON.2019.8919665
4. Raeis, M., Tizghadam, A., Leon-Garcia, A.: Queue-learning: a reinforcement learning approach for providing quality of service (2021)
5. Kumar, R., Lewis, M.E., Topaloglu, H.: Dynamic service rate control for a single server queue with Markov modulated arrivals. Naval Res. Logistics (NRL) 60(8), 661–677 (2013)
6. Lee, N., Kulkarni, V.G.: Optimal arrival rate and service rate control of multi-server queues. Queueing Syst. 76(1), 37–50 (2013). https://doi.org/10.1007/s11134-012-9341-7
7. Mandel, A., Laptin, V.: Myopic channel switching strategies for stationary mode: threshold calculation algorithms. In: Vishnevskiy, V.M., Kozyrev, D.V. (eds.) DCCN 2018. CCIS, vol. 919, pp. 410–420. Springer, Cham (2018). https://doi.org/10.1007/978-3-319-99447-5_35
8. Mandel, A.S., Laptin, V.A.: Channel switching threshold strategies for multi-channel controllable queuing systems. In: Vishnevskiy, V.M., Samouylov, K.E., Kozyrev, D.V. (eds.) DCCN 2020. CCIS, vol. 1337, pp. 259–270. Springer, Cham (2020). https://doi.org/10.1007/978-3-030-66242-4_21
9. Granin, S., Laptin, V., Mandel, A.: Inventory control with returns and controlled Markov queueing systems. In: Vishnevskiy, V.M., Samouylov, K.E., Kozyrev, D.V. (eds.) Distributed Computer and Communication Networks: Control, Computation, Communications, DCCN 2022. Lecture Notes in Computer Science, vol. 13766, pp. 361–370. Springer, Cham (2022). https://doi.org/10.1007/978-3-031-23207-7_28
10. Saaty, T.: Elements of Queueing Theory. Mc Grau Hill Company Inc, New York (1961)
11. Schaul, T. Quan, J., Antonoglou, I., Silver, D.: Prioritized experience replay. CoRR, pp. 1–23 (2015)
12. Haykin, S. Neural Networks. A Comprehensive Foundation, 2nd Edition. Prentice Hall, Upper Saddle River (2020)

Fuzzy Classification Model Based on Genetic Algorithm with Practical Example

Olga Kochueva(✉) [ID]

National University of Oil and Gas "Gubkin University", 65 Leninsky Prospekt,
Moscow 119991, Russia
olgakoch@mail.ru

Abstract. The paper presents a new classification model based on a symbolic regression method, fuzzy inference system and genetic algorithm. For complex practical problems, building a unified predictive model for various states of a system or a process encounters a lot of difficulties, but the task can be divided into 2 stages: a) to obtain a classification of system states; b) to build models with good predictive qualities for each class. The fuzzy approach makes it possible to specify states (sets of parameters) which can be assigned to more than one class. A feature of the presented model is the use of symbolic regression to identify input variables that form the basis of the classification model, and to clarify the interaction of parameters. The paper also presents an example of the practical application of the model, which shows the advantage of the proposed approach over traditional methods of building models.

Keywords: Machine learning · Symbolic regression · Genetic algorithm · Fuzzy classification model · Knowledge extraction

1 Introduction and Literature Review

Classification problem is one of the main components of knowledge extraction. Nowadays, with the development of measurement technology, a large amount of data is available, and it is necessary to use a special strategy for specifying parameters or their combinations as the feature variables of the model. In empirical classification problems, the number of the feature variables is unknown, moreover the definition of a set of variables for classification poses its own challenges especially when there is an effect of reciprocal influence of several parameters.

Fuzzy classification model can be a powerful instrument and provide an insight for practical problems when crisp classification reaches its limits. One example of such a situation is the classification of emissions of harmful pollutants, where a crisp approach can define a certain amount of a pollutant as safe, while its increase by 0.01% can exceed the limit and be classified as dangerous. The use of fuzzy logic makes it easy to overcome this disadvantage - the set of initial data can be assigned to two classes at once, for each of them, the degree of membership can be determined.

V. M. Vishnevskiy et al. (Eds.): DCCN 2022, CCIS 1748, pp. 257–268, 2023.
https://doi.org/10.1007/978-3-031-30648-8_21

The application of classification problem is extremely wide and fuzzy app-roach has given good results in a number of real world applications (examples can be found in [1–4]). An overview of algorithms for fuzzy classification models is given in [5]. In most of the algorithms proposed by the authors (see [6], for example), two stages can be distinguished - feature extraction step and infer-ence step. Optimization algorithms are used to determine the distribution of fuzzy sets for each feature variable in the first step. In the second step, the opti-mization procedure helps to determine weights for each rule to obtain the best classification result.

The most popular optimization algorithms for multivariate functions applied to the solution of the classification problem are

- Particle Swarm Optimization technique (PSO) [2];
- Grey Wolf Optimization algorithm (GWO) [3];
- Artificial Bee Colony (ABC) [7];
- various modifications of Genetic Algorithm (GA) [6, 8–10]. The main ideas to use genetic programming (GP) methodology in fuzzy inference system (FIS) design are presented in [11].

In [3], to solve the flood classification problem, the idea of an iterative fuzzy clustering model based on a combination of subjective and objective weights of the parameters was presented.

A number of difficulties in building rule-based classification models like

- the exponential increase in the number of fuzzy rules with the number of input variables,
- worsening interpretability of rules when they involve many antecedent condi-tions,
- multiple objectives in optimizing fuzzy rule-based system

were formulated in [9]. Therefore, the authors consider an optimization problem with three objectives: to maximize the number of correctly classified training patterns, minimize the number of fuzzy rules, and minimize the total number of antecedent conditions, they use GA to solve optimization problem.

In [10], a classification rule base mining algorithm combining the GA and intuitionistic fuzzy-rough set for large-scale fuzzy information system has been presented. The feature of this paper is that it has proposed the definitions and measurement metrics of completeness, interaction and compatibility describing the whole rule base.

In [12], a neuro-fuzzy model named fuzzy broad learning system based on the Takagi-Sugeno fuzzy system [13] was proposed. Its performance was tested for several datasets for regression and classification problems.

The interest of most researchers is focused directly on solving the classifica-tion problem, but this paper considers a set of algorithms where classification serves as an auxiliary tool that makes it possible to single out separate classes for the system or the process under study, after which its own model can be built for each class. The similar idea is developed in [14]. The authors present piecewise regression model and use the maximum likelihood method to estimate break-points and regression parameters simultaneously. Breakpoint detection is based on classifying data into groups of similar observations. This approach allows to obtain a composite model with good predictive quality.

2   Methodology

Let $x = \{x_1, x_2, ..., x_m\}$ be the set of input variables, y - the dependent variable. A training dataset $\{X, Y\}$, where $X \in \mathbb{R}^{n \times m}$, n - number of input samples.

The ultimate goal is to build a predictive model, the input and output variables have a wide range of values, but often the model does not work well on the entire data set. In this case, it makes sense to define multiple classes, build predictive models for each class, and use classification model to assign a set of input variables to a specific class. In the case of using a fuzzy classification model, the calculation algorithm is shown in Fig. 1.

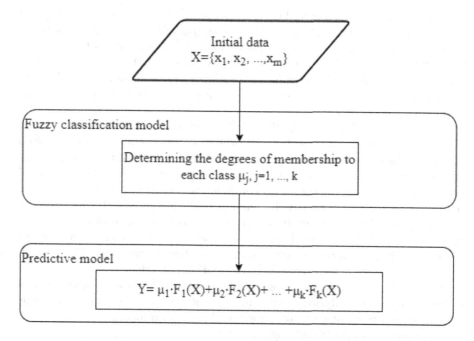

Fig. 1. Combination of classification and predictive models.

A predictive model can be obtained using various methodologies: linear regression, non-linear function approximation, neural networks and other machine learning algorithms. The advantage of symbolic regression (SR) is that it allows to generate models in the form of analytic equations, the researcher does not need to determine the structure of the model in advance, and the interaction between variables is transparent. In [15], using the SR methodology, an explicit approximation of the Colebrook equation was presented. It should be noted that obtaining a formula that defines the dependence in an explicit form, instead of an implicit calculation procedure, is a very popular issue in many practical problems. In [16], the authors discuss industrial applications of symbolic regression and its potential applications in materials science.

The idea of SR is based on a genetic algorithm [17,18], a model is built as a sequence of chromosomes. Chromosomes are generated randomly from a set of genes, as a gene can be used a predictor, a number, an arithmetic operation or a function. At the first step a population of functions $F_l, l = 1, .., N$ (N – the size of population) is randomly generated, then for each individual F_l the fitness function FF_l is calculated as follows

$$FF_l = -\sum_{i=1}^{n}(F_l(x_{1,i}, x_{2,i}, ..., x_{m,i}) - y_i)^2 \qquad (1)$$

The sum of squared differences between the observed dependent variable y_i and the value predicted by function F_l for i-th sample of the training set, is taken with a negative sign, since a larger value of the fitness function is favorable for the individual. According to the evolutionary principle, the fittest individuals survive, the function FF_l determines the viability of specific individual F_l in the population, that is, the ability of the function F_l to correctly predict the value of the target variable y.

Then, from the functions F_l, those individuals that will become parents are selected, the offsprings are created using the crossover operation. The next step in the GP algorithm is mutation (transformation of a chromosome that accidentally changes one or more of its genes). This operation avoids falling to a local extremum and makes it possible to continue the search in a wide area. Then, the fitness function FF for new individuals is calculated and the next generation is formed. The described procedure is repeated until one of the following conditions is met: the change in the best value of the fitness function becomes less than a given tolerance; a predetermined number of generations is obtained, or the maximum time to complete the calculation is reached.

Thus, using the SR methodology, it is possible to obtain a predictive model in the form of an analytical equation, and the new idea is to use its terms (chromosomes) as feature variables for the classification model.

Consider the stages of building a model:

1. Determine the required number of classes k. Mark up the data set according to the selected number of classes.
2. For each class i, using the SR method (the implementation in MATLAB [19,20]), obtain a formula in the form

$$F_i(x_1, x_2, ..., x_m) = f_{i,1}(x_1, x_2, ..., x_m) + f_{i,2}(x_1, x_2, ..., x_m) + ...$$
$$+ f_{i,m}(x_1, x_2, ..., x_m) \qquad (2)$$

A special case of the function $f_{i,j}$ can be a function of one input variable, for example $f_{i,j}(x_1, x_2, ..., x_m) = C \cdot x_m$.

Thus, for the i-th class, an individual F_i with the best value of the fitness function is determined.

3. Analyze the chromosome-functions $f_{i,j}$ and select those that will be input variables for classification, since their values have the greatest differences for data belonging to different classes.

The further procedure will be described for the case of two classes, since the practical problem that initiated this study required the consideration of just two classes.

For the samples of the training set belonging to the first class, calculate the set of the values $f_{i,j}^{(1)}(x_1^{(1)}, x_2^{(1)}, ..., x_m^{(1)})$ and for the second class $f_{i,j}^{(2)}(x_1^{(2)}, x_2^{(2)}, ..., x_m^{(2)})$. If the value sets $f_{i,j}^{(1)}$ and $f_{i,j}^{(2)}$ do not intersect for at least one pair (i, j), a crisp classification can be applied and the function $f_{i,j}$ is a feature variable. In the case of intersection of sets, it is possible to build a fuzzy classification model and it is necessary to choose which functions $f_{i,j}$ will be input variables. Therefore, we need a numerical criterion to determine how well chromosome $f_{i,j}$ values classify.

Let a median of values $f_{i,j}^{(1)}$ for training set for the first class is greater than a median of values $f_{i,j}^{(2)}$ for the second class, let $s_{i,j} \in [min(f_{i,j}^{(1)}), max(f_{i,j}^{(2)})]$, $K_1(s_{i,j})$ – the number of training set samples belonging to the first class for which $f_{i,j}^{(1)} < s_{i,j}$, $K_2(s_{i,j})$ – the number of training set samples belonging to the second class for which $f_{i,j}^{(2)} > s_{i,j}$, n_i – the number of samples in i-th class. The function

$$g_{i,j}(s_{i,j}) = K_1(s_{i,j})/n_1 + K_2(s_{i,j})/n_2 \tag{3}$$

tends to 0 if there is a clear difference between the sets of values $f_{i,j}^{(1)}$ and $f_{i,j}^{(2)}$. Thus, for two classes, the problem turns to finding the minimum of function of one variable (3) on a fixed interval, the algorithm based on golden section search can be used. The value $g_{i,j}$ determines to what extent the chromosome $f_{i,j}$ is suitable for the set of FIS input variables. Smaller values of $g_{i,j}$ indicate that $f_{i,j}$ should be included in the set of input variables.

4. Determine the parameters of membership functions and form a system of rules for a FIS. For two classes, two terms ("LOW" and "HIGH") are used for each input variable as shown in Fig. 2. For the term "LOW" a Z-shaped linear or nonlinear membership function can be used. Each of them requires two parameters – the first parameter is the maximum value of the argument when the function is 1 ($2S_j - A_j$ in Fig. 2), and the second parameter is the minimum value of the argument when the function is 0 (A_j in Fig. 2). Since the membership functions are assumed to be symmetrical and the point S_j is

determined when searching for the minimum of function (3), the membership
function for the term "LOW" is defined as

$$
Zmf(x, S_j, A_j) = \begin{cases} 1, & x \le 2S_j - A_j \\ 1 - \frac{1}{2}\left(\frac{x-S_j}{A_j-S_j}+1\right)^2, & 2S_j - A_j < x \le S_j \\ \frac{1}{2}\left(\frac{x-A_j}{A_j-S_j}\right)^2, & S_j \le x < A_j \\ 0, & x \ge A_j \end{cases}, \quad (4)
$$

Then it remains for each input variable of the FIS to determine the parameter A_j.

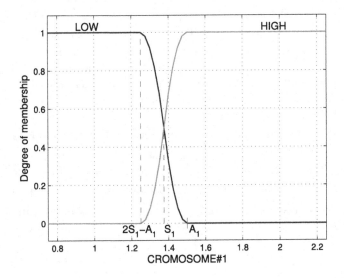

Fig. 2. Membership functions for terms "LOW" and "HIGH" and their parameters.

The rules for the FIS are formulated according to the "IF-THEN" scheme.
An example of three rules is shown below:
IF CHROMOSOME#1 is LOW THEN CLASS is FIRST;
IF CHROMOSOME#2 is LOW THEN CLASS is SECOND;
IF CHROMOSOME#3 is HIGH THEN CLASS is FIRST;
Each rule is assigned a weight w_j corresponding to its contribution to the
correct decision.
It should be noted that the use of chromosome-functions as feature variables
makes it possible to obtain a FIS with a simple structure, where the number
of rules for each class is equal to the number of feature variables, while the
interaction of the parameters is taken into account.
The effectiveness of a classification model is determined by metrics such as
Precision, *Recall*, and F_1-score, that can be calculated as follows:

$$
Precision = \frac{T1}{T1 + F1}, Recall = \frac{T1}{T1 + F2}, F_1 = 2\frac{Precision \cdot Recall}{Precision + Recall}, \quad (5)
$$

where $T1$—the number of first class predictions that actually belong to the first class; $F1$—the number of first class predictions that actually belong to the second class; $F2$—the number of second class predictions that actually belong to the first class.

The parameters of membership functions A_j and the weights w_j are determined from the solution of the constrained optimization problem. The goal is to maximize the value of F_1-score (5), with the control parameters $A_j, w_j, (j = \overline{1, J})$ for each input variable (chromosome-function) and corresponding rule. The constraints are:

$$s_{i,j} \leq A_j \leq max(f_{i,j}(x_1, x_2, ..., x_m)),$$
$$0 \leq w_j \leq 1. \tag{6}$$

To solve the multidimensional optimization problem, a genetic algorithm implemented as a standard function in MATLAB can be used. Thus, the optimal set of FIS parameters is obtained, it gives the highest value for F_1-score (5) for the training set, and then, using fuzzy inference, for any set of input data $(x_1, x_2, ..., x_m)$, the degree of membership to the first and second classes can be calculated.

5. To improve the quality metrics of the resulting model, new chromosome function $f_{i,j}$ should be added to the number of FIS input variables, and it is necessary to determine the membership functions parameters for the new feature variable, modify the system of rules for FIS, and calculate the quality metrics. The choice of new functions should be carried out in accordance with the values of function (3).

The general scheme for constructing a fuzzy classification model using the components of the SR formulae as feature variables is shown in Fig. 3.

3 Example and Discussion

The described model was tested in the classification of carbon monoxide emissions from gas turbines, the data for the study is presented in the open data repository [21]. The dataset was collected for 5 years, it contains 36,733 instances of 11 sensor measures aggregated over one hour, including three external environmental parameters (air temperature (AT), air humidity (AH), atmosphere pressure (AP)), six indicators of the gas turbine technological process (air filter difference pressure (AFDP), gas turbine exhaust pressure (GTEP), turbine inlet temperature (TIT), turbine after temperature (TAT), turbine energy yield (TEY), compressor discharge pressure (CDP)), and two target variables (emissions of carbon monoxide (CO) and the total emissions of nitrogen monoxide and nitrogen dioxide (NOx)). Description of the dataset is given in [22], some more details are given in [23].

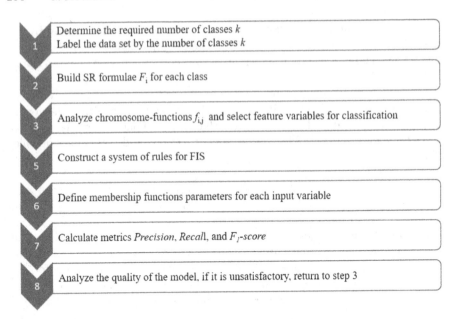

Fig. 3. The general scheme for building a fuzzy classification model using the components of the SR formulae.

CO emissions can increase dramatically up to 20–40 times the median value, so it is important to indicate the situation that leads to extraordinary emissions. So the dataset was split into 2 classes: "Standard" - for emissions less than $5\,\mathrm{mg/m^3}$ (about 90% of data) and "Extreme" for the higher values. For each class, a separate predictive model was built using the SR methodology. An example of the formula obtained for the data set for the "Extreme" class and the chromosomes from which the input variables for the classification model are selected is shown in Fig. 4. To build a model the input variables were standardized using the Z-score.

CHROMOSOME#1   CHROMOSOME#2   CHROMOSOME#3

$$F_i = 225.59 \cdot \boxed{GTEP^2 \cdot TAT} - 20.67 \cdot \boxed{TAT \cdot exp(-TIT)} - 20.67 \cdot \boxed{AFDP \cdot TIT} + 7.88 \times$$

$$\boxed{GTEP \cdot exp(-TIT^2)} - 67.05 \cdot AT \cdot AFDP \cdot TIT - 24.68 \cdot AT \cdot TIT \cdot exp(-TAT) -$$

CHROMOSOME#4

$$68.81 \boxed{AFDP \cdot GTEP \cdot TIT \cdot exp(-GTEP)} + 7.25.$$

CHROMOSOME#7

Fig. 4. Predictive formula for the "Extreme" class and chromosomes to be analyzed for feature selection.

The examples of chromosomes selected as inputs to FIS (CHROMOSOME#1 and CHROMOSOME#2) are given in Fig. 5. The red dashed lines correspond to the values s_j determined when optimizing the function (3).

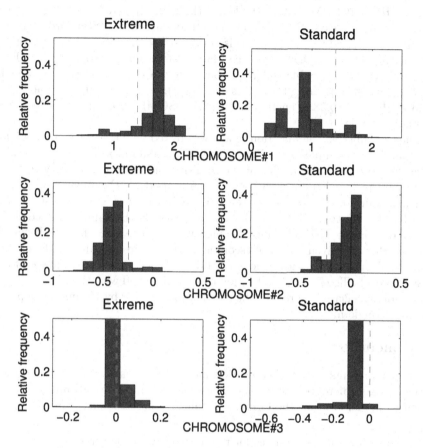

Fig. 5. Histograms of chromosomes #1 and #2 selected as inputs to FIS and chromosome #3, which is not relevant for classification. (Color figure online)

For CHROMOSOME#1, to the left of the point $s_{1,1}$ (red dashed line) there are 87% of the values of $GTEP^2 \cdot TAT$ for the data belonging to the "Standard" class for the training set, and to the right of this point, there are more than 90% of the values of $GTEP^2 \cdot TAT$ for the data belonging to the "Extreme" class.

For CHROMOSOME#2, 89% of the values of $TAT \cdot exp(-TIT)$ for data belonging to the "Standard" class for the training sample are located to the right of the point $s_{1,2}$, and 90% of the values of $TAT \cdot exp(-TIT)$ for data belonging to the "Extreme" class are to the left of this point. So CHROMOSOME#1 and CHROMOSOME#2 were chosen as input variables for FIS. It is worth noting that CHROMOSOME#1 represents the interaction of the variables $GTEP$ and

TAT, while CHROMOSOME#2 shows the interaction of the variables TIT and TAT. Consideration of the combination of these parameters makes it possible to improve the metrics of the model in comparison with the use of individual variables $GTEP$, TAT, TIT as input variables of the FIS.

For CHROMOSOME#3, to the left of the point $s_{1,3}$ (red dashed line) there are 95% of the values $AFDP{\cdot}TIT$ for data belonging to the "Standard" class for the training set, and to the right of this point, 46% of the values $AFDP{\cdot}TIT$ for data belonging to the "Extreme" class are located. Thus, CHROMOSOME#3 does not give a noticeable difference between the classes, as can be seen in Fig. 5 and when comparing the values of $g_{i,j}(s_{i,j})$: $g_{1,1}(s_{1,1}) = 0.223, g_{1,2}(s_{1,2}) = 0.206, g_{1,3}(s_{1,3}) = 0.583$. Therefore, CHROMOSOME#3 cannot be selected as an input variable for FIS.

The CO emission classification model was tested on a test set including 9968 "Standard" class samples and 1053 "Extreme" class samples. To obtain the value of the F_1-score to check quality of the resulting model, it is necessary to bring the fuzzy sets to crisp ones, so α-cut $= 0.5$ is used for the calculated degrees of membership for the "Standard" and "Extreme" classes. For the test set, F_1-score for the "Extreme" class is 0.82, which exceeds the value obtained using the Random Forest (RF) classification algorithm for the same dataset. It should be noted that the given metric F_1-score is not the main indicator of the quality of the resulting model, since it does not take into account the advantage of the fuzzy approach. The fuzzy classification model, combined with the SR prediction models for each class, can reduce the mean absolute error by 10 times, compared to using a single model.

4   Conclusion

The paper presents a new classification model based on a symbolic regression method, fuzzy inference system and genetic algorithm. The advantages of the presented model are that the formulas obtained using symbolic regression

- can be used as predictive models;
- form a set of feature variables for the fuzzy classification model;
- clarify the interaction of input parameters;
- make it possible to obtain a FIS with a simple structure, where the number of rules is equal to the number of feature variables, while the interaction of the parameters is taken into account.

The genetic algorithm is used in the framework of symbolic regression and as a tool for solving the problem of multivariate optimization to determine the parameters of the FIS. The model has been applied to real data and the results are superior to the models presented in [22].

References

1. Xu, S., Feng, N., Liu, K., Liang, Y., Liu, X.: A weighted fuzzy process neural network model and its application in mixed-process signal classification. Expert Syst. Appl. **172**, 114642 (2021)

2. Marimuthu S., Mohamed Mansoor Roomi S.: Particle swarm optimized fuzzy model for the classification of banana ripeness. IEEE Sens. J. **17**(15), 4903–4915 (2017)

3. Zou, Q., Liao, L., Ding, Y., Qin, H.: Flood classification based on a fuzzy clustering iteration model with combined weight and an immune grey wolf optimizer algorithm. Water **11**(1), 80 (2019)

4. Saini, J., Dutta, M., Marques, G.: ADFIST: adaptive dynamic fuzzy inference system tree driven by optimized knowledge base for indoor air quality assessment. Sensors **22**, 1008 (2022). https://doi.org/10.3390/s22031008

5. Ducange, P., Fazzolari, M., Marcelloni, F.: An overview of recent distributed algorithms for learning fuzzy models in big data classification. J. Big Data **7**(1), 1–29 (2020). https://doi.org/10.1186/s40537-020-00298-6

6. Guo, N.R., Li, T.-H.S.: Construction of a neuron-fuzzy classification model based on feature-extraction approach. Expert Syst. Appl. **38**(1), 682–691 (2017)

7. Feng, T.-C., Chiang, T.-Y., Li, T.-H.S.: Enhanced hierarchical fuzzy model using evolutionary GA with modified ABC algorithm for classification problem. In: ICCSS 2015 - Proceedings: 2015 International Conference on Informative and Cybernetics for Computational Social Systems, Chengdu, Sichuan, China, vol. 7281146, pp. 40–44 (2015)

8. Feng, T.-C., Li, T.-H.S., Kuo, P.-H.: Variable coded hierarchical fuzzy classification model using DNA coding and evolutionary programming. Appl. Math. Model. **39**(23–24), 7401–7419 (2015)

9. Kalia, H., Dehuri, S., Ghosh, A., Cho, S.-B.: Surrogate-assisted multi-objective genetic algorithms for fuzzy rule-based classification. Int. J. Fuzzy Syst. **20**(6), 1938–1955 (2018). https://doi.org/10.1007/s40815-018-0478-3

10. Zhang, C.: Classification rule mining algorithm combining intuitionistic fuzzy rough sets and genetic algorithm. Int. J. Fuzzy Syst. **22**(5), 1694–1715 (2020). https://doi.org/10.1007/s40815-020-00849-2

11. Ojha, V., Abraham, A., Snasel, V.: Heuristic design of fuzzy inference systems: a review of three decades of research. Eng. Appl. Artif. Intell. **85**, 845–864 (2019)

12. Feng, S., Chen, C.L.P.: Fuzzy broad learning system: a novel neuro-fuzzy model for regression and classification. IEEE Trans. Cybern. **50**(2), 414–424 (2020). https://doi.org/10.1109/TCYB.2018.2857815

13. Takagi, T., Sugeno, M.: Fuzzy identification of systems and its applications to modeling and control. IEEE Trans. Syst. Man Cybern. **15**, 116–132 (1985)

14. Lu, K.-P., Chang, S.-T.: A fuzzy classification approach to piecewise regression models. Appl. Soft Comput. J. **69**, 671–688 (2018)

15. Praks, P., Brkić, D.: Symbolic regression-based genetic approximations of the colebrook equation for flow friction. Water **10**, 1175 (2018)

16. Wang, Y., Wagner, N., Rondinelli, J.M.: Symbolic regression in materials science. MRS Commun. **9**(3), 793–805 (2019). https://doi.org/10.1557/mrc.2019.85

17. Koza, J.R.: Genetic programming as a means for programming computers by natural selection. Stat. Comput. **4**(2), 87–112 (1994)

18. Mitchell, M.: An Introduction to Genetic Algorithms. MIT Press, Cambridge (1996)

19. Searson, D.P., Leahy, D.E., Willis, M.J.: GPTIPS: an open source genetic programming toolbox for multigene symbolic regression. In: Proceedings of the International MultiConference of Engineers and Computer Scientists IMECS 2010, pp. 77–80. IMECS, Hong Kong (2010)

20. Searson, D.P.: GPTIPS 2: an open-source software platform for symbolic data mining. In: Gandomi, A.H., Alavi, A.H., Ryan, C. (eds.) Handbook of Genetic Programming Applications, pp. 551–573. Springer, Cham (2015). https://doi.org/10.1007/978-3-319-20883-1_22

21. Dua, D., Graff, C.: UCI machine learning repository. http://archive.ics.uci.edu/ml. Accessed 10 May 2022

22. Kaya, H., Tüfekci, P., Uzun, E.: Predicting CO and NOx emissions from gas turbines: novel data and a benchmark PEMS. Turk. J. Electr. Eng. Comput. Sci. **27**(6), 4783–4796 (2019). https://doi.org/10.3906/elk-1807-87

23. Kochueva, O., Nikolskii, K.: Data analysis and symbolic regression models for predicting CO and NOx emissions from gas turbines. Computation **9**(139), 1–16 (2021). https://doi.org/10.3390/computation9120139

Preparing Traffic to Analyze the Dynamics of Its States by Method of Partial Correlations

D. N. Nikol'skii[1]([⊠]) and A. E. Krasnov[2]

[1] Smart Wallet LLC (SWiP), Nezhinskaya str., 1, bldg. 4, 159, Moscow, Russia
nikolskydn@mail.ru
[2] Russian State Social University, Wilhelm Peak str., 4, bldg. 1, Moscow, Russia

Abstract. The aggregation procedures network traffic data packets for the analysis of its states by the partial correlations method of aggregated data packets are described. An algorithm for the formation of partial correlation functions based on the flows of characteristics of aggregated data packets, including the number of packets and their information load, is presented. An extension of the method's application area of partial correlations for describing the dynamics of telecommunication network's states is proposed.

Keywords: Network traffic · Time series flow · Partial correlations · DDoS attacks · State pattern

1 Introduction

DDoS attacks are complex types of attacks [1] and special methods of network traffic analysis are required to identify (detect and classify) them. Such methods based on network traffic data packets aggregation and subsequent partial correlations calculation of adjacent aggregates were developed and explored by the authors in [2,3]. The high efficiency identification of various types of DDoS attacks based on the non-stationary states description of network traffic by the so-called patterns—histograms of the values of partial correlations of adjacent aggregates is shown [3]. At the same time, in these works, no attention was paid to the description of the procedures for the formation of the aggregates themselves. Papers [4–6] are devoted to the formation of network traffic aggregates for various purposes.

The purpose of this work is a detailed description of the procedures for aggregating network traffic data packets for analyzing its states by the partial correlations method of aggregated data packets, a brief description of the algorithm for forming partial correlations, as well as an extension of the scope of the partial correlation method for describing the dynamics of telecommunication network's states.

The specific tasks of the work are: collection of network traffic—description of the transformation of the parameters of the headers of its data packets;

preparation of network traffic - formation time series of its aggregated packets, considering bit flags from the data transmission protocol TCP header.

2 Network Traffic Collection

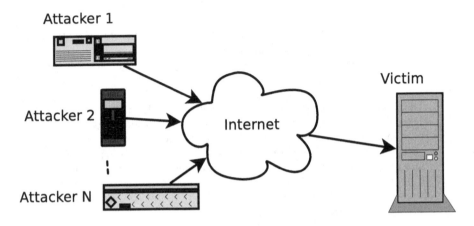

Fig. 1. Experimental stand for picking up network traffic.

The network traffic primary signs were represented by the TCP/IP data transfer protocol [7,8]. Bit flags from the TCP network packet header were selected for the study. The change in the statistical characteristics of these flags over time can be used to describe the unique network traffic states: its normal and abnormal states. To study the traffic, a stand was created (see Fig. 1), using which attacks were carried out on the victim node from distributed nodes. The some Frontend server of some Web-service was chosen as a victim node. A Qlogic 10 Gb/s network card was used as hardware support. The packages were parsed with the tshark utility. As a result, a csv file was obtained with network packets stored on the victim node.

As an example, in Table 1 first 10 packets on the victim host when performing a Slowloris attack are shown. The columns are:

– `frame.time` is the packet saving time, the start of the report is taken as the time when the first packet fell;
– `ip.len` is the total length of the packet in bytes;
– `tcp.flags` is TCP flags.

3 Network Traffic Cooking

To build time series (a aggregated data stream packages with different flag indices), Table 1 was pivoted. The new table columns are the flags values from the TCP headers tcp.flags. The values in these columns are the values sum from

Table 1. Captured network packets

Number	frame.time, sec	tcp.flags	ip.len, byte
0	0.00000	0x10	52
1	0.00004	0x04	40
2	0.00494	0x02	60
3	0.00497	0x12	60
4	0.00528	0x10	52
5	0.01012	0x02	60
6	0.01014	0x12	60
7	0.01021	0x18	286
8	0.01023	0x10	52
9	0.01047	0x10	52

the ip.len field. The data pivoting result with source packets for a Slowloris attack are presented in Table 2.

Table 2. Packages after pivoting

frame.time \ tcp.flags	0x02	0x04	0x10	0x11	0x12	0x18
0 days 00:00:00	0	0	52	0	0	0
0 days 00:00:00.000040	0	40	0	0	0	0
0 days 00:00:00.004940	60	0	0	0	0	0
0 days 00:00:00.004970	0	0	0	0	60	0
0 days 00:00:00.005280	0	0	52	0	0	0
0 days 00:00:00.010120	60	0	0	0	0	0
0 days 00:00:00.010140	0	0	0	0	60	0
0 days 00:00:00.010210	0	0	0	0	0	286
0 days 00:00:00.010230	0	0	52	0	0	0
0 days 00:00:00.010470	0	0	52	0	0	0

Next, the received data aggregation and the formation of a time stream were performed: changes in the aggregated data packets count and their total length over time. As a result, for each j-th flag index ($j = 1, 2, \ldots, J$) of each aggregate of the interval ΔT, the number $N_{\Delta T(t)}^{(j)}$ of packets and their total length or information capacity $I_{\Delta T}^{(j)}(t)$ ($j = 1, 2, \ldots, J$) at the current time t.

The pictures below show the first 30 seconds of the process. So, in Fig. 2, aggregation is performed using the function of counting the number of packets, and in Fig. 3 using the function of counting the total length of packets. The size ΔT of the aggregation window is equal to one second.

It should be noted that in Fig. 2 the lines for different flag states have merged. The reason for this is that the same number of packets for different flags are registered at different times.

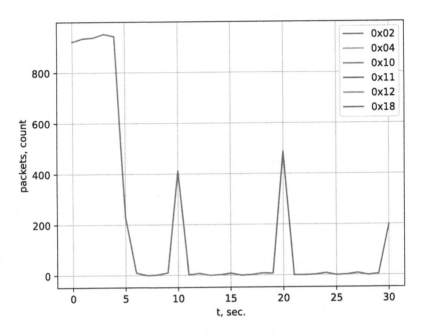

Fig. 2. Changes in the number of packets over time.

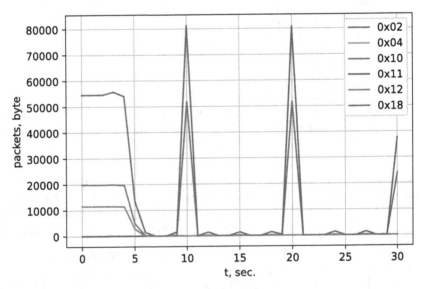

Fig. 3. Changes in the total length of packets over time.

4   Algorithm for Forming Partial Correlations

The above technology was based on the formation of a stream of time series by aggregating such network traffic characteristics as the number of packets and their information capacity.

Each time series was associated with the number $N_{\Delta T}^{(j)}(t)$ of aggregated packets and their information capacity $I_{\Delta T}^{(j)}(t)$ ($j = 1, 2, \ldots, J$), and the number of time series was determined by the number J of possible flag states.

In accordance with [3], further analysis of the flow of J time series is carried out on the basis of their patterns or histograms of the values of partial correlation functions $h(t_k, t_{k-1})$, formed by averaging the real part of the evolution operator $\mathbf{S}(t_k, t_{k-1})$ time series flow by statistical operator $\mathbf{Q}(t_k, t_{k-1}) = [q_{jm}(t_k, t_{k-1})]$:

$$h(t_k, t_{k-1}) = \mathrm{tr}\left[\mathbf{Q}(t_k, t_{k-1})\mathrm{Re}\mathbf{S}(t_k, t_{k-1})\right] =$$

$$= \sum_j \sum_m q_{jm}(t_k, t_{k-1})\left[x_{\Delta T}^{(j)}(t_k)x_{\Delta T}^{(m)}(t_{k-1}) + y_{\Delta T}^{(j)}(t_k)y_{\Delta T}^{(m)}(t_{k-1})\right];$$

$$q_{jm}(t_k, t_{k-1}) = \begin{cases} \dfrac{I_{\Delta T}^{(j)}(t_k)I_{\Delta T}^{(m)}(t_{k-1})}{\sum_j I_{\Delta T}^{(j)}(t_k)\sum_m I_{\Delta T}^{(m)}(t_{k-1})}, & j \neq m, \\[3ex] \dfrac{I_{\Delta T}^{(j)}(t_k)I_{\Delta T}^{(j)}(t_{k-1})}{\sum_j I_{\Delta T}^{(j)}(t_k)I_{\Delta T}^{(j)}(t_{k-1})}, \\[3ex] \mathrm{tr}\mathbf{Q}(t_k, t_{k-1}) = \sum_j q_{jj}(t_k, t_{k-1}) = 1. \end{cases} \tag{1}$$

The diagonal matrix elements of the statistical operator $\mathbf{Q}(t_k, t_{k-1})$ from (1) correspond to the weights or values of the flag indices of the same name of adjacent aggregates. The off-diagonal matrix elements determine the statistical weights of the links of their heterogeneous flag indices.

Dynamic variables $x_{\Delta T}^{(j)}(t_k)$ and $y_{\Delta T}^{(j)}(t_k)$ determine the unit phase vector $\mathbf{z}_{\Delta T}^{(j)}(t_k)$ of network traffic state:

$$\mathbf{z}_{\Delta T}(t_k) = \begin{vmatrix} \mathbf{x}_{\Delta T}(t_k) \\ \mathrm{i}\mathbf{y}_{\Delta T}(t_k) \end{vmatrix}, \quad \mathbf{z}_{\Delta T}^{+}(t_k) = \mathbf{x}_{\Delta T}^{T}(t_k) - \mathrm{i}\mathbf{y}_{\Delta T}^{T}(t_k),$$

$$\left\| \mathbf{z}_{\Delta T}^{(j)}(t_k) \right\| = \sqrt{\mathbf{z}_{\Delta T}^{+}(t_k)\mathbf{z}_{\Delta T}(t_k)} =$$

$$= \sqrt{\mathbf{x}_{\Delta T}^{T}(t_k)\mathbf{x}_{\Delta T}(t_k) + \mathbf{y}_{\Delta T}^{T}(t_k)\mathbf{y}_{\Delta T}(t_k)} = 1,$$

$$\mathbf{x}_{\Delta T}(t_k) = \frac{\mathbf{X}_{\Delta T}(t_k)}{\sqrt{\sum_j \left(\left(\mathbf{X}_{\Delta T}^{(j)}(t_k) \right)^2 + \left(\mathbf{Y}_{\Delta T}^{(j)}(t_k) \right)^2 \right)}}, \tag{2}$$

$$\mathbf{y}_{\Delta T}(t_k) = \frac{\mathbf{Y}_{\Delta T}(t_k)}{\sqrt{\sum_j \left(\left(\mathbf{X}_{\Delta T}^{(j)}(t_k) \right)^2 + \left(\mathbf{Y}_{\Delta T}^{(j)}(t_k) \right)^2 \right)}},$$

$$\mathbf{X}_{\Delta T}^{T}(t_k) = \left[X_{\Delta T}^{(1)}(t_k), X_{\Delta T}^{(2)}(t_k), \ldots, X_{\Delta T}^{(J)}(t_k) \right]$$

$$\mathbf{Y}_{\Delta T}^{T}(t_k) = \left[Y_{\Delta T}^{(1)}(t_k), Y_{\Delta T}^{(2)}(t_k), \ldots, Y_{\Delta T}^{(J)}(t_k) \right],$$

$$k = 0, 1, \ldots, K.$$

The dynamic model of the change in the phase vector $\mathbf{z}_{\Delta T}(t_k)$ moving along a single hypersphere is described by the evolution operator $\mathbf{S}(t_s, t_p)$ (shift operator):

$$\mathbf{z}_{\Delta T}(t_s) = \mathbf{S}(t_s, t_p)\mathbf{z}_{\Delta T}(t_p), \quad p < s, \quad s = 1, 2, \ldots, K. \tag{3}$$

The given non-Hermitian evolution operator for every t_s, t_p-th discrete time instants is defined as the projection operator [3]:

$$\mathbf{S}(t_s, t_p) = \mathbf{z}_{\Delta T}(t_s)\mathbf{z}_{\Delta T}^{+}(t_p), \quad p < s, \quad s = 1, 2, \ldots, K, \tag{4}$$

where the generalized coordinates $X_{\Delta T}^{(j)}(t_k)$ and the generalized velocities $X_{\Delta T}^{(j)}(t_k)$ of the phase vector $\mathbf{z}_{\Delta T}(t_k)$ form a discrete model of the time series flow in the form of a set of analytical signals:

$$X_{\Delta T}^{(j)}(t_k) + \mathrm{i}Y_{\Delta T}^{(j)}(t_k), \quad j = 1, 2, \ldots, J, \quad s = 1, 2, \ldots, K,$$

In [3], it was proposed to set $X_{\Delta T}^{(j)}(t_k) = \sqrt{N_{\Delta T}^{(j)}(t_k)}$ and $Y_{\Delta T}^{(j)}(t_k)$ to form by discrete Hilbert transform of signals $X_{\Delta T}^{(j)}(t_k)$.

Descriptions of the dynamic behavior of network traffic based on K values of partial correlations $h(t_k, t_{k-1})$ can be obtained in the following ways:

1. form a network traffic state pattern in the form of an empirical distribution (normalized histogram) $w[h]$ of $h(t_k, t_{k-1})$ values for $k = 1, 2, \ldots, K$, defined by expression (1);

2. form a network traffic state pattern in the form of an empirical distribution (normalized histogram) $w[H]$ of $H(t_k, t_{k-1})$ values for $k = 1, 2, \ldots, K$, defined by expression (1) and [9]:

$$H(t_k, t_{k-1}) = \left(-(1 - 3h(t_k, t_{k-1})) + \sqrt{1 + (1 - 3h(t_k, t_{k-1})^2)} \right)^{\frac{1}{3}} +$$
$$+ \left(-(1 - 3h(t_k, t_{k-1})) - \sqrt{1 + (1 - 3h(t_k, t_{k-1})^2)} \right)^{\frac{1}{3}} . \tag{5}$$

If the physical meaning of the partial correlation h is that it is associated with the evolution operator (or shift operator) of the phase vector of the network traffic state, then the physical meaning of the quantity H is due to the Hamiltonian operator of a dynamical system, which is a telecommunications data transmission network [9].

Note also that in accordance with (1) $0.0 \leq h(t_k, t_{k-1}) \leq 1.0$, and in accordance with (5) $-0.596 \leq H(t_k, t_{k-1}) \leq 1.0$. Thus, the distribution $w[H]$ has a higher resolution than the distribution $w[h]$.

3. form a network traffic state pattern in the form of an empirical discrete power spectrum $W[m]$ of a periodically extended discrete sequence $H(t_k, t_{k-1})$ from (5):

$$W[m] = \left| \sum_{k=1}^{K} H(t_k, t_{k-1}) \exp\left(-i\, 2\pi \frac{mk}{K} \right) \right|^2 . \tag{6}$$

The first two methods, compared with the third, have the advantage that at the stage of detecting network traffic anomalies using the Wald statistical analysis method, they can use a small number of aggregated packets, which is necessary to quickly determine the moment of anomaly onset [9]. The implementation of the third method requires the entire sample of K aggregated packages. At the same time, at the learning stage, all methods require a sufficiently large value of $K \approx 10^3 \div 10^4$.

For clarity, Fig. 4 shows the distributions of partial correlation values $w[h|r]$ from (1) to describe various states of network traffic: normal state – normal ($r = 0$); attack states tcpcon – TCP Connection Flood ($r = 1$), slowloris – Slow Loris ($r = 2$), httpget - HTTP Get Flood ($r = 3$). All distributions are constructed for $\Delta T = 50$ ms and $K = 10^4$ [3].

As shown in [9], the second method, using $w[H|r]$ patterns, is more selective in identifying various types of network attacks ($r = 0, 1, 2, \ldots, R$). Thus, experiments carried out on the stand (see Fig. 1) made it possible to detect and identify the following types ($R = 24$) of network attacks with a detection probability of 95% and a type 2 error probability of 5%:

- attacks with incorrectly assembled or incorrect fragments of IP packets;
- attacks by IP packets with incorrect values in the headers;
- TCP SYN flood attacks;
- TCP RST flood attacks;
- TCP FIN flood attacks;

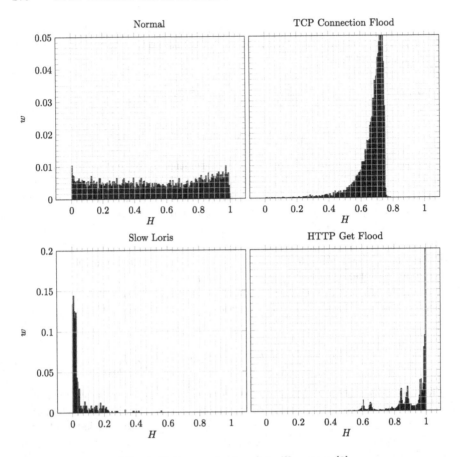

Fig. 4. Patterns of network traffic states [3].

- ICMP request flood attacks;
- ICMP reply flood attacks;
- TCP connection flood attacks;
- TCP Slow Read attacks;
- attacks with the installation of slow sessions such as SlowLoris;
- UDP flood attacks on arbitrary ports;
- UDP Fragmented Flood attacks;
- port brute force attacks;
- random traffic attacks from certain countries/regions/cities (according to the GeoIP database);
- DNS Amplification;
- NTP Amplification;
- SSDP Amplification;
- attacks with incorrect GET or POST requests over the HTTP protocol (HTTP GET flood, HTTP POST flood);
- attacks by random requests (Random) via the DNS protocol;

- attacks on non-existent domains;
- attacks over the SMTP protocol with sending random emails to non-existent addresses;
- attacks with incorrect SSL/TLS packets;
- attacks with the installation of slow sessions;
- attacks on the vulnerability of encryption protocols by incorrect installation SSL/TLS Renegotiation and session maintenance.

5   Expanding the Scope Partial Correlation Method

It is easy to see that the algorithm given in the form of formulas (1)–(5) describes the partial correlation functions. This functions applied for the flow of time series based on the aggregated packets $N_{\Delta T}^{(j)}(t_k)$ and $I_{\Delta T}^{(j)}(t_k)$ $(j = 1, 2, \ldots, J)$. If we consider the variable j as the number of some information transmission channel, then it is obvious that the algorithm is completely independent of the physical essence of this channel. Therefore, the algorithm can be applied, and completely without any changes. Therefore, it can be applied, and completely without any changes, to the analysis of the activity of a telecommunications network, in particular, to the description of the dynamics of its states.

The review [10] shows that two main types of dynamic processes in networks can be distinguished: processes that change the network topology, i.e. the number of vertices and the nature of network connections (topology dynamics), and processes that change the characteristics of network elements (vertex states, edge weights, etc.) with the topology unchanged (state dynamics).

The description of the dynamics of the topology of networks is considered in numerous works, the most famous of which are [11–13].

The description of the dynamics of the states of the nodes of the telecommunication network with their constant connections is considered, for example, in [14]. It should be noted that there are few works devoted to the consideration of the joint dynamics of nodes and links of a telecommunication network [15].

Therefore, if we consider a telecommunications network with an unchanged topology, i.e. with a fixed set of its nodes (for example, routers and servers) and links (channels) between them, it will be practically useful to describe the dynamics of the states of a given network, taking into account changes in the characteristics transmitted over these links: the number of packets and their information load.

At the same time, as one of the approaches to such a description, one can consider the approach to the description based on patterns of correlation functions of network states. Recall that this approach was borrowed from the semiclassical theory of electromagnetic fields based on the apparatus of field correlation functions [3]. Here, the fundamental point is that the approach is based on the rejection of the temporal averaging of field correlations, and the consideration of partial correlations connecting only "adjacent" time moments, which makes it possible to describe non-stationary fields. It is these "fields" that form information data flows spreading through the telecommunications network. It is these "information fields" that form the data flows spreading through the telecommunications network.

In this formulation, the problem may look as follows. There is a telecommunications network, which corresponds to a weighted undirected multigraph with U nodes and J pairs of edges (see Fig. 5). Each j-th pair of edges ($j = 1, 2, \ldots, J$) corresponds to the number of $N_{\Delta T}^{(j)}(t_k)$ aggregated (in the interval ΔT) packets with the number of $I_{\Delta T}^{(j)}(t_k)$ information duplex transmitted at k-th time points between nodes incident to a given pair of edges. In this setting, the matrix elements $q_{jm}(t_k, t_{k-1})$ of the statistical operator $\mathbf{Q}(t_k, t_{k-1})$ have the physical meaning of the relative weights of the information load of the edges of the same name ($j \neq m$) and significance connections of unlike edges at adjacent times ($j = m$).

Based on the algorithm described above, it is possible to calculate both the partial correlation function $h(t_k, t_{k-1})$ of adjacent aggregates and the family of partial correlation functions $h(t_k, t_{k-l})$ for L different levels of adjacency ($l = 1, 2, \ldots, L$) aggregated packages.

For clarity, Table 3 shows an example of formation the sequence of partial correlation functions of adjacent aggregated packages for different levels of adjacency $(1, 2)$, by which it is possible to construct patterns of telecommunication network states.

Attention should be paid to the fact that for telecommunication networks with a large number of connections between nodes, the task of calculating correlation functions is quite laborious. Thus, in the problem considered in Sects. 2 and 3, time series for $J = 64$ were formed. In the problem of describing the dynamics of a network telecommunications states the number of time series

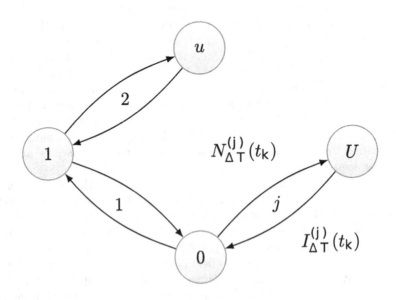

Fig. 5. Telecommunication network graph at the t_k time.

Table 3. Sequence of formation of partial correlations.

t	t_1	t_2	t_3	...	t_k	...	t_K
t_1		$h(t_2,t_1)$					
t_2			$h(t_3,t_2)$				
t_3			$h(t_3,t_1)$				
...							
t_{k-2}							
t_{k-1}					$h(t_k,t_{k-1})$		
t_k					$h(t_k,t_{k-2})$		
...							
t_{K-2}							
t_{K-1}							$h(t_K,t_{K-1})$
t_K							$h(t_K,t_{K-2})$

significantly exceeds this number. For example, for a network of L nodes, the maximum number of pair connections is $J_{max} = U(U-1)/2$. So, for $U = 100$, $J_{max} = 4950$.

To compile the corresponding time series that characterize the dynamics of the states of a telecommunications network, a more complex technology is required than that described in Sects. 2 and 3. For this purpose, it is possible to use, for example, the NS system developed at Berkeley University in collaboration with Xerox [16]. To describe the topology of a network, NS offers two basic concepts: a node and a link, which have many attributes that allow you to describe real equipment.

Another useful tool for the analysis and visualization of dynamic graphs is the Gephi software package [17], which has an open source code for visualization and analysis of related data (`gephi.org`). Gephi has built-in tools for analyzing graphs that model dynamic systems. In addition, Gephi allows you to visualize the dynamics of a graph by adding a player that shows the dynamics of changing relationships between graph objects over time.

6   Conclusion

The procedures described in the paper allow one to form discrete dynamic models of network traffic aggregate flows during information exchange both between individual nodes of an information system and a telecommunications network as a whole. In this case, the following steps are important:

- Selection of information about the number of packets and their information capacity from the primary features of the TCP/IP data transfer protocol;
- Formation of the number of packets and their information capacity for different information channels (flag states in the case of data transfer between

individual nodes and duplex communication channels for a telecommunications network);
- Aggregation of data packets and the formation of time series for a certain time interval and obtaining the integrated number of packets and their load capacity (the criterion for determining the integration time in the case of data transmission between individual nodes for identifying DDoS attacks);
- Formation in all information channels of generalized coordinates from the number of aggregated data packages and generalized rates of their birth and destruction;
- Formation of a single phase vector of the state of network traffic by normalizing its generalized coordinates and speeds;
- Formation of a statistical network traffic operator based on its load data in various information channels for adjacent units;
- Calculation of partial correlation functions of adjacent aggregates of network traffic;
- Formation of network traffic patterns in the form of normalized histograms of the values of the correlation functions of its adjacent aggregated data packets.

In the future, the authors intend to create a research bench to study various modifications of the algorithm considered in the article to search for more advanced solutions. In particular, it is of considerable interest to calculate a more accurate approximation of the value of H by the values of h. This will increase the resolution in the description of the network traffic state pattern.

References

1. Bhattacharyya, D., Kalita, J.: DDoS Attacks: Evolution, Detection, Prevention, Reaction, and Tolerance. CRC Press Taylor & Francis Group, Boca Raton (2016)
2. Krasnov, A.E., Nikol'skii, D.N.: Statistical distributions of partial correlators of network traffic aggregated packets for distinguishing DDoS attacks. In: Vishnevskiy, V.M., Samouylov, K.E., Kozyrev, D.V. (eds.) DCCN 2019. LNCS, vol. 11965, pp. 365–378. Springer, Cham (2019). https://doi.org/10.1007/978-3-030-36614-8_28
3. Krasnov, A.E., Nikol'skii, D.N.: Formation of one-dimensional distributions of values of correlators of network traffic aggregates. Russ. Phys. J. **63**, 563–573 (2020)
4. Konoplev, V.V., Zaharov, D.Y., Boyarskij, M.N., Nazirov, R.R.: Skhema adaptivnogo agregirovaniya dlya klasterizacii dannyh setevogo trafika (in Russia). ISSLEDOVANO V ROSSII, pp. 2555–2568 (2003)
5. Get'man, A.I., Ivannikov, V.P., Markin, Y.V., Padaryan, V.A., Tihonov A.Y.: Model' predstavleniya dannyh pri provedenii glubokogo analiza setevogo trafika (in Russia). Trudy ISP RAN. **27**(4), 5–20 (2015)
6. Efimov, A.Y.: Problemy obrabotki statistiki setevogo trafika dlya obnaruzheniya vtorzhenij v sushchestvuyushchih informacionnyh sistemah (in Russia). Programmnye produkty i sistemy / Software & Systems **1**, 17–21 (2016)
7. Postel, J. (ed.): Internet protocol - DARPA Internet Program Protocol Specification, rfc791 (Internet Standard). USC/Information Sciences Institute, Los Angeles (1981)

8. Postel, J. (ed.): Transmission control protocol - DARPA Internet Program Protocol Specification, rfc793 (Internet Standard). USC/Information Sciences Institute, Los Angeles (1981)
9. Krasnov, A.E., Nadezhdin, E.N., Nikol'skii, D.N., et al.: Detecting DDoS attacks using the analysis of network traffic dynamics and characteristics interrelation. Vestnik udmurtskogo universiteta: Matematika, Mekhanika, Komp'yuternye nauki. **28**(3), 407–418 (2018)
10. Kuznecov, O.P.: Slozhnye seti i rasprostranenie aktivnosti (in Russia). Avtomatika i telemekhanika, **12**, 3–26 (2015)
11. Watts, D.J., Strogatz, S.H.: Collective dynamics of "small-world" networks. Nature **393**, 440–442 (1998)
12. Barabasi, A., Albert, R.: Emergence of scaling in random networks. Science **286**, 509–512 (1999)
13. Bollobas, B., Riordan, O.: Robustness and vulnerability of scale-free random graphs. Internet Math. **1**(1), 1–35 (2003)
14. Harush, U., Barzel, B.: Dynamic patterns of information flow in complex networks. Nat. Commun. **8**, 2181 (2017)
15. Karyotis, V., Sta, E., Papavassiliou, S.: Evolutionary Dynamics of Complex Communications Networks, p. 31. CRC Press, Boca Raton (2013)
16. Kevin, F., Kannan, V.: The ns Manual - The VINT Project (2006)
17. Bastian, M., Heymann, S., Jacomy, M.: Gephi : an open source software for exploring and manipulating networks. In: Third International AAAI Conference on Weblogs and Social Media. AAAI Publications (2019)

On Approximation
of the Time-Probabilistic Measures
# of a Resource Loss System
with the Waiting Buffer

A.V. Daraseliya[1]([⊠]) [iD], E. S. Sopin[1,2] [iD], and S. Ya. Shorgin[1,2] [iD]

[1] Peoples' Friendship University of Russia (RUDN University),
Moscow, Russian Federation
{daraselia-av,sopin-es}@rudn.ru, sshorgin@ipiran.ru
[2] Institute of Informatics Problems, Federal Research Center Computer Science
and Control of Russian Academy of Sciences, Moscow, Russian Federation

Abstract. In this paper we consider a multi-server model in terms of a resource loss system with the waiting buffer and the multi-type of resources. A customer accepted for servicing occupies a random amount of resources with described distribution functions. Based on the assumptions of a Poisson arrival process and exponential service times, we analytically find the system of equilibrium equations. We proposed an approximation of the model with the single type of resources. We analytically find the system of equilibrium equations, solving which we get the stationary probabilities for the simpler model. For various probability distributions, we evaluated metrics of interest such as the loss probability of the system, the average waiting time, the average number of customers, and the average resource requirements of blocked customers.

Keywords: queuing system · resource loss system · multi-service network · queuing system with resources · random amount of resources · arithmetic probability distribution

1 Introduction

Resource loss systems (LS), also known as resource queuing systems (QS) in the research literature, have been one of the topical areas of research for a long time. The early studies [1] investigated systems with the allocation of a certain random amount of resources of a single type to each arriving customer. Further in [2] the author proposed a finite LS with arbitrary distribution functions of resource and service time, and in [3,4] for dependent random variables. Subsequently, resource

The research was funded by the Russian Science Foundation, project no.22-79-10128 (recipients E.S. Sopin and A.V. Daraseliya, Sects. 2–3). The reported study has been partially supported by RFBR, project 20-07-01052 (recipients S.Ya. Shorgin, Sects. 1, 4).

LS continued to develop actively due to the possibility of their application for modeling a fairly wide range of technical devices, information and computing systems, as well as wireless networks, an overview of the modeling of which is presented in [5]. In particular, in several studies, resource LSs were used as one of the tools for modeling LTE [19,20] and 5G [21–23] cellular networks.

For the reason that the analytical calculation and modeling the respective random processes are rather complicated, most of the research results were obtained under simplifying assumptions, such as exponentially distributed service time, the simplest incoming flow, the simplest LS configurations and so on [5–8,10]. In [9] the authors reviewed resource LS with general service discipline, in particular with the recurrent service discipline. The studies [10–15] propose multiphase resource LS with a non-Poisson arriving flow and an unlimited amount of allocated resources on each of the phases. However, in our study, we will consider the Poisson arriving flow and exponential service as a simplifying assumption. Another broad sub-field of research, which will not be explored in this paper, is resource LS with signals [17–19], which takes into consideration the mobility of subscribers.

In particular, one of the effective methods is the use of well-known Erlang network algorithms for the analysis of resource LS. In [16] the authors draws parallels between multi-service QS and resource LS, which makes it possible to obtain the loss probability in the resource LS by narrowing down to the use well-known methods developed for QS. However, the contributors investigated LS without the waiting buffer where each customer only occupies a certain amount of resources. In this paper, we propose the system of equilibrium equations for the resource LS with the waiting buffer and calculation of some technical characteristics of the system.

2 Mathematical Model

2.1 Description of the Model

Consider a multiserver resource loss system with $N < \infty$ servers and waiting buffer of size $V < \infty$, see Fig. 1. Assume that the system has a single Poisson flow of customers with arrival rate λ, and duration of customer servicing are are exponentially distributed with parameter μ. Serving process of each customer requires a server and a random number r_i of resources, $1 \leq i \leq N$, $0 < r_i \leq R$. The distribution of resource requirements are given by probability mass function (pmf) $\{p_{r_i}\}$, where p_{r_i} is the probability that a customer requires r_i resources.

The system operates as follows.

1. An arriving customer requiring r_i, $1 \leq i \leq N$, resources is accepted for service if at the moment of arrival there is a free server and the number of occupied resources is no more than $R - r_i$, $1 \leq i \leq N$.
2. An arriving customer requiring r_i, $1 \leq i \leq N$, resources takes up space in the waiting buffer if there is a room for it and at the moment of arrival there is no free server or the number of available resources is greater than $R - r_i$,

$1 \leq i \leq N$, . This customer is assigned the amount of resource requirements $s_j, 0 < j \leq V$, in the waiting buffer.

3. Assume that customers can arrive for service on the servers only at the time of departure of the customer, and only one at a time. A customer randomly selected from the waiting buffer is trying to get up for service. It's accepted for service if the number of occupied resources is no more than $R - s_j, 0 < j \leq V$, and remains in the waiting buffer otherwise.

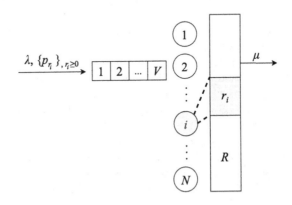

Fig. 1. Loss system with resources and waiting buffer

The system behavior can be described by a random process

$$X(t) = \left(\xi(t), \left(\gamma_1(t), \gamma_2(t), \dots \gamma_{\xi(t)}(t)\right), \eta(t), \left(\beta_1(t), \beta_2(t), \dots, \beta_{\eta(t)}(t)\right)\right), \quad (1)$$

where $\xi(t)$ is the number of customers on the servers, $\xi(t) \leq N$, $\eta(t)$ is the number of customers in the waiting buffer, $\eta(t) \leq V$, $\gamma_i(t)$ is the amount of resources allocated at time moment t to the i-th customer on the servers, $\sum_{i=1}^{\xi(t)} \gamma_i(t) \leq R$, and $\beta_i(t)$ is the amount of requirements for i-th customer in the waiting buffer. The state space of $X(t)$ for the system is described by the set

$$S = \bigcup_{n=0}^{N} S_n, \quad (2)$$

where

$$S_0 = \{(0,0,0,0)\}, \quad (3)$$

$$S_n = \left\{(n, (r_1, \dots, r_n), k, (s_1, \dots, s_k)) : \sum_{i=1}^{n} r_i \leq R, \ k \in [0, V]\right\}, n \in [1, N]. \quad (4)$$

Let us define $P_{n,k}\left((r_1, \dots, r_n), (s_1, \dots, s_k)\right)$ as the stationary probability that there are n customers on the servers that totally occupy $\sum_{i=1}^{n} r_i$ resources and k customers in the waiting buffer.

$$P_{0,0} = \lim_{t\to\infty} \{\xi(t) = 0, \eta(t) = 0\} \tag{5}$$

$$P_{n,k}\left((r_1,...,r_n),(s_1,...,s_k)\right) = \tag{6}$$
$$= \lim_{t\to\infty} \{\xi(t) = n, \gamma_i(t) = r_i, i = 1,...,n, \eta(t) = k, \beta_j(t) = s_j, j = 1,...,k\}$$

Then the system of equilibrium equations can be written as follows:

$$\lambda P_{0,0}(0) = \mu \sum_{j=1}^{R} P_{1,0}(j), \tag{7}$$

$$(\lambda + n\mu)\, P_{n,0}(r_1,...,r_n) = \lambda P_{n-1,0}(r_1,...,r_{n-1})p_{r_n} + \tag{8}$$
$$+ \mu \sum_{i=0}^{n} \sum_{r=1}^{R-r_1-...-r_n} P_{n+1,0}(r_1,...,r_i,r,r_{i+1},...,r_n) +$$
$$+ \mu \sum_{i=0}^{n-1} \sum_{r=1}^{R-r_1-...-r_n} P_{n,0}\left((r_1,...,r_i,r,r_{i+1},...,r_{n-1}),(r_n)\right), \quad n \in [1,N],$$

$$(\lambda + N\mu)P_{N,0}(r_1,...,r_n) = \lambda P_{N-1,0}(r_1,...,r_{N-1})p_{r_N} + \tag{9}$$
$$+\mu \sum_{i=0}^{N-1} \sum_{r=1}^{R-r_1-...-r_{N-1}} P_{N,1}\left((r_1,...,r_i,r,r_{i+1},...,r_{N-1}),(r_N)\right)$$

$$(\lambda + n\mu)\, P_{n,k}\left((r_1,...,r_n),(s_1,...,s_k)\right) = \tag{10}$$
$$= \lambda P_{n-1,k}\left((r_1,...,r_{n-1}),(s_1,...,s_k)\right) p_{r_n} +$$
$$+ \lambda\theta(r_1 + ... + r_n + s_k - R)P_{n,k-1}\left((r_1,...,r_n),(s_1,...,s_{k-1})\right) p_{s_k} +$$
$$+ \mu \sum_{i=0}^{n-1} \sum_{r=1}^{R-r_1-...-r_n} \sum_{j=1}^{k} \frac{1}{k}\theta(r_1 + ... + r_n + s_j - R)\times$$
$$\times P_{n+1,k}\left((r_1,...,r_i,r,r_{i+1},...,r_n),(s_1,...,s_k)\right) +$$
$$+ \mu \sum_{i=0}^{n-1} \sum_{r=1}^{R-r_1-...-r_{n-1}} \frac{1}{k+1}\times$$
$$\times \sum_{j=0}^{k} P_{n,k+1}\left((r_1,...,r_i,r,r_{i+1},...,r_{n-1}),(s_1,...,s_j,r_n,s_{j+1},...,s_k)\right),$$
$$n \in [1,N], k \in [1,V],$$

$$\left(\lambda \sum_{j=1}^{R-r_1-\ldots-r_n} p_j + n\mu\right) P_{n,V}\left((r_1,\ldots,r_n),(s_1,\ldots,s_V)\right) = \tag{11}$$

$$= \lambda P_{n-1,V}\left((r_1,\ldots,r_{n-1}),(s_1,\ldots,s_V)\right)p_{r_n} +$$
$$+ \lambda\theta(r_1+\ldots+r_n+s_V-R)P_{n,V-1}\left((r_1,\ldots,r_n),(s_1,\ldots,s_{V-1})\right)p_{s_V} +$$
$$+ \mu\sum_{i=0}^{n}\sum_{r=1}^{R-r_1-\ldots-r_n}\frac{1}{V}\sum_{j=1}^{V}\theta(r_1+\ldots+r_n-s_j-R)\times$$
$$\times P_{n+1,V}\left((r_1,\ldots,r_i,r,r_{i+1},\ldots,r_n),(s_1,\ldots,s_k)\right),\quad n\in[1,N),$$

$$(\lambda+N\mu)P_{N,k}\left((r_1,\ldots,r_N),(s_1,\ldots,s_k)\right) = \tag{12}$$
$$= \lambda P_{N-1,k}\left((r_1,\ldots,r_{N-1}),(s_1,\ldots,s_k)\right)p_{r_N} +$$
$$+ \lambda P_{N,k-1}\left((r_1,\ldots,r_N),(s_1,\ldots,s_{k-1})\right)p_{s_k} +$$
$$+ \mu\sum_{i=0}^{N-1}\sum_{j=0}^{R-r_1-\ldots-r_{N-1}}\frac{1}{k+1}\times$$
$$\times\sum_{j=0}^{k}P_{N,k+1}\left((r_1,\ldots,r_i,r,r_{i+1},\ldots,r_{N-1}),(s_1,\ldots,s_j,r_N,s_{j+1},\ldots,s_k)\right),$$
$$k\in[1,V),$$

$$N\mu P_{N,V}\left((r_1,\ldots,r_N),(s_1,\ldots,s_V)\right) = \tag{13}$$
$$= \lambda\sum_{j=0}^{r}P_{N-1,V}\left((r_1,\ldots,r_{N-1}),(s_1,\ldots,s_V)\right)p_{r_N} +$$
$$+ \lambda P_{N,V-1}\left((r_1,\ldots,r_N),(s_1,\ldots,s_{N-1})p_{s_V}\right),$$

where $\sum_{i=1}^{n}r_i \le R$, and $\theta(x)$ is the Heaviside step function, i.e. $\theta(x)=1_{x>0}$.

Now let's estimate the number of states in the system. Assume that $N = R$ and $p_i > 0, i = 1,2,\ldots,R$. The number of ordered pairs $(n,(r_1,\ldots,r_n))$: $\sum_{i=1}^{n}r_i \le R$ is equal to the value of multiset coefficients $\left(\!\binom{n+1}{R-n}\!\right)$, then total number of all ordered pairs is $\sum_{n=1}^{R}\left(\!\binom{n+1}{R-n}\!\right) = 2^R - 1$. For any $n > 0$, there are $(1+R+R^2+\ldots+R^V)$ ways to fill the waiting buffer, then the number of states in the system is

$$|S| = 1 + \left(2^R - 1\right)\frac{R^V - 1}{R-1} = 1 + \frac{(2^R-1)(R^{V+1}-1)}{R-1}. \tag{14}$$

Due to the exponential growth in the number of states (14), we propose an approximation of the model with the following simplifying propositions: (i) we do not monitor the amount of resources for each customer on service, but monitor for the total amount of resources allocated at the time moment, (ii) we do not monitor the amount of resource requirements of customers that are in the waiting buffer. We assume as a simplification that the amount of resource requirements of the customer from the waiting buffer is the same as that of arriving customers.

2.2 Simplification Approach

Consider a model similar to the previous one, described in the Subsect. 2.1, except the fact that we assumed there is the single type of resources and the amount of resource requirements s in the waiting buffer is not taken into account, see Fig. 2. Then in contrast to the original model, serving process of each customer requires a server and a random number r of resources, $r \in [0, R]$. The distribution of resource requirements are given by pmf $\{p_r\}_{r \in [1,R]}$, where p_r is the probability that a customer requires r resources.

The system operates as follows.

1. An arriving customer requiring r resources is accepted for service if at the moment of arrival there is a free server and the number of occupied resources is no more than $R - r$.
2. An arriving customer requiring r resources takes up space in the waiting buffer if there is a room for it and at the moment of arrival there is no free server or the number of available resources is greater than $R - r$.
3. If there are n customers on the servers and they totally occupy r resources, then at the departure of a customer j resource units are released with probability $\frac{p_{r-j}^{(n-1)} p_j}{p_r^{(n)}}$, and one random customer from the waiting buffer is trying to enter the service.

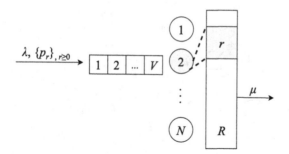

Fig. 2. Simplified LS with resources and waiting buffer

The system behavior can be described by a random process

$$X(t) = (\xi(t), \gamma(t), \eta(t)), \tag{15}$$

where $\xi(t)$ is the number of customers on the servers, $\eta(t)$ is the number of customers in the waiting buffer, and $\gamma(t)$ is the total amount of resources allocated at the time moment t to the customers on the servers.

The state space of the system is described by the set

$$S = \bigcup_{n=0}^{N} S_n, \tag{16}$$

where

$$S_0 = \{(0,0,0)\}, \quad S_n = \left\{(n,r,k) : p_r^{(n)} > 0, k \in [0,V]\right\}, n \in [1,N]. \tag{17}$$

Here $\{p_r^{(n)}\}_{r \geq 0}$ is n-fold convolution of pmf $\{p_r\}_{1 \leq r \leq R}$, $p_r^{(n)}$ is the probability that n customers occupy r resource units, and is calculated as follows

$$p_j^{(n)} = \sum_{i=1}^{j-n+1} p_{j-i}^{(n-1)} p_i, \quad j \in [1,R], n \geq 2, \tag{18}$$

where $p_j^{(1)} = p_j$, $p_0^{(0)} = 1$ and $p_j^{(0)} = 0$, $j \geq 0$.

Let us denote the stationary probability that there are n customers on the servers that totally occupy r resources and k customers in the waiting buffer by $P_{n,k}(r)$. Then the system of equilibrium equations for the simplified model can be written as follows:

$$\lambda P_{0,0}(0) = \mu \sum_{j=1}^{R} P_{1,0}(j), \tag{19}$$

$$(\lambda + n\mu) P_{n,0}(r) = \lambda \sum_{j=0}^{r-1} P_{n-1,0}(j) p_{r-j} + \tag{20}$$

$$+ (n+1)\mu \sum_{j=r+1}^{R} P_{n+1,0}(j) \frac{p_{j-r} p_r^{(n)}}{p_j^{(n+1)}} +$$

$$+ n\mu \sum_{j=n}^{R} P_{n,1}(j) \sum_{i=1}^{j-n+1} \frac{p_i p_{j-i}^{(n-1)}}{p_j^{(n)}} p_{r-j+i}, \quad n \in [1,N),$$

$$(\lambda + N\mu) P_{N,0}(r) = \lambda \sum_{j=N-1}^{r-1} P_{N-1,0}(j) p_{r-j} + \tag{21}$$

$$+ N\mu \sum_{j=N}^{R} P_{N,1}(j) \sum_{i=1}^{j-N+1} \frac{p_i p_{j-i}^{(N-1)}}{p_j^{(N)}} p_{r-j+i},$$

$$(\lambda + n\mu)\, P_{n,k}(r) = \lambda \left(\sum_{j=n-1}^{r-1} P_{n-1,k}(j)p_{r-j} + \sum_{j=R-r+1}^{R} P_{n,k-1}(r)p_j \right) + \quad (22)$$

$$+ (n+1)\mu \sum_{j=r+1}^{R} P_{n+1,k}(j)\frac{p_{j-r}p_r^{(n)}}{p_j^{(n+1)}} \sum_{i=R-r+1}^{R} p_i +$$

$$+ n\mu \sum_{j=n}^{R} P_{n,k+1}(j) \sum_{i=1}^{j-n+1} \frac{p_i p_{j-i}^{(n-1)}}{p_j^{(n)}} p_{r-j+i}, \quad n \in [1,N], k \in [1,V],$$

$$\left(\lambda \sum_{j=1}^{R-r} p_j + n\mu \right) P_{n,V}(r) = \quad (23)$$

$$= \lambda \left(\sum_{j=n-1}^{r-1} P_{n-1,V}(j)p_{r-j} + \sum_{j=R-r+1}^{R} P_{n,V-1}(r)p_j \right) +$$

$$+ (n+1)\mu \sum_{j=r+1}^{R} P_{n+1,V}(j)\frac{p_{j-r}p_r^{(n)}}{p_j^{(n+1)}} \sum_{i=R-r+1}^{R} p_i, \quad n \in [1,N],$$

$$(\lambda + N\mu)P_{N,k}(r) = \lambda \sum_{j=N-1}^{r-1} P_{N-1,k}(j)p_{r-j} + \lambda P_{N,k-1}(r) + \quad (24)$$

$$+ N\mu \sum_{j=N}^{R} P_{N,k+1}(j) \sum_{i=1}^{j-N+1} \frac{p_i p_{j-i}^{(N-1)}}{p_j^{(N)}} p_{r-j+i}, \quad k \in [1,V],$$

$$N\mu P_{N,V}(r) = \lambda \sum_{j=0}^{r} P_{N-1,V}(j)p_{r-j} + \lambda P_{N,V-1}(r), \quad (25)$$

where $r \in [1,R]$.

By solving the system (19)–(25), we numerically obtained the stationary probabilities $P_{n,k}(r)$, $n \in [0,N]$, $k \in [0,K]$, $r \in [0,R]$.

2.3 Performance Metrics of Interest

Now, we proceed with performance metrics of interest. With the stationary distribution, one can obtain the loss probability π.

$$\pi = \sum_{r=N}^{R} P_{N,V}(r) + \sum_{n=1}^{N-1} \sum_{r=n}^{R} P_{n,V}(r) \sum_{i=R-r+1}^{R} p_i. \quad (26)$$

The average number $\bar{N}$ of customers in the waiting buffer is

$$\bar{N} = \sum_{n=1}^{N} \sum_{r=n}^{R} \sum_{k=1}^{V} k P_{n,k}(r). \tag{27}$$

The average waiting time W takes the following form:

$$W = \frac{\bar{N}}{\lambda(1 - \pi)}. \tag{28}$$

Resource requirements of blocked customers can be found using the following formula

$$p_b(i) = \frac{p_i}{\pi} \left(\sum_{n=1}^{N-1} \sum_{r=R-j+2}^{R} P_{n,V}(r) + \sum_{r=N}^{R} (i+1) P_{N,V}(r). \right) \tag{29}$$

Then the average resource requirements of blocked customers is

$$E[p_b] = \sum_{i=1}^{R} i p_b(i) \tag{30}$$

3   Numerical Results

In this section, we present our numerical results for the previously obtained system efficiency characteristics.

For numerical analysis, we used the simplified model described in Subsect. 2.2. We consider a system witch consists of $N = 20$ servers and waiting buffer of size $V = 10$. The other parameters used for numerical calculations was summarized and provided in Table 1.

Table 1. Default model parameters.

Parameter	Value
Number of servers, N	20
Number of resource units, R	30
Waiting buffer size, V	10
Customer arrival rate, λ	10
Customer servicing intensity, μ	1

Now let's proceed directly to the analysis of the performance metrics of interest and the impact of the system parameters on it. Figure 3 shows the effect of the session arrival rate λ, on the loss probability, π, for several values of waiting buffer size, V. It confirms the logic of our model that with an increase in offered system load the loss probability will gradually increase. At low system

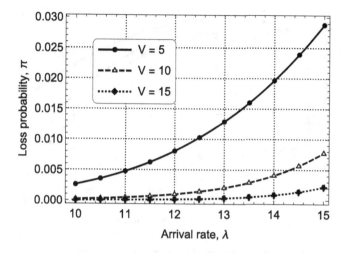

Fig. 3. Loss probability

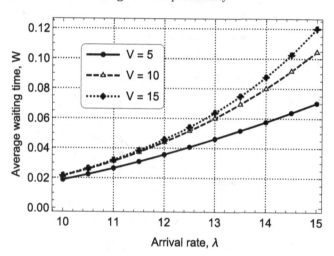

Fig. 4. Average waiting time

loads, the length of the wait buffer does not greatly affect the loss probability, since all customers have time to be processed on the server. With an increase in load, an increase in the waiting buffer size is also required, however, with a gradual increase in V, the loss probability graphs will gradually approach each other until they overlap each other for certain λ values. The dependence of the sessions arrival rate λ on the average waiting time W on Fig. 4 is also confirm the logic of our model that for the considered input parameters, with an increase in the waiting buffer size, the waiting time increases, since those customers that previously did not manage to be served and were lost now queue up in the waiting buffer.

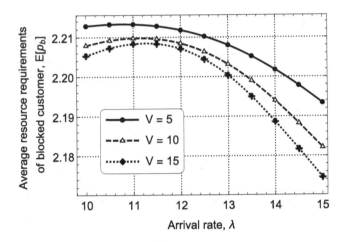

Fig. 5. Average resource requirements of blocked customers

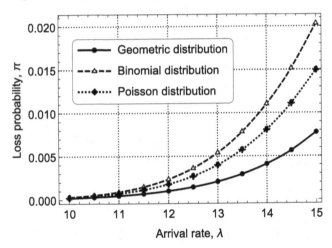

Fig. 6. Loss probability

Figure 5 shows the dependence of the average resource requirements of blocked customers on the arrival rate. The presence of an extremum on the function graph can be explained as follows. With small system loads the impact of the left sum in (26), in which any customers are blocked when the waiting buffer and servers are busy, on the loss probability is much higher than that of the right sum in (26), in which only "heavy" customers are lost. This has such an effect that the average resource requirements of blocked customers is slightly higher. And vice versa, after the peak $V = 15$, the system resources are almost completely occupied, so the loss probability of blocking increases, and the average resource requirement decreases.

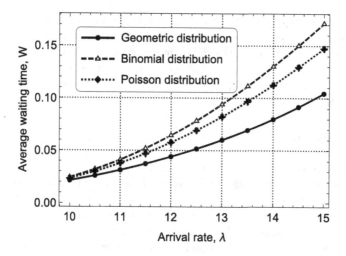

Fig. 7. Average waiting time

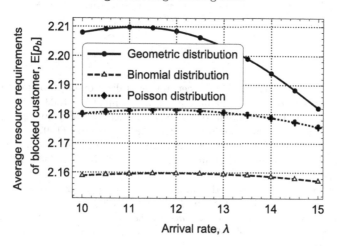

Fig. 8. Average resource requirements of blocked customers

During the modeling of wireless networks, one should take into account the features of various schemes for distributing radio resources corresponding to different technologies. Thus, the amount of required resources is determined by a given probability distribution, which may differ. In this study, we use three different distributions. We assume that resource requirements of customers have geometric distribution with parameter 0.5, binomial distribution with parameter 0.25 and Poisson distribution with parameter 1. Figure 6 shows an effect on loss probabilities depending on probability distributions. It can be noted that the losses of customers in the system increase as the variance of distribution decreases. This can be explained by the fact that as the variance increases, the average resource requirements of blocked customers increases, see Fig. 8. The dependence of the average waiting time W on the probability distributions (see Fig. 7) is generally similar to the dependence of the loss probability π (Fig. 6). By analogy with Fig. 5, in Fig. 8 we also see the presence of an extremum of the function. Note that the most pronounced inflection of the function is observed for the geometric distribution.

4 Conclusion

In our paper we consider a model of a multi-server loss system with the multi-type of resources, waiting buffer and derived the system of equilibrium equations for the stationary probabilities of the system. We proposed a simplified model with the single-type of resources, for which we also obtained stationary probabilities. Besides, we obtained the loss probability of the system, the average waiting time, the average number of customers, and the average resource requirements of blocked customers. Numerical analysis was carried out for three different distributions. The accuracy of this model will also be further investigated using a simulation model.

References

1. Romm, E.L., Skitovitch, V.V.: On certain generalization of problem of Erlang. In: Automation and Remote Control, vol. 32, no. 6, pp. 1000–1003 (1971)
2. Tikhonenko, O.M.: Determination of the characteristics of queueing systems with limited memory. Autom. Remote. Control. **58**(6), 969–973 (1997)
3. Tikhonenko, O.M., Klimovich, K.G.: Analysis of queuing systems for random-length arrivals with limited cumulative volume. Probl. Peredachi Inform. **37**(1), 78–88 (2001)
4. Tikhonenko, O.M.: Generalized erlang problem for queueing systems with bounded total size. Probl. Peredachi Inform. **41**(3), 64–75 (2005)
5. Gorbunova, A.V., Naumov, V.A., Gaidamaka, Y.V., Samouylov, K.E.: Resource queuing systems as models of wireless communication systems. Inf. Appl. **12**(3), 48–55 (2018)
6. Naumov, V.A., Samuilov, K.E., Samuilov, A.K.: On the total amount of resources occupied by serviced customers. Autom. Remote. Control. **77**(8), 1419–1427 (2016). https://doi.org/10.1134/S0005117916080087
7. Basharin, G.P., Samouylov, K.E., Yarkina, N.V., Gudkova, I.A.: A new stage in mathematical teletraffic theory. Autom. Remote. Control. **70**(12), 1954–1964 (2009)
8. Naumov, V., Samouylov, K.: Analysis of multi-resource loss system with state-dependent arrival and service rates. Probab. Eng. Inf. Sci. **31**(4), 413–419 (2017)
9. Gorbunova, A.V., Naumov, V.A., Gaidamaka, Y.V., Samouylov, K.E.: Resource queuing systems with general service discipline. Inform. Primen. **13**(1), 99–107 (2019)
10. Galileyskaya, A.A., Lisovskaya, E.Y., Moiseeva, S.P., Gaidamaka Y.V.: On sequential data processing model, that implements the backup storage. Sovremennye informacionnye tehnologii i IT-obrazovanie = Modern Inf. Technol. IT-Educ. **15**(3), 579–587 (2019)

11. Lisovskaya, E., Moiseeva, S., Pagano, M.: Infinite–server tandem queue with renewal arrivals and random capacity of customers. In: Vishnevskiy, V.M., Samouylov, K.E., Kozyrev, D.V. (eds.) DCCN 2017. CCIS, vol. 700, pp. 201–216. Springer, Cham (2017). https://doi.org/10.1007/978-3-319-66836-9_17

12. Moiseev, A., Moiseeva, S., Lisovskaya, E.: Infinite-server queueing tandem with MMPP arrivals and random capacity of customers. In: 31st European Conference on Modelling and Simulation Proceedings - Budapest, Hungary, pp. 673–679 (2017)

13. Kononov, I.A., Lisovskaya, E.Y.: Issledovanie beskonechnolineynoy SMO MAP|GI|∞ s zayavkami sluchaynogo ob"ema [Study of the infinite server queue MAP|GI|∞ with customers of random volume]. In: 15th Conference (International) named after A. F. Terpugov "Information Technologies and Mathematical Modelling" Proceedings. Tomsk: Tomsk State University, pp. 67–71 (2016)

14. Lisovskaya, E.Y., Moiseeva, S.P.: Asimptoticheskiy analiz nemarkovskoy beskonechnolineynoy sistemy obsluzhivaniya trebovaniy sluchaynogo ob"ema s vkhodyashchim rekurrentnym potokom [Asymptotical analysis of a non-Markovian queueing system with renewal input process and random capacity of customers]. Vestnik Tomskogo gosudarstvennogo universiteta. Upravlenie, vychislitel'naya tekhnika i informatika [Tomsk State University J. Control Computer Science], pp. 30–38 (2017)

15. Lisovskaya, E., Moiseeva, S., Pagano, M., Potatueva, V.: Study of the MMPP|GI|∞ queueing system with random customers' capacities. Informatika i ee Primeneniya - Inform. Appl. 11(4), 111–119 (2017)

16. Naumov, V.A., Samouylov, K.E.: On relationship between queuing systems with resources and Erlang networks. Inf. Appl. 10(3), 9–14 (2016)

17. Ageev, K., Sopin, E., Konstantin, S.: Simulation of the limited resources queuing system with signals. In: 10th International Congress on Ultra Modern Telecommunications and Control Systems and Workshops (ICUMT), Moscow, Russia, pp. 1–5 (2018)

18. Samuvlov, A., Moltchanov, D., Krupko, A., Kovalchukov, R., Moskaleva, F., Gaidamaka, Y.: Performance analysis of mixture of unicast and multicast sessions in 5G NR systems. In: 2018 10th International Congress on Ultra Modern Telecommunications and Control Systems and Workshops (ICUMT), Moscow, Russia, pp. 1–7 (2018)

19. Sopin, E., Vikhrova, O., Samouylov, K.: LTE network model with signals and random resource requirements 2017. In: 9th International Congress on Ultra Modern Telecommunications and Control Systems and Workshops (ICUMT), pp. 101–106 (2017)

20. Sopin, E.S., Ageev, K.A., Markova, E.V., Vikhrova, O.G., Gaidamaka, Y.V.: Performance analysis of M2M traffic in LTE network using queuing systems with random resource requirements. Autom. Control. Comput. Sci. 52(5), 345–353 (2018). https://doi.org/10.3103/S0146411618050127

21. Lu, X., et al.: Integrated use of licensed- and unlicensed-band MMWave radio technology in 5G and beyond. IEEE Access 7, 24376–24391 (2019)

22. Daraseliya, A.V.E., Sopin, E.S., Moltchanov, D.A., Samouylov, K.E.E.: Analysis of 5G NR base stations offloading by means of NR-U technology. Inf. Appl. 15(3), 98–111 (2021)

23. Naumov, V., Beschastnyi, V., Ostrikova, D., Gaidamaka, Y.: 5G new radio system performance analysis using limited resource queuing systems with varying requirements. In: Vishnevskiy, V.M., Samouylov, K.E., Kozyrev, D.V. (eds.) DCCN 2019. LNCS, vol. 11965, pp. 3–14. Springer, Cham (2019). https://doi.org/10.1007/978-3-030-36614-8_1

A Unified Regenerative Stability Analysis of Some Non-conventional Queueing Models

Stepan Rogozin[1,2(✉)] and Evsey Morozov[1,2,3]

[1] Institute of Applied Mathematical Research, Karelian Research Centre RAS,
185910 Petrozavodsk, Russia
ppexa@mail.ru, emorozov@karelia.ru
[2] Institute of Mathematics and Information Technologies,
Petrozavodsk State University, 185910 Petrozavodsk, Russia
[3] Moscow Center for Fundamental and Applied Mathematics,
Moscow State University, 119991 Moscow, Russia

Abstract. In this research we demonstrate how the regenerative methodology allows to deduce stability conditions in some particular queueing models by a unified way. First we consider the retrial system with classical retrial discipline in which the retrial times becomes exponential when the orbit size exceeds a high threshold. Then we study the buffered system with a state-dependent arrival rate, and finally, a buffered batch-arrival and batch-service system with random serving capacity and batch-size-dependent service. We present short transparent regenerative proofs of the stability conditions of these systems which have been obtained in previous works by various methods. Some numerical examples are given.

Keywords: Regenerative method · stability analysis · modified classic retrials · batch arrival · batch service · state-dependent input

1 Introduction

In this research we study the stability of a few $M/G/c$-type queueing systems which are analysed by a unified regenerative approach [1]. This approach can be applied if the basic processes that describe the dynamics of the system are regenerative, and the purpose of the analysis is to find conditions when these processes are *positive recurrent* [1,6]. This approach is a very intuitive and powerful tool for analyzing the performance and stability of queueing processes, including a wide class of non-Markov models. The aim of the regenerative stability analysis is based on a characterization of the limiting *remaining regeneration time*. More precisely, if this time does not go to infinity in probability, then the mean regeneration period length is finite, and thus the process is positive recurrent [1,7–9].

We consider the basic processes describing the dynamics of the corresponding system at the departure and arbitrary time instances. This approach has been

V. M. Vishnevskiy et al. (Eds.): DCCN 2022, CCIS 1748, pp. 296–310, 2023.
https://doi.org/10.1007/978-3-031-30648-8_24

successfully applied for the first time in the paper [2] to reprove the stability condition of a multiclass multiserver $M/G/c$ system with *classical retrials*. In present research we extend this analysis to some other regenerative (non-conventional) queueing models.

The first model we consider is a modification of a single-class multiserver $M/G/c$-type system with classical retrials considered in [3] in which the distribution of the retrial times switches from phase-type to the exponential provided the orbit size exceeds a predefined (deterministic) threshold. A motivation and performance analysis of this model can also be found in [3].

In the second model, the input has a state-dependent rate and, in addition to the 'basic' servers, there are a set of the so-called 'assistant' servers. Such a queueing model with *Markovian arrival process* is considered in the paper [4]. This model is motivated by the queueing systems managed by the ticket technology which is widely used in the service industries and the government offices.

The final model, considered in work [5], is a batch-arrival and batch-service system with random serving capacity and batch-size-dependent service. This model has applications in the group screening policy to detect viruses and in the area of telecommunications where IP packets often arrive by groups and are transmitted by batches as well.

As we mentioned above, these different systems are studied by various authors by different methods. For instance, two first models are investigated by Matrix-analytic method while the analysis of the last model is based on standard Markovian technique. The main contribution of the present research is that we obtain short and intuitive proofs based on an unified method of regenerative stability analysis [1].

The rest of this paper is organized as follows. Section 2 describes the analysis of the first model with switching between the retrial time distributions. In Sect. 3, the analysis of a state-dependent system with the assistance servers is given. Finally, in Sect. 4 we present the stability analysis of a batch-arrival and batch-service system with random serving capacity and batch-size-dependent service. Some numerical results based on the stochastic simulation are presented in Sect. 5 which illustrate the theoretical results.

2   Exponential Retrial Times for Large Orbits

We first describe a basic retrial model which is used below as a useful tool of the analysis. This is a single-class retrial $M/G/c$ queueing system with classical retrials. We denote by $\{t_n\}$ the arrival instants and let τ be the generic exponential interarrival time with rate $\lambda = 1/\mathsf{E}\tau$. Denote by $\{S_n\}$ the iid service times with the generic time S with rate $\mu = 1/\mathsf{E}S$. A customer meeting server busy joins an infinite capacity orbit. Also we assume that the retrial time has exponential distribution with rate γ. Such a (basic) multiclass retrial model has been studied in the paper [2]. Now we modify this basic model by the following way. We assume that if $Q(t)$, the orbit size at instant t^-, exceeds some fixed threshold K, that is $Q(t) \geq K$, then the retrial times of all orbital customers become exponential, while if $Q(t) < K$ then retrial times follow some general distribution. The latter distribution can be in particular phase-type and this case is considered in the paper [3] which has attracted our attention to this system. We emphasize that the regenerative stability analysis we apply in this research is extended to an arbitrary retrial distribution when the orbits size is below the threshold.

First we describe a regenerative structure of the system, which remains unchanged for other systems we consider in this paper. Denote by $\{t_k\}$ the instants of the arrival Poisson process. Then the regeneration instants $\{T_n\}$ of the process $\{Q(t), t \geq 0\}$ are recursively defined as

$$T_{n+1} = \inf_k(t_k > T_n : Q(t_k) = 0, \ n \geq 0), \tag{1}$$

($T_0 := 0$). The distance between two regeneration points (which are called the generic regeneration period lengths) are denoted by T. We recall, that if $\mathsf{E}T < \infty$ then the regenerative process $\{Q(t)\}$ is called *positive recurrent* (or stable), and, because the input process is Poisson, this implies the existence of the stationary process $Q(t) \Rightarrow Q, t \to \infty$ (where $\Rightarrow$ means convergence in distribution) [1]. Denote by $\rho = \lambda/\mu$. Now we are ready to prove the following statement.

Theorem 1. *If condition*

$$\rho < c, \tag{2}$$

holds then the (initially idle) system under consideration is positive recurrent, $\mathsf{E}T < \infty$.

Proof. The regenerative method used in [2] is based on a balance between the received and the remaining work, which is written for the departure instants d_k, $k \geq 1$. We further develop this approach, and denote by $V(t)$ the work which bring the customers arriving in the system in interval $[0, t]$. Also denote by $W(t)$ the remaining workload at instant t, by $I_i(t)$ the idle time of server i in time interval $[0, t]$, and then $I(t) = \sum_{i=1}^c I_i(t)$ is the total idle time of the servers in this interval. Then we obtain the following balance equation

$$W(d_k) = V(d_k) - cd_k + I(d_k), \ k \geq 1. \tag{3}$$

Denote by $Q(d_k) = Q_k$, then using the proof by contradiction, we assume that the orbit size *increases in probability*, that is

$$Q_k \Rightarrow \infty, \ k \to \infty. \tag{4}$$

It is easy to obtain, using the Strong Law of Large Numbers (SLLN) and the renewal theory, that with probability (w.p.1)

$$\lim_{k \to \infty} \frac{V(d_k)}{d_k} = \rho. \tag{5}$$

The most challenging point is to show that, under assumption (4), the mean total idle time $\mathsf{E}I_k$ of the servers in the interval $[0, d_k]$ satisfies

$$\lim_{k \to \infty} \frac{\mathsf{E}I_k}{k} = 0. \tag{6}$$

First we denote by $\Delta_k = I(d_{k+1}) - I(d_k)$ the idle time of all servers between the kth and $(k + 1)$th departures. The main observation is that, provided $Q_k \geq n$, the total mean idle time of the servers after the kth departure is upper by the constant

$$C_n := \frac{c}{\lambda + n\gamma} \to 0,$$

as $n \to \infty$. For an arbitrary $\varepsilon > 0$, take n_1 such that for all $n \geq n_1$, $C_n \leq \varepsilon$, and for a fixed constant L take n_2 such that, for all $n \geq n_2$

$$P(Q(d_n) \leq L) \leq \varepsilon.$$

Take now $n_0 = \max(n_1, n_2)$ and let $n > n_0$. Then

$$\mathsf{E}I_n = \mathsf{E}\sum_{k=1}^{n_0} \Delta_k + \mathsf{E} \sum_{k=n_0+1}^{n} \Delta_k,$$

and we obtain

$$\mathsf{E} \sum_{k=n_0+1}^{n} \Delta_k = \sum_{k=n_0+1}^{n} \mathsf{E}(\Delta_k|Q_k > L)\mathsf{P}(Q_k > L)$$
$$+ \sum_{k=n_0+1}^{n} \mathsf{E}(\Delta_k|Q_k \leq L)\mathsf{P}(Q_k \leq L)$$
$$\leq \varepsilon(n - n_0)(1 + c/\lambda).$$

Because ε is arbitrary, then (6) follows. It is known (see [1]) that then there exists a subsequence $d_{n_k} \to \infty$, $k \to \infty$ such that, w.p.1,

$$\lim_{k \to \infty} I_{n_k}/d_{n_k} = 0.$$

Then we write the balance Eq. (3) for d_{n_k}, and the rest of the proof remains the same as for the basic model in the paper [2] and using the renewal theory we arrive in the limit, by contradiction, to the stability condition (2).

3 A State-Dependent System with the Assistance Servers

Now we consider a $M/G/c$-type queuing model with infinite buffer in which input rate is state-dependent, which has been considered in the paper [4] for the first time. First we assume that an arriving customer leaves the system with a probability q_i if meets no free servers and the queue size equals i. It is assumed that there are R states of the system and the input rates are ordered as

$$\lambda_1 \leq \lambda_2 \leq \cdots \leq \lambda_R. \tag{7}$$

It follows from [4] that the minimal input rate λ_1 corresponds to a saturated regime when the orbit size approaches infinity. (The authors thank Prof. A. Dudin for this comment.) We make this statement more precise and assume that the input has the minimal rate λ_1 as long as the queue size exceeds a finite threshold D. (We call this regime saturated or overloaded.) We denote by $\{S_n^{(1)}\}$ the general (iid) service times of the nth arrival customer, accepted in the system, with generic time $S^{(1)}$ and rate $\mu_1 = 1/\mathsf{E}S^{(1)}$. We also assume that there are M additional assistance servers (assistants), and after finishing service by a main server, and with the given probability $p < 1$, the customer asks an assistant to help. It is worth mentioning that this customer continues to block the basic server while obtains the help. Otherwise he waits until an assistant becomes idle and *does not block* the server, and thus other customers can use this server. Then this customer is served by a free assistant during a (generic) time $S^{(2)}$ with rate $\mu_2 = 1/\mathsf{E}S^{(2)}$. After finishing this additional service time, the customer returns to the main server to start a new trial with a generic service time $S^{(1)}$, and so on. The regeneration instants of this system are defined in (1). We denote by γ_n the probability that the number of busy assistants equals n, provided the system is overloaded. Assume that the limit $\lim_{i\to\infty} q_i = q$ exists, in which case $1 - q$ represents the probability to join the system regardless of the queue size. Such a queueing model, with a Markovian arrival process, is considered and motivated in the paper [4]. We reprove the following statement from this paper. Denote

$$\rho_1 = \frac{\lambda_1}{\mu_1}\frac{(1-q)}{(1-p)}. \tag{8}$$

Theorem 2. *If condition*

$$\rho_1 < \sum_{n=1}^{\min(c,M)} \gamma_n(c-n) = c - \sum_{n=1}^{\min(c,M)} \gamma_n n, \tag{9}$$

holds then the (initially idle) system under consideration is positive recurrent, $\mathsf{E}T < \infty$.

Proof. Denote, in interval $[0, t]$, by $V(t)$ the total work generated by the customers accepted in the system, $B_1(t)$ the total busy time of main servers when they *work* (this time *does not include the blocking time*), $I(t)$ the total idle time (of all servers), $B_2(t)$ the total blocking time (when the customers are with the assistance servers). Also denote by $W(t)$ the remaining workload at instant t and $S_i(t)$ the remaining service time in server i at instant t^-. We can represent the busy time of main servers $B_1(t)$ as

$$B_1(t) = ct - I(t) - B_2(t), \quad t \geq 0. \tag{10}$$

Then the following balance equation holds:

$$V(t) = W(t) + ct - I(t) - B_2(t), \quad n \geq 1. \tag{11}$$

We assume that the number of customers

$$Q(t) \Rightarrow \infty, \quad t \to \infty. \tag{12}$$

We note that the system has infinite capacity buffer and the service discipline is non-idling: the servers can not be idle while there are waiting customers. Then it is clear that, as $t \to \infty$,

$$P(Q(t) > c) \le P(S_i(t) > 0) \to 1, \quad i = 1, \ldots, c, \tag{13}$$

and by (13),

$$P(S_i(t) = 0) = 1 - P(S_i(t) > 0) \le 1 - P(Q(t) > c) \to 0.$$

Hence, we obtain

$$\lim_{t \to \infty} \frac{1}{t} \mathsf{E} I(t) = \lim_{t \to \infty} \frac{1}{t} \mathsf{E}\Big[\sum_{i=1}^{c} \int_0^t \mathbf{1}(S_i(u) = 0) du \Big] \tag{14}$$

$$= \lim_{t \to \infty} \frac{1}{t} \sum_{i=1}^{c} \int_0^t P(S_i(u) = 0) du = 0.$$

Let the indicator $\mathbf{1}_k = 1$ if kth arriving customer joins the system, and $\mathbf{1}_k = 0$ otherwise. Denote by ξ_k the number of the returns of customer k for the new service (after a help). Note that $\{\xi_k\}$ is an iid sequence with generic element ξ. Then each new accepted arrival will have in average

$$\mathsf{E}\xi = \sum_{k=1}^{\infty} k p^{k-1}(1-p) = \frac{1}{1-p},$$

returns for new service. We could threat these repeated attempts as the *new arrivals*. By the basic assumption (12), the following convergence in distribution holds:

$$\mathbf{1}_k \xi_k \Rightarrow \mathbf{1}\xi, \quad k \to \infty,$$

where $\mathbf{1}_k \Rightarrow \mathbf{1}$ and $\mathsf{E}\mathbf{1} = 1 - q$. Then

$$\lim_{k \to \infty} \mathsf{E}\mathbf{1}_k = \lim_{k \to \infty} (1 - q_k) = 1 - q.$$

Now using independence $\mathbf{1}_k$ and ξ_k we can apply the convergence in mean resulting in

$$\lim_{k \to \infty} \mathsf{E}(\mathbf{1}_k \xi_k) = \mathsf{E}\xi \lim_{k \to \infty} \mathsf{E}\mathbf{1}_k = \frac{1-q}{1-p}. \tag{15}$$

Thus (15) expresses the mean number of return to server which arbitrary customer makes (after help) in the saturated regime. Denote by

$$T_D(t) = \int_0^t \mathbf{1}(Q(u) \le D) du,$$

the total time, in the interval $[0, t]$, when the queue size exceeds the threshold D and hence, during time $t - T_D$ the input process is Poisson with rate λ_1. By (12),

$$\mathsf{E}T_D(t) = o(t), \quad t \to \infty. \tag{16}$$

Denote by $A(t)$ the total number of the accepted customers in interval $[0, t]$. We may represent $A(t)$ as

$$A(t) = A_0(T_D(t)) + A_1(t - T_D(t)), \tag{17}$$

where the 1st term denotes the number of arrivals during total time $T_D(t)$, when $Q(t) \leq D$, while the 2nd terms denotes the number of arrivals during total time $t - T_D(t)$ when $Q(t) > D$. As we mention above $\{A_1(t)\}$ is a renewal process with rate λ_1, and

$$\lim_{t \to \infty} \frac{EA_1(t)}{t} = \lambda_1, \ \lim_{t \to \infty} \frac{A_1(t)}{t} = \lambda_1 \ \text{w.p.1.}$$

On the other hand, by (16),

$$EA_0(T_D(t)) = \int_0^t EA_0(u)P(T_D(t) \in du) \leq \lambda_R \int_0^t uP(T_D(t) \in du)$$
$$= \lambda_R ET_D(t) = o(t), \ t \to \infty, \tag{18}$$

where the inequality follows by (7). Then it follows that

$$\frac{1}{t}EA(t) \sim \frac{1}{t}EA_1(t - T_D(t)),$$

where $a \sim b$ means that $a/b \to 1$. It also follows from the renewal theory that

$$EV(t) \sim EV_1(t - T_D(t)), \ t \to \infty, \tag{19}$$

where $V_1(t - T_D(t))$ means the work arrived during time periods, within $[0, t]$, when $Q(t) \geq D$ and the input is Poisson with rate λ_1. Denote by $\hat{S}$ the total service time generated by an arbitrary customer (including repeated attempts). Note that $\{V_1(t)\}$ is a process with regenerative increments (over exponential interarrival interval with parameter λ_1), and in particular in our setting, for $t > 0$,

$$\lim_{t \to \infty} \frac{EV_1(t)}{t} = \lambda_1 E\hat{S} = \lambda_1 ES^{(1)}\frac{(1-q)}{(1-p)} = \rho_1. \tag{20}$$

(For detail see Chap. 1 in [1]. Indeed, because the input is Poisson, this process is stationary and $EV_1(t)/t = \rho_1, \ t > 0$.) Then we obtain, also see (16),

$$\frac{1}{t}EV_1(t - T_D(t)) = \frac{1}{t}\int_0^t \frac{EV_1(t-u)}{t-u} \cdot (t-u)P(T_D(t) \in du)$$
$$\sim \frac{1}{t}\int_0^t \rho_1(t-u)P(T_D(t) \in du)$$
$$= \rho_1 - \rho_1\frac{1}{t}ET_D(t) \to \rho_1, \ t \to \infty. \tag{21}$$

Now we denote by $a(t)$ the number of busy assistants at instant t^-. Then, on the event (12), for each k,

$$\frac{1}{t}\int_0^t \mathbf{1}(a(u) = k)du \Rightarrow \gamma_k,$$

and because the process $\{B_2(t)/t\}$ is uniformly integrable (since $B_2(t)/t \leq \min(c, M)$) we obtain the following limit in mean:

$$\lim_{t\to\infty}\frac{1}{t}\mathsf{E}B_2(t) = \sum_{k=1}^{\min(c,M)} k\lim_{t\to\infty}\frac{1}{t}\mathsf{E}\int_0^t \mathbf{1}(a(u) = k)du = \sum_{k=1}^{\min(c,M)} k\gamma_k, \quad (22)$$

because, by definition,

$$\lim_{t\to\infty}\frac{1}{t}\mathsf{E}\int_0^t \mathbf{1}(a(u) = k)du = \gamma_k.$$

Combining it with (19) and (21), we obtain from (11) the following inequality

$$\lim_{t\to\infty}\frac{1}{t}\mathsf{E}W(t) = \rho_1 - c + \sum_{k=1}^{\min(c,M)} k\gamma_k \geq 0, \quad (23)$$

which implies

$$\rho_1 \geq \sum_{k=1}^{\min(c,M)} \gamma_k(c - k) = c - \sum_{k=1}^{\min(c,M)} k\gamma_k. \quad (24)$$

This contradicts (9), and thus assumption (12) does not hold, that is

$$Q(t) \not\to \infty, \quad t \to \infty. \quad (25)$$

Then, using a standard approach and an unloading procedure (see [2]), we obtain the system is positive recurrent, that is $\mathsf{E}T < \infty$.

4 A Batch-Arrival and Batch-Size-Dependent Service

Now we consider the queueing system studied in the paper [5]. The batches of customers arrive in the infinite capacity buffer by a Poisson process with rate λ. Sizes of bathes are iid with generic size X with a finite mean $\mathsf{E}X$. The server has a random service capacity Y with a finite mean $\mathsf{E}Y$. We assume that Y has size i with the given probability $\mathsf{P}_i = \mathsf{P}(Y = i)$ in which case the service time of this batch is S_i, $i = 1, \ldots, B$. Denote by $\{d_n\}$ the departure instants of the batches (when the server is ready to take a new batch to be served), and let $Q_n = Q(d_n)$, $n \geq 1$. For a fixed i, the service times are iid and the maximal capacity B is assumed to be finite in [5]. If, at the instant d_n (and if the system is not empty), the queue size Q_n, turns out to be less than the sampled service capacity Y_n, then the server takes all present customers $\tilde{Y}_n$, and we denote $\tilde{Y} = \min(Y_n, Q_n)$. The following stability condition has been obtained in [5]

$$\lambda EX \sum_{i=1}^{B} P_i ES_i < EY. \tag{26}$$

Again we apply the regenerative method to establish the same stability condition in a simpler way. Now we assume that the queue size

$$Q_n \Rightarrow \infty, \quad n \to \infty. \tag{27}$$

Under this assumption

$$\tilde{Y}_n = \min(Y_n, Q_n) \Rightarrow Y, \quad n \to \infty. \tag{28}$$

We note that $\tilde{Y}_n \leq Y_n$, and since $EY < \infty$, then the sequence $\{Y_n\}$ is *uniformly integrable* and we obtain

$$\lim_{n \to \infty} E\tilde{Y}_n = EY. \tag{29}$$

Denote by $\{S_k\}$ the iid service times of the batch in the overloaded regime, with a generic element S. Then the following stochastic equality holds:

$$S =_{st} \sum_{i=1}^{B} 1_i S_i, \tag{30}$$

where indicator $1_i = 1$ if the capacity of the system is i, implying

$$ES = \sum_{i=1}^{B} ES_i P_i. \tag{31}$$

Note that, by construction, $\{Q_n\}$ is a Markov chain, and it turns out to be that the mean increment of the queue size (on the event (27)) is asymptotically negative. More exactly, as $n \to \infty$,

$$E\Delta_n = E(Q_{n+1} - Q_n) \to \lambda EX ES - EY < 0,$$

because, by the Little's law, $\lambda EX ES$ is the mean number of arrivals during (generic) service time S, and EY is the mean number of departed customers at any departure instant in the *saturated regime*. As a result, the negative drift of the Markov chain $\{Q_n\}$ implies $Q_n \not\Rightarrow \infty$, that is

$$\inf_k P(Q_{n_k} \leq D) \geq \varepsilon,$$

for some non-random $D, \varepsilon > 0$ and a subsequence $n_k \to \infty$. Because the input is Poisson then it now follows by the regeneration arguments that this Markov chain is positive recurrent, for detail see [1]. The proof is completed.

Remark. Unlike verifying aperiodicity and irreducibility of $\{Q_n\}$ which is required in classic stability analysis of a numerable Markov chain, in this regenerative analysis, it is enough to show the state $Q_n = 0$ is reached with a positive probability from any state, and thus this Markov chain is a regenerative process. Then the obtained above negative drift implies that this chain turns out to be positive recurrent that is $ET < \infty$.

5 Simulation

In this section we present some numerical results of the system considered in Sect. 2. The purpose of this simulation is to show, for a few distributions of retrial times and values of the threshold K, that the system remains positive recurrent under stability condition (2).

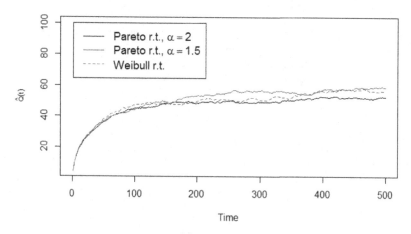

Fig. 1. The mean orbit size for different retrial time distribution vs. time; $\rho = 2.7$; $K = \infty$.

For this purpose we apply the discrete event simulation a system with $c = 3$ identical servers and one class of the retrial customers with service time S. We assume that the service time has Pareto distribution

$$F(x) = 1 - \left(\frac{x_0}{x}\right)^{\alpha}, \quad x \geq x_0. \tag{32}$$

(Below we use Pareto distribution (32) for the retrial time as well.) In all our examples parameters of the Pareto service times are $x_0 = 0.05$, $\alpha = 2$, implying

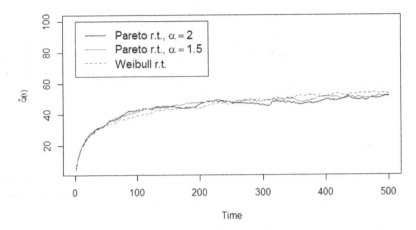

Fig. 2. The mean orbit size for different retrial time distribution vs. time; $\rho = 2.7$; $K = 50$.

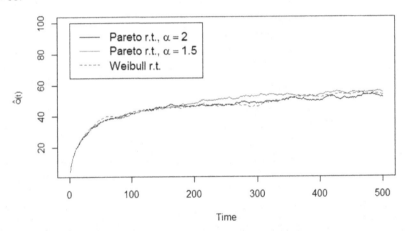

Fig. 3. The mean orbit size for different retrial time distribution vs. time; $\rho = 2.7$; $K = 40$.

$\mathsf{E}S = 0.1$. We also assume that the arrivals follow Poisson input with rate $\lambda = 27$ for stable system, implying $\rho = 2.7$, and $\lambda = 30$ for unstable system, implying $\rho = 3.0$.

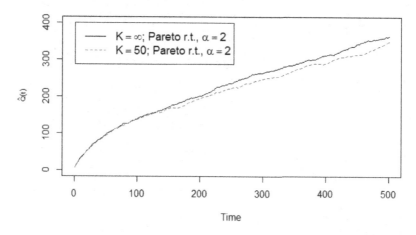

Fig. 4. The mean orbit size for different threshold values vs. time; $\rho = 3.0$.

Moreover we consider three different retrial time distributions but with the same mean 0.1. The first case is Pareto distribution with parameters $x_0 = 0.05$, $\alpha = 2$ (as the service time). The second case is Pareto distribution with parameters $x_0 = 1/30$, $\alpha = 1.5$, and the third case is Weibull distribution

$$F(x) = 1 - \exp\left\{ -\left(\frac{x}{b}\right)^a \right\}, \ x \geq 0, \tag{33}$$

with the shape parameter $a = 0.9$ and the scale parameter $b = 0.095$. Recall that the retrial times become exponential if the orbit size $Q(t) \geq K$.

Figure 1, 2 and 3 present the sample mean orbit size $\widehat{Q}(t)$ based on 1500 sample paths of the orbit size $Q(t)$, provided the system is stable, $\rho = 2.7$. At that the plot on Fig. 1 represents a system with no switching retrial distribution which corresponds to $K = \infty$. This result demonstrates an intuitive result that the system with classic retrials and general retrial time distribution also remains stable under condition (2). This result however has not a full complete proof, particular cases see in [10–12]. On the other hand, Fig. 2, 3 correspond to the cases with thresholds $K = 50$ and $K = 40$, respectively. Figure 4 illustrates the dynamics of orbit for $\rho = 3.0$ implying that the sample mean orbit size $\widehat{Q}(t)$ increases unlimitedly. Thus the simulation results are consistent with the theoretical findings.

Now we present some numerical results related to the system considered in Sect. 4 which verify the positive recurrence under stability condition (26).

We assume that the size X of the arriving batches has mean $\mathsf{E}X = 5$ and is either binomial

$$\mathsf{P}(X = i) = \binom{n-1}{i-1} \cdot p^{i-1}(1-p)^{n-i}, \ i = 1, \ldots, n,$$

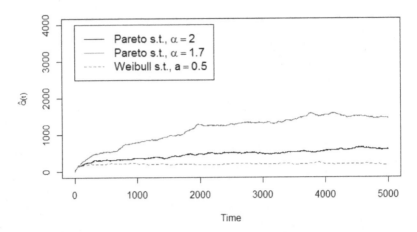

Fig. 5. The (sample) mean queue size for different service time distribution vs. time; $\rho = 0.9$; binomial batch size X.

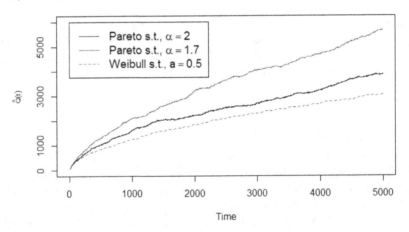

Fig. 6. The (sample) mean queue size for different service time distribution vs. time; $\rho = 1.0$; binomial batch size X.

with parameters $p = 4/19$ and $n = 20$, or geometric

$$P(X = i) = (1 - p)^{i-1}p, \quad i = 1, 2, \ldots$$

with $p = 0.2$. The service batch size distribution P_i is binomial

$$P_i = P(Y = i) = \binom{B - 1}{i - 1} \cdot p^{i-1}(1 - p)^{B-i}, \quad i = 1, \ldots, B,$$

with parameters $p = 9/19$ and $B = 20$, implying the mean $EY = 10$.

We also consider, for each batch size i, three service time distribution with the same means $ES_i = 0.02\,i$, where $i = 1, \ldots, B$. More exactly, we consider

Pareto distributions (32) with parameters $x_0 = 0.01\,i$, $\alpha = 2$, and with $x_0 = 0.0082\,i$, $\alpha = 1.7$. Moreover we use the Weibull service time distribution (33) with the parameters

$$a = 0.5,\ b = \frac{0.02\,i}{\Gamma(3)},\ \ i = 1, \ldots, B,$$

where Γ is the Gamma function. Under these parameters the traffic intensity is

$$\rho = \lambda \frac{\mathsf{E}X}{\mathsf{E}Y} \sum_{i=1}^{B} \mathsf{P}_i\, 0.02\,i = \lambda \mathsf{E}X\, 0.02 = 0.1\,\lambda.$$

We consider $\rho = 0.9$ and $\rho = 1.0$. Figure 5 presents the sample mean queue size $\widehat{Q}(t)$ based on 500 sample paths of the queue size for binomial arrival batch size and $\rho = 0.9$, implying positive recurrence. Figure 6 illustrates an unstable system ($\rho = 1.0$) where the sample mean queue size $\widehat{Q}(t)$ increases unlimitedly. The results for geometric arrival batch sizes are quite similar and omitted.

6   Conclusion

We consider several queueing models and analyze stability using the regenerative approach. The first system is a single-class multiserver $M/G/c$-type system with classical retrials and the threshold-dependent retrial time distribution. The second model is a multiserver system with a set of the assistance servers, and the last model is a batch-arrival system with batch-size-dependent service.

The regenerative method allows to obtain the new and short proofs of the known stability conditions of these systems. Some numerical examples, obtained by simulation, are included as well which confirm the theoretical results.

Acknowledgement. The paper was published with the financial support of the Ministry of Education and Science of the Russian Federation as part of the program of the Moscow Center for Fundamental and Applied Mathematics under the agreement № 075-15-2022-284.

References

1. Morozov, E., Steyaert, B.: Stability Analysis of Regenerative Queueing Models. Springer, Cham (2021). https://doi.org/10.1007/978-3-030-82438-9
2. Morozov, E. V., Rogozin, S. S.: Stability condition of multiclass classical retrials: a revised regenerative proof. Proceedings of the CITDS 2022 : 2022 IEEE 2nd Conference on Information Technology and Data Science, Hungary, Debrecen (2022)
3. Klimenok, V.I., Dudin, A.N., Vishnevsky, V.M., Semenova, O.V.: Retrial BMAP/PH/N queueing system with a threshold-dependent inter-retrial time distribution. Mathematics **10**(2), 269 (2022). https://doi.org/10.3390/math10020269
4. Dudin, A., Dudina, O., Dudin, S., Gaidamaka, Y.: Self-service system with rating dependent arrivals. Mathematics **10**(3), 297 (2022). https://doi.org/10.3390/math10030297

5. Pradhan, S., Gupta, U.C.: Stationary distribution of an infinite-buffer batch-arrival and batch-service queue with random serving capacity and batch-size-dependent service. Int. J. Oper. Res. **40**(1), 1–31 (2021)
6. Asmussen, S.: Applied Probability and Queues. Wiley, Hoboken (1987)
7. Morozov, E.: The tightness in the ergodic analysis of regenerative queueing processes. Queue. Syst. **27**, 179–203 (1997)
8. Morozov, E.: Weak regeneration in modeling of queueing processes. Queue. Syst. **46**, 295–315 (2004)
9. Feller, W.: An Introduction to Probability Theory, vol. 2. Wiley, New York (1971)
10. Morozov, E., Nekrasova, R.: Stability conditions of a multiclass system with NBU retrials. In: Phung-Duc, T., Kasahara, S., Wittevrongel, S. (eds.) QTNA 2019. LNCS, vol. 11688, pp. 51–63. Springer, Cham (2019). https://doi.org/10.1007/978-3-030-27181-7_4
11. Chakravarthy, S.R.: Analysis of MAP/PH/c retrial queue with phase type retrials - simulation approach. Mod. Prob. Methods Anal. Telecommun. Netw. **3**, 7–49 (2013)
12. Shin, Y. W., Moon D. H.: Approximation of PH/PH/c retrial queue with PH-retrial time. Asia-Pac. J. Oper. Res. **31**, 1440010 (2014) https://doi.org/10.1142/S0217595914400107

Speed-Up Simulation for Reliability Analysis of Wiener Degradation Process with Random Failure Threshold

Alexandra Borodina[1,2(✉)] [iD]

[1] Institute of Applied Mathematical Research of the Karelian Research Centre of RAS, Petrozavodsk, Russia
borodina@krc.karelia.ru
[2] Petrozavodsk State University, Petrozavodsk, Russia
musen@cs.pertsu.ru

Abstract. This paper proposes an alternative approach to the analysis and modeling of the reliability of degradation systems with rare failures. We assume that the degradation measure of the multiphase system is described by the Wiener process and instantaneous failure occurs when the process reaches a given level. Let us consider a model of a multiphase system, where the degree of degradation is described by a Wiener process and an instantaneous failure occurs when the process reaches a given level. The failure threshold value can be either a constant or a random variable that takes values on the positive semiaxis or on the fixed interval. Evaluation of the failure probability and other characteristics of the system is crucial for optimal control, however, for complex multi-phase systems with multiple failure thresholds, it is often impossible to find an analytical solution.

For the process of regenerative degradation, instead of the standard Monte-Carlo procedure, it is proposed to use an accelerated version of the construction of regeneration cycles. The splitting of the Wiener process trajectory occurs at the moment of transition to the next phase of the degradation. This approach is useful when failures are rare.

Keywords: Degradation process · Wiener process · Failure probability · Reliability analysis · Lifetime · Speed-up simulation · Regenerative simulation · Splitting method

1 Introduction

The effective methods for analysis of degradation and shock systems is a key challenge in the development and implementation of modern highly reliable

The study was carried out under state order to the Karelian Research Centre of the Russian Academy of Sciences (Institute of Applied Mathematical Research KarRC RAS).

technologies. The nature of these systems are not monotonic, so that appropriate models should adequately describe the behavior of systems with multiphase degradation when the degradation rate changes significantly and simultaneously under the influence of some random environment. In the overview of currently popular models, one can see manifold combinations of processes with independent increments like the Wiener or Gamma process with different system failure thresholds and imitation of a random environment (for istance, see [1–3]). More complex models of two-stage or multi-stage degradation include Gamma-Gamma, Wiener-Wiener, Gamma-Wiener degradation models are also considered increasingly.

The models become more complex, but analytical methods are less available. In this regard, we propose to look for more efficient alternatives for degradation simulation than the naive Monte Carlo method, which is nevertheless still popular in a large number of modern works, despite its known inefficiency for rare failures [4,5].

Previously in the papers [6–8] were proposed and tested variance reduction methods (conditional Monte-Carlo, Importance sampling) and regenerative splitting algorithms for a degradation process model describing the thickness of an anticorrosion coating for heavy-tailed and light-tailed phase distributions. For a more detailed consideration of modern speed-up simulation methods, one can refer to the sources [9,10].

In this paper, we propose an accelerated simulation method based on splitting and regenerative techniques for the Wiener degradation process. In addition, the system is assumed to be multi-phase with a random failure threshold.

The outline of the paper is as follows. Firstly we describe the degradation model in detail, define the main metrics of Wiener degradation measure and phase distributions. In Sect. 3 the regenerative structure of the degradation process is discussed and highlighted the main analytical results for the random value of the failure threshold. Finally, in Sect. 4 the used speed-up approach was presented for designing the regeneration cycles of degradation process and estimating system reliability. Also Sect. 4 shows the results of a comparative analysis between speed-up and naive methods for exponential and Beta distributed failure threshold.

2   Preliminaries

Earlier in [6–8] the degradation model with gradual and instantaneous failures was considered, where the degradation process $X := \{X(t), t \geq 0\}$ with a finite state space $E = \{0, 1, \ldots, L, \ldots, M, \ldots, K; F\}$ was described as an accumulation process with independent but not necessarily identically distributed increments. Following the existing notation, now we will consider a more generalized model of the degradation process in this paper. Let, as before, the set E describe the degradation stages (phases) of the system and consider the two-threshold policy $(K, L), K \geq L$, which means that the system is restored in stage K (preventive repair stage) and then proceeds to stage L (*stage of return*).

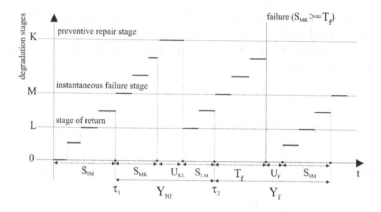

Fig. 1. Change of degradation phases for two types of regeneration cycles.

As it is shown in Fig. 1, the process X starting in state $X(0) = 0$ goes through $K - 1$ intermediate stages of degradation. Between stages 0 and K there is the stage $M \geq L$ (*instantaneous failure stage*) from which failure is possible.

Each stage has its own duration, so let T_i be the transition time from i to $i+1$ stage. Random variables (r.v.) T_i are independent but not necessarily identically distributed.

Note that

$$S_{i,j} := \sum_{k=i}^{j-1} T_k, \quad 0 \leq i \leq K - 1, \, j > i,$$

is the transition time from state j to state i.

In this paper, unlike the previous model (for instance, see [6]), it is assumed that the degradation measure of system between stages 0 and K is described by the Wiener process with linear drift,

$$W(t) = w_0 + \mu t + \sigma B(t), \quad 0 \leq t \leq \min\{T_c, S_{M,K}\}, \tag{1}$$

where w_0 is the initial value of the degradation performance, μ is a drift parameter, σ is a diffusion parameter, $B(t)$ is a standard Brownian motion. Thus, $W(t)$ is the *degradation measure* to characterize the system failure phase F. Without loss of generality, we will further consider the case $\sigma = 1$.

In the proposed model, the acceleration of the degradation process occurs due to a decrease in the residence time T_i in the phases. To do this, consider the case when the independent times T_j are distributed exponentially with parameters λ_j (hereinafter, this fact will be denoted by $T_j \sim Exp(\lambda_j)$) with density

$$f_{T_j}(x) = \lambda_j e^{-\lambda_j x}, \, x \geq 0,$$

where the heterogeneity is ensured by the condition

$$\lambda_0 < \cdots < \lambda_{K-1}. \tag{2}$$

The probability density function (pdf) of $S_{i,i}$ can be obtained through the Fourier transform for the characteristic function $\varphi_{S_{i,j}}(t)$.

Consider exponential distribution as a special case of gamma distribution with characteristic function

$$\varphi_{T_j}(t) = (1 - \frac{it}{\lambda_j})^{-1}, \tag{3}$$

then the independence of T_i entails

$$\varphi_{S_{i,j}}(t) = \mathbb{E}e^{it(T_i + \cdots + T_{j-1})} = \mathbb{E}e^{itT_i} \cdots \mathbb{E}e^{itT_{j-1}} = \prod_{k=i}^{j-1}(1 - \frac{it}{\lambda_k})^{-1}. \tag{4}$$

Further, by induction (for more details see [11]), one can show that

$$\varphi_{S_{i,j}}(t) = \sum_{k=i}^{j-1}\varphi_{T_k}(t) \prod_{l=i, l \neq k}^{j-1} \frac{\lambda_l}{\lambda_l - \lambda_k}. \tag{5}$$

Denoting

$$\prod_{l=i, l \neq k}^{j-1}(\lambda_l - \lambda_k) = A_k, \tag{6}$$

from (5) we have

$$\varphi_{S_{i,j}}(t) = \prod_{k=i}^{j-1}\lambda_k \sum_{k=i}^{j-1}\varphi_{T_k}(t)\frac{1}{\lambda_k A_k}. \tag{7}$$

Now, through the Fourier transform, we obtain the pdf

$$f_{S_{i,j}}(x) = \frac{1}{2\pi} \int_{-\infty}^{\infty} e^{-itx}\varphi_{S_{i,j}}(t)dt$$

$$= \prod_{k=i}^{j-1}\lambda_k \sum_{k=i}^{j-1}f_{T_k}(x)\frac{1}{\lambda_k A_k} = \prod_{k=i}^{j-1}\lambda_k \sum_{k=i}^{j-1}\frac{e^{-\lambda_k x}}{A_k}. \tag{8}$$

Instantaneous failure event $\{W(t) \geq c\}$ associated with the achievement by the process of a certain (generally random) threshold c at the first failure-point T_c:

$$T_c = \inf\{t|W(t) \geq c, t > 0\}. \tag{9}$$

This means that the process enters the *final state F*.

It is well known (see [4]) that for constant threshold c the random value T_c follows a transformed inverse Gaussian distribution $T_c \sim IG(c/\mu, c^2)$ with pdf

$$f_{T_c}(t) = \frac{c - w_0}{\sqrt{2\pi}t^{3/2}}e^{-\frac{(c - w_0 - \mu t)^2}{2t}}, \tag{10}$$

and cumulative distribution function (cdf)

$$F_{T_c}(t) = \Phi\left(\frac{w_0 - c + \mu t}{\sqrt{t}}\right) + e^{2\mu(c - w_0)}\Phi\left(\frac{w_0 - c - \mu t}{\sqrt{t}}\right), \tag{11}$$

where $\Phi(x) = \dfrac{1}{\sqrt{2\pi}} \displaystyle\int_{-\infty}^{x} e^{-t^2/2} dt$ is a standard normal distribution function. From the property of Wiener process the expectation of the first hitting time is

$$\mathbb{E}[T_c] = \frac{c - w_0}{\mu},$$

where threshold c assumed constant.

It is clear from the Fig. 1, that after the stage M either *instantaneous failure* may happen during a random period T_c, or transition to the *preventive repair stage* occurs during the time

$$S_{M,K} = \sum_{i=M}^{K-1} T_i.$$

Preventive repair of the system at stage K takes random time $U_{K,L}$ and then the system returns to the stage L and then again the degradation performance measurement of the system follows a Wiener process (1) with initial value $w_0 = W(S_{0,L})$.

If a instantaneous failure occurs, then the system after disaster recovery in random time U_F returns to phase 0, where $W(0) = 0$.

In real systems, the degradation process tends to accelerate over time or degradation rate is influenced by random working environment. In degradation models, as a rule, this behavior is achieved by changing failure threshold c or drift parameter μ from phase to phase. From the point of view of reliability problems, the threshold c is most likely not a constant, then further cases will be considered random cases of c. For the experiments, it is supposed to use the exponential distribution case and the case of the Beta distribution on the fixed interval $[c_1, c_2]$.

For further analysis, we will use the following cdf notation:

$$F_{T_i}(t) = \mathbb{P}(T_i \leq t); \quad F_f(t) = \mathbb{P}(f \leq t);$$
$$F_{U_F}(t) = \mathbb{P}(U_F \leq t); \quad F_{U_{K,L}}(t) = \mathbb{P}(U_{K,L} \leq t);$$
$$F_{S_{i,j}}(t) = \mathbb{P}(S_{i,j} \leq t) = F_{S_{i,j-1}} * F_{T_j}(t) = \int_{0-}^{t} F_{S_{i,j-1}}(t-v) dF_{T_j}(v);$$
$$F_{S_{i,i+1}}(t) = F_{T_i}(t), \quad F_{S_{i,i}}(t) \equiv 0, \tag{12}$$

where $*$ means convolution, and similar indices for the pdf $f_{(.)}$ of distributions with condition that they exist.

3   Regenerative Structure of Degradation Process

Since the degradation process X is regenerative, then the simulation built on independent regeneration cycles will have the properties necessary for the quality of the estimation, such as unbiasedness and consistency. For reliability research, we are interested in the hitting times of the stage M, after which a failure may occur.

The basic process X is regenerative with moments

$$\tau_{n+1} = \inf\{Z_i > \tau_n : X(Z_i^+) = M\}, \; n \geq 0, \; \tau_0 := 0,$$

where Z_i is the the hitting time of the stage $i \geq 1$, and cycle lengths $Y_i = \tau_{i+1} - \tau_i$, $i \geq 1$ are independent and identically distributed. We might as well define cycles in terms of reaching any other phase of degradation, not necessarily M. We choose to follow phase M because after this point, the system is likely to fail. In the future, this will help save computational resources on the simulation procedure up to stage M.

In general, the regenerative method advantages are based on the time-averaged properties of the process as a consequence of the special case of the Central Limit Theorem and the Renewal Reward Theorem.

Figure 1 illustrates the regenerative structure and the piecewise change of $X(t)$ degradation phases. As shown in Fig. 2 the underlying Wiener degradation measure process $W(t)$ either can reach the preventive repair phase K during the cycle without failure, or failure will occur in a period T_c.

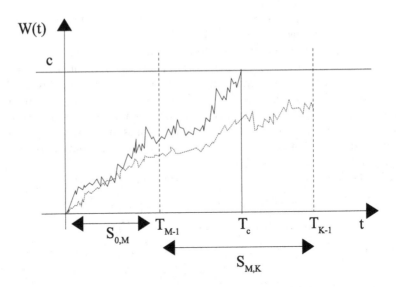

Fig. 2. Sample paths for the degradation measure process in two types of cycle.

Since the degradation measure W corresponds the Wiener law, then the process W satisfies the Gaussian stationarity:

$$W(t) - W(s) \sim \mathbb{N}(0, t - s), \quad 0 \leq s < t < \infty, \; \sigma = 1,$$

with continuous sample paths and

$$cov(W(s), W(t)) = \min(s, t).$$

Moreover W is a time-homogeneous strong Markov process, so for any finite stopping time τ and for all $t > 0$, holds:

$$(W(\tau + t)|W(u), u \geq \tau) \sim (W(\tau + t)|W(\tau)).$$

Thus, any change-phase moment of the degradation process will be Markovian point, and regenerative splitting can be used, as was previously proposed in [7,8] for a simpler degradation model.

Next, we will consider the most significant characteristics of regeneration cycles that are important for the reliability analysis of the degradation systems.

The typical length of a regeneration cycle is

$$Y = Y_F \cdot I_{\{T_c \leq S_{M,K}\}} + Y_{NF} \cdot I_{\{T_c > S_{M,K}\}}, \tag{13}$$

where for failed cycle length is $Y_F = T_c + U_F + S_{0,M}$, $Y_{NF} = S_{M,K} + U_{K,L} + S_{L,M}$ otherwise and indicator function $I_{\{A\}} = 1$ if the event A happened.

An indicative characteristic of the system reliability is the so-called *lifetime*, which can be written as

$$T = \min\{T_c, S_{M,K}\}. \tag{14}$$

The reliability of a system is related to the *reliability function* $R(t)$, which is defined as the probability that at time t the system is still operational. Due to the independence of random variables (r.v.) T_c and $S_{M,K}$ we can get

$$\begin{aligned} R(t) &= \mathbb{P}(T > t) = \mathbb{P}(\min\{T_c, S_{M,K}\} > t) \\ &= \mathbb{P}(T_c > t, S_{M,K} > t) = \overline{F_{T_c}}(t)\overline{F_{S_{M,K}}}(t), \end{aligned} \tag{15}$$

where $\overline{F_\xi}(t) = 1 - F_\xi(t)$ denotes tail distribution function of r.v. ξ and cdf and pdf in (11) and (10) for constant threshold f respectively.

Based on the regenerative structure of process X, it is possible to calculate the probability of a instantaneous failure

$$\begin{aligned} p_F &= \mathbb{P}(S_{M,K} \geq T_c) = \int_0^\infty \overline{F_{S_{M,K}}}(t) f_{T_c}(t) dt \\ &= \int_0^\infty \frac{c - w_0}{\sqrt{2\pi} t^{3/2}} e^{-\frac{(c - w_0 - \mu t)^2}{2t}} \overline{F_{S_{M,K}}}(t) dt, \end{aligned} \tag{16}$$

as well as other characteristics of the system functioning, such as the average length of the regeneration cycle is

$$\begin{aligned} \mathbb{E}[Y] &= \mathbb{E}[\min\{T_c, S_{M,K}\}] \\ &+ (\mathbb{E}[U_F] + \mathbb{E}[S_{0,M}])p_F \\ &+ (\mathbb{E}[U_{K,L}] + \mathbb{E}[S_{L,M}])(1 - p_F), \end{aligned} \tag{17}$$

From (13) it turns out that the average conditional length of cycle with failure is

$$\begin{aligned} \mathbb{E}[Y_F] &= \mathbb{E}[Y|T_c \leq S_{M,K}] \\ &= \frac{1}{p_F} \int_0^\infty z f_{T_c}(z) \overline{F_{S_{M,K}}}(z) dz + \mathbb{E}[U_F] + \mathbb{E}[S_{0,M}], \end{aligned} \tag{18}$$

and the average conditional length of the cycle without failure is calculated in the same way

$$\mathbb{E}[Y_{NF}] = \mathbb{E}[Y|S_{M,K} < T_c]$$
$$= \frac{1}{1 - p_F} \int_0^\infty z \, f_{S_{M,K}}(z) \, \overline{F_{T_c}}(z) \, dz + \mathbb{E}[U_{K,L}] + \mathbb{E}[S_{L,M}]. \quad (19)$$

Then the average lifetime of the system is:

$$\mathbb{E}[T] = \int_0^\infty R(t)dt = \int_0^\infty \overline{F_{T_c}}(t) \overline{F_{S_{M,K}}}(t)dt$$
$$= \int_0^\infty \left[\Phi\left(\frac{c - w_0 - \mu t}{\sqrt{t}}\right) - e^{2\mu(c - w_0)} \Phi\left(\frac{w_0 - c - \mu t}{\sqrt{t}}\right) \right] \overline{F_{S_{M,K}}}(t)dt. \quad (20)$$

Finally note that (8) implying

$$\overline{F_{S_{M,K}}}(t) = \mathbb{P}(S_{M,K} \geq t) = \int_t^\infty f_{S_{M,K}}(z)dz = \prod_{j=M}^{K-1} \lambda_j \sum_{j=M}^{K-1} \frac{e^{-\lambda_j t}}{\lambda_j A_j}, \quad (21)$$

where $A_j = \prod_{l=M, l \neq j}^{K-1} (\lambda_l - \lambda_j)$.

Next, we consider cases for random values of the threshold f. Recall that this option is more natural than the constant one.

3.1   The Case of Nonnegative Random Failure Threshold

For nonnegative random threshold c with pdf $f_c(t)$ from expression (10) we obtain

$$f_{T_c}(t) = \int_0^\infty f_{T_c}(t|c = z)f_c(z)dz = \int_0^\infty \frac{z - w_0}{\sqrt{2\pi}t^{3/2}} e^{-\frac{(z - w_0 - \mu t)^2}{2t}} f_c(z)dz. \quad (22)$$

Then, more precisely, due to (11) and (8) the reliability function in (15) becomes

$$R(t) = \int_t^\infty f_{S_{M,K}}(z)dz \int_0^\infty (1 - F_{T_c}(t|c = z))f_c(z)dz$$
$$= \prod_{j=M}^{K-1} \lambda_j \sum_{j=M}^{K-1} \frac{e^{-\lambda_j t}}{\lambda_j A_j} \int_0^\infty \left[\Phi\left(\frac{z - w_0 - \mu t}{\sqrt{t}}\right) - \right.$$
$$\left. - e^{2\mu(z - w_0)} \Phi\left(\frac{w_0 - z - \mu t}{\sqrt{t}}\right) \right] f_c(z)dz. \quad (23)$$

Finally the expression for the probability of failure corresponds to

$$p_F = \int_0^\infty \overline{F_{S_{M,K}}}(t) \int_0^\infty f_{T_c}(t|c = z)f_c(z)dzdt$$
$$= \prod_{j=M}^{K-1} \lambda_j \iint_0^\infty \sum_{j=M}^{K-1} \frac{e^{-\lambda_j t}}{\lambda_j A_j} \frac{z - w_0}{\sqrt{2\pi}t^{3/2}} e^{-\frac{(z - w_0 - \mu t)^2}{2t}} f_c(z)dzdt. \quad (24)$$

By setting the appropriate distribution for threshold like exponential $f \sim Exp(\nu)$ or the Webull type it will be possible to obtain the values for the formulas above numerically only, due to their complexity for analytical calculations.

3.2   The Case of Random Threshold from Finite Interval

Another option for setting the threshold for failures assumes that the random level for the degradation measure belongs to a certain interval with fixed boundaries $[c_1, c_2]$. Most often, Uniform and Beta distributions are used for these purposes. For generalized Beta distributed random value c denote pdf

$$f_c(t) = \frac{1}{B(\alpha, \beta)} \frac{(t - c_1)^{\alpha-1}(c_2 - t)^{\beta-1}}{(c_2 - c_1)^{\alpha+\beta-1}},$$

where for Beta-function we have $B(\alpha, \beta) = \dfrac{\Gamma(\alpha)\Gamma(\beta)}{\Gamma(\alpha + \beta)}$.

Then for main reliability measures of degradation implies that

$$f_{T_c}(t) = \int_{c_1}^{c_2} \frac{z - w_0}{\sqrt{2\pi}t^{3/2}} e^{-\frac{(z - w_0 - \mu t)^2}{2t}} f_c(z) dz, \tag{25}$$

and

$$R(t) = \prod_{j=M}^{K-1} \lambda_j \sum_{j=M}^{K-1} \frac{e^{-\lambda_j t}}{\lambda_j A_j} \int_{c_1}^{c_2} \left[\Phi\left(\frac{z - w_0 - \mu t}{\sqrt{t}} \right) - \right.$$
$$\left. - e^{2\mu(z - w_0)} \Phi\left(\frac{w_0 - z - \mu t}{\sqrt{t}} \right) \right] f_c(z) dz, \tag{26}$$

and (16) becomes

$$p_F = \prod_{j=M}^{K-1} \lambda_j \int_0^\infty \int_{c_1}^{c_2} \sum_{j=M}^{K-1} \frac{e^{-\lambda_j t}}{\lambda_j A_j} \frac{z - w_0}{\sqrt{2\pi}t^{3/2}} e^{-\frac{(z - w_0 - \mu t)^2}{2t}} f_c(z) dz dt. \tag{27}$$

In fact, to describe real systems where degradation models are required, even more complex constructions are considered. For instance see [3,12], with two or more degradation thresholds.

For example, for our model, we could assume a change of failure threshold at each moment $T_i, i \in [M - 1, K - 1]$, when the degradation phase changes. So, for the new multy-threshold degradation model with constant or random thresholds $[c_M, \ldots, c_{K-1}]$ the lifetime is

$$T = \inf\{t | W(t) \geq Y(t)\}, \quad Y(t) = \sum_{k=M}^{K-1} c_i I_{\{T_{k-1} < t \leq T_k\}}.$$

The analytical solution for the reliability function in this case is quite cumbersome, may involve the calculation of multiple integrals, and cannot be represented explicitly.

Nevertheless, for the regenerative simulation algorithm proposed below, such a complication is practically not difficult and does not differ too much from the algorithm for one threshold (constant or random).

4   Reliability Simulation Procedures

Due to the complexity of the analytical results in the problems of finding the values of the reliability function, simulation modeling by the naive Monte-Carlo method is usually used (for instance, see [4,5,12–14] and many others). Previous papers have already discussed the apparent inefficiency of the Monte-Carlo method for highly reliable systems where failures are rare events.

Previous papers [6–8] have already discussed the apparent inefficiency of the Monte-Carlo method for highly reliable systems where failures are rare events. Besides, the variance reducing methods for probability of failure estimation were also discussed, and the splitting procedure was successfully tested to speed up the system characteristics estimation.

Within this work, we propose a similar approach, but for constructing an empirical reliability function of a generalized model with a Wiener process of changing the degradation measure. As for model in [8], the most advantageous splitting option for the process we are interested is *fully branching* case.

Splitting procedure starts at stage M with the start of R_{M-1} trajectories, the splitting levels l_i, $M \le i \le K - 1$ are fixed and strictly correspond to the degradation phases. At each level, we split the trajectory of the process $W(t)$ into R_i copies. So, each original trajectory generates

$$D = R_M \cdots R_{K-1}$$

dependable subpaths called *group of cycles*. We will refer to these groups as

$$X_1^*, \ldots, X_{R_{M-1}}^*.$$

Then splitting procedure give us the set of cycles in the first group:

$$X_1^* = \{G^{(M)}, \ldots, G^{(N^*)}\}.$$

By successively applying it to all trajectories starting from the point T_{M-1} and completing the trajectories to a whole cycle (with length Y_F or Y_{NF}), we obtain a full set of cycles much faster than by standard Monte-Carlo simulation.

We do not need to build entire regeneration cycles for failure probability estimation. However, to obtain all other metrics, including the reliability function, this is necessary.

In order not to duplicate a similar description of the algorithm proposed in the paper [8], we will consider only the key steps that are essential in the context of the new model. The main idea is to turn a branching structure into a sequential set of standard regeneration cycles, while maintaining time intervals. In the framework of the given in [8] notation:

g_i - generation of $l_i, i \in [M, K - 1]$ threshold;

n_i - the number of trajectories starting from the level l_{i-1} and reaching the level l_i, $i \in [M + 1, K - 1]$, $n_M = R_{M-1}$;

n_i - the number of trajectories starting from the level l_{i-1} and reaching the level l_i, $i \in [M + 1, K - 1]$, $n_M = R_{M-1}$;

N^* - the maximum stage number reached by the trajectories before the failure (if it occurs) $N^* = \max\{i : n_i > 0, i \in [M, K-1]\}$;

$t_{k_i}^{(i)}$ - the splitting moment of the bunch number $k_i \in [1, n_i]$ at the level $l_i, i \in [M+1, K-1]$;

$H_{k_i,j}^{(i)}(t)$ - accumulated time spent in degradation stages at time t by jth process copy that started from l_i threshold at the moment $t_{k_i}^{(i)}$;

$\eta_{k_i,j}^{(i)}$ - the instantaneous failure moment for jth copy of the process after the hitting l_i threshold;

$Z_{k_i,j}^{(i+1)}$ - the moment of the next event, namely, either a transition to the next stage or a failure occur at the level l_i,

for every level l_i holds

$$H_{k_i,j}^{(i)}(t_{k_i}^{(i)}) = \sum_{p=M}^{i-1} T_p, \quad i \geq M+1; \tag{28}$$

$$\eta_{k_i,j}^{(i)} = \min\{t > t_{k_i}^{(i)} : W(H_{k_i,j}^{(i)}(t)) \geq c\}; \tag{29}$$

$$t_{k_i,j}^{(i+1)} = t_{k_i}^{(i)} + T_{i,j}, \quad T_{i,j} =_{st} T_i \sim F_{T_i}; \tag{30}$$

$$Z_{k_i,j}^{(i+1)} = \min\{\eta_{k_i,j}^{(i)}, t_{k_i,j}^{(i+1)}\}. \tag{31}$$

It is considered that the trajectory *belongs to the generation* g_i if it started from a level l_i and either reached the level l_{i+1} or the failure occurred, so:

$$g_i = \{H_{k_i,j}^{(i)}, j \geq M\}, \tag{32}$$

$$H_{k_i,j}^{(i)} = \{H_{k_i,j}^{(i)}(t), t_{k_i}^{(i)} \leq t \leq Z_{k_i,j}^{(i+1)}\}. \tag{33}$$

All of D cycles in each group $X_i^*, i \in [1, R_{M-1}]$ constructed from trajectories of Wiener process $W(t)$, that either met failure threshold at one of the degradation phases (cycles with the failure Y_F), or otherwise reached stage K (cycles without failure Y_{NF}).

Then, the total number of the failures in the ith group is

$$B_i = \sum_{j=(i-1)\cdot D+1}^{i \cdot D} I^{(j)}, i = 1, \ldots, R_{M-1},$$

where $I^{(j)} = 1$ for the cycle with failure ($I^{(j)} = 0$, otherwise). Sequence $\{I^{(j)}, j \geq 1\}$ is discrete D-dependent regenerative with constant cycle length D and regeneration instants are $\{i \cdot D\}, i \in [1, R_{M-1}]$.

The regenerative structure gives the following unbiased and strongly consistent estimator $\widehat{p}_F$:

$$\widehat{p}_F = \frac{\sum_{j=1}^{R_{M-1}} B_j}{R_{M-1} \cdot D} \to \frac{\mathbb{E} \sum_{j=1}^{D} I^{(j)}}{D} = p_F \tag{34}$$

as $R_{M-1} \to \infty$ w.p.1. and provides a basis for building $100(1 - \delta)\%$ confidence interval for p_F based on the *regenerative variant of Central Limit Theorem*

$$\left[\widehat{p}_F \pm \frac{z(\delta)\sqrt{v_n}}{\sqrt{n}} \right] \tag{35}$$

where quantile $z(\delta)$ satisfies $\mathbb{P}[N(0,1) \leq z(\delta)] = 1 - \delta/2$, and

$$v_n = \frac{n^{-1} \sum_{i=1}^{n} [B_i - \widehat{p}_F D]^2}{D^2} \tag{36}$$

is a weakly consistent estimator of $\sigma^2 = \mathbb{E}[B_1 - p_F D]^2 / D^2$ if $\mathbb{E}(B_1 - \gamma \alpha_1)^2 < \infty$.

Note that, if we want to calculate the system reliability $R(t)$ the special array $Z_a rr$ can be formed from the moments $Z_{k_i,j}^{(i+1)}$ during the splitting procedure. When the splitting is stopped the reliability at the moment t estimated as $\widehat{R}_D(t) = \dfrac{D - n(t)}{D}$, where $n(t)$ is the points number in sample $Z_a rr$ before time t.

Numerical comparison of two simulation procedures: Monte-Carlo (MC) and regernerative splitting (RS) was made with respect to the standard evaluation quality criteria: relative error RE

$$RE[\widehat{p}_F] = \frac{\sqrt{Var[\widehat{p}_F]}}{\mathbb{E}[\widehat{p}_F]},$$

and simulation time over the equal number n of regeneration cycles.

All numerical tests were executed on ultrabook HP ENVY Intel(R) Core(TM) i3 7100U 2.4 GHz processor with 4 GB of RAM, running Windows 10. The algorithms were developed in C++ language. For the numerical calculation of the failure probability value by formulas (24) and (27), the Python 3 language libraries were used.

Define the phase-rate parameters $K = 15$, $M = 5$, $L = 2$, $\mu_F = 1.5$, $\mu_{K,L} = 2$ for exponential U_F and $U_{K,L}$, choose $w_0 = 0$, $\mu = 1$ For $T_j \sim Exp(\lambda_j)$ we will vary number of regeneration cycles n and sequence of values

$$\lambda_j = \lambda_{K-1} - (K - j - 1)s, \ j \in [0, K - 2],$$

where λ_{K-1} will be initialized before starting the splitting procedure, and the other values will be shifted by step s, thus the condition

$$\lambda_0 < \cdots < \lambda_{K-1}, \ \nu < \lambda_j, \ j = [0, K - 1],$$

for increasing the degradation rate is guaranteed.

Consider $c \sim Exp(\nu)$, $\nu = 0.001$, then the time for constructing of n degradation cycles for MC and RS shown in Table 1, where the p_F values are calculated according to the formula (24) numerically.

For the Beta-distributed threshold c with $\alpha = 1$, $\beta = 1$, $c_1 = 0.1$, $c_2 = 2$ the comparison of the simulation time for n cycles given in Table 2. The numarical value of failure probability given by formula (27). As expected for current model,

Table 1. Time (sec.) comparison: MC vs. RS $c \sim Exp(\nu)$

λ_{K-1}, s	n	p_F	t_{MC}	t_{RS}
1000, 50	10^4	$8.53 \cdot 10^{-5}$	0.056	0.019
10000, 500	10^4	$2.42 \cdot 10^{-6}$	0.815	0.037
10000, 50	10^5	$1.53 \cdot 10^{-7}$	351.0	27.31
50000, 5000	10^7	$1.18 \cdot 10^{-8}$	659.5	75.10

Table 2. Time (sec.) comparison: MC vs. RS $c \sim Beta(c_1, c_2)$

λ_{K-1}, s	n	p_F	t_{MC}	t_{RS}
1000, 50	10^4	$6.40 \cdot 10^{-4}$	0.041	0.023
5000, 50	10^4	$1.40 \cdot 10^{-5}$	0.541	0.031
10000, 500	10^5	$1.49 \cdot 10^{-6}$	287.4	25.01
10000, 100	10^7	$1.27 \cdot 10^{-7}$	415.0	68.05

the splitting method RS allows to build the same number of cycles faster than MC. Note that in the case of RS, the trajectories are partially duplicated and dependent.

Now in Tables 3 and 4 we can see a comparison of the relative error for both methods.

Table 3. RE comparison: MC vs. RS $c \sim Exp(\nu)$

λ_{K-1}, s	n	RE_{MC}	RE_{RS}
1000, 50	10^4	0.61	3.71
10000, 500	10^4	0.36	3.15
10000, 50	10^5	0.35	4.35
50000, 5000	10^7	0.24	4.02

The relative estimation error for the Monte-Carlo method is slightly better, but the splitting method also gives an acceptable estimation error.

It is important to note that the estimation accuracy is worse for the exponential failure threshold than for the Beta distribution.

In general, the results confirm a similar relationship between the estimation accuracy and the speed of constructing estimates between the two methods MC and RS for another model in [8].

It can be summarized that for estimating the parameters of low-failure systems, the splitting technique makes it possible to estimate the probability of a rare failure and other characteristics of the system by constructing complete cycles much faster than the naive Monte-Carlo method.

Table 4. RE comparison: MC vs. RS $c \sim Beta(c_1, c_2)$

λ_{K-1}, s	n	RE_{MC}	RE_{RS}
1000, 50	10^4	0.15	1.02
5000, 50	10^4	0.13	1.95
10000, 500	10^5	0.21	2.30
10000, 100	10^7	0.78	2.85

5   Conclusion

The proposed simulation algorithm for speed-up regenerative simulation allows building estimates of the main reliability characteristics of the system when analytical results cannot be obtained explicitly.

For systems with a large number of parameters, thresholds and distributions, proposed technique is also useful, especially in the case of systems with rare instantaneous failures.

For the Wiener model of the degradation process, analytical results were proposed for the case of multiple degradation phases and a random failure threshold.

For the exponential and Beta distributed failure threshold and the same number of cycles, we compared the accuracy of the estimates constructed by the standard Monte-Carlo and by splitting methods.

The experimental results showed that the proposed method in the region of not very rare probabilities shows less accuracy than the standard Monte-Carlo method. Nevertheless, it will not be possible to compare both methods on rarer failures of order 10^{-20} due to the unacceptably long running time of the Monte-Carlo method.

References

1. Jensen, H., Papadimitriou, C.: Reliability analysis of dynamical systems. In: Substructure Coupling for Dynamic Analysis, pp. 69–111 (2019)
2. Shahraki, A., Yadav, O., Liao, H.: A review on degradation modelling and its engineering applications. Int. J. Performabil. Eng. 13(3), 299–314 (2017)
3. Dong, Q., Cui, L.: Reliability analysis of a system with two-stage degradation using Wiener processes with piecewise linear drift. IMA J. Manag. Math. 32(1), 3–29 (2021). https://doi.org/10.1093/imaman/dpaa009
4. Dong, Q., Cui, L.: A study on stochastic degradation process models under different types of failure thresholds. Reliabil. Eng. Syst. Saf. 181, 202–212 (2019)
5. Yousefi, N., Coit,D., Song, S., Feng, Q.: Optimization of on-condition thresholds for a system of degrading components with competing dependent failure processes. Reliabil. Eng. Syst. Saf. 192 (2019). https://doi.org/10.1016/j.ress.2019.106547
6. Borodina, A., Lukashenko, O.: A variance reduction technique for the failure probability estimation. In: Proceedings of the First International Workshop on Stochastic Modeling and Applied Research of Technology (SMARTY), pp. 71–76 (2018). http://ceur-ws.org/Vol-2278/#paper08

7. Borodina, A., Lukashenko, O., Morozov, E.: On estimation of rare event probability in degradation process with light-tailed stages. In: Proceedings the 22-th International Conference on Distributed Computer and Communication Networks: Control, Computation, Communications (DCCN 2019), pp. 477–483 (2019)
8. Borodina, A., Tishenko, V.: On splitting scenarios for effective simulation of reliability models. Commun. Comput. Inf. Sci. **1337**, 353–366 (2020). https://doi.org/10.1007/978-3-030-66242-4_28
9. Rubinstein, R.Y., Kroese, D.P.: Simulation and the Monte Carlo method. John Wiley & Sons Inc., Hoboken (2017)
10. Rubinstein, R.Y., Ridder, A., Vaisman, R.: Fast Sequential Monte Carlo Methods for Counting and Optimization. John Wiley & Sons Inc, Hoboken (2014)
11. Bibinger, M.: Notes on the sum and maximum of independent exponentially distributed random variables with different scale parameters. arXiv preprint arXiv:1307.3945 (2013)
12. Gao, H., Cui, L., Kong, D.: Reliability analysis for a Wiener degradation process model under changing failure thresholds. Reliabil. Eng. Syst. Saf. **171**, 1–8 (2018)
13. Li, J., Zhang, Y., Wang, Z., Fu, H., Xiao, L.: Reliability analysis of the products subject to competing failure processes with unbalanced data. Eksploatacja i Niezawodnosc Maintenance and Reliability **18**(1), 98–109 (2016). https://doi.org/10.17531/ein.2016.1.13
14. Sun, Q., Ye, Z., Zhu, X.: Managing component degradation in series systems for balancing degradation through reallocation and maintenance. IISE Trans. **52**(7), 797–810 (2020)

Optimization and Stability of Some Discrete-Time Risk Models

Ekaterina V. Bulinskaya[✉][iD]

Lomonosov Moscow State University, Moscow 119991, Russia
ebulinsk@mech.math.msu.su

Abstract. An appropriate mathematical model is needed to study a real process or system. Dealing with such probability theory applications as communication, reliability, queueing, inventory, finance, insurance and many others, a researcher usually considers input-output models. Modern period in actuarial sciences is characterized by consideration of complex systems, interplay of cost and reliability approaches and wide use of computers, therefore it is desirable to treat discrete-time models. They can give approximation to corresponding continuous-time ones. Moreover, in some cases they provide a better description of a real-life situation. So, we consider three discrete-time models and investigate their optimal performance and stability.

Keywords: Stability · Reliability · Optimization · Discrete Time · Risk Models

1 Introduction

The models used in applications of probability theory are for the most part of input-output type. They are described by specifying the planning horizon, input and output processes and objective function, see, e.g., [6]. Giving different interpretations to input and output processes one can pass from one applied research domain to another. Thus, the results obtained in one domain can be used in others. Since insurance is the oldest domain of application, we are going to formulate all the problems in terms of insurance company performance. According to the choice of objective function, there exist two main approaches (cost and reliability ones), see, e.g., [7]. We consider three discrete-time models. The first one, illustrating the cost approach, describes the performance of insurance company using proportional reinsurance and bank loans. The second model treats functioning of the company employing investment in non-risky asset for a fixed period. Algorithms for calculation of finite-time and ultimate ruin probabilities are obtained. Thus, reliability approach is used in this case. The third model deals with the functioning of a company using investment in risky and non-risky assets combined with quota-share reinsurance. Optimal investment policy is established in the framework of cost approach. The survival probability is also studied.

Supported by the Russian Foundation for Basic Research, project 20-01-00487.

Discrete-time models became popular during the last decades. On the one hand, they give an approximation to corresponding continuous-time models (see, e.g., [21]), on the other, sometimes provide a better description of a real-life situation. One of the first papers concerning discrete models in ruin theory [24] appeared in 1988. The paper [26] gives review of discrete-time models considered until 2009. A review of recent results on discrete-time models is supplied, as well, in Sect. 5 of [6]. It is important to mention that for obtaining the optimal behavior of a risk model one can use such decisions as investment, bank loan, dividends and reinsurance. Necessary definitions and useful results one can find in textbooks [2, 23, 25, 28], see also papers [1, 4]- [17, 19, 20, 22, 29].

2 Discrete-Time Model with Reinsurance and Bank Loans

2.1 Model Description

Suppose that the claims arriving to insurance company are described by a sequence of independent identically distributed (i.i.d.) non-negative random variables (r.v.'s) $\{X_i, i \geq 1\}$. Here X_i is the claim amount during the i-th period (year, month or day). Let $F(x)$ be its distribution function (d.f.) having density $\varphi(x)$ and finite expectation. The company uses proportional reinsurance with quota α, that is, pays only αX if the claim amount is X, and bank loans. If a loan is taken at the beginning of period (before the claim arrival) the rate is k whereas the loan after the claim arrival is taken at the rate r with $r > k$. Our aim is to choose the loans in such a way that the additional payments entailed by loans are minimized. Denote by M the premium acquired by direct insurer (after reinsurance) during each period. If x is the initial capital then $f_1(x)$, the minimal expected additional costs during one period, are given by

$$f_1(x) = \min_{y \geq x}[k(y - x) + rE(\alpha X - y)^+], \tag{1}$$

here y is the company capital after the loan and $(\alpha X - y)^+ = \max(\alpha X - y, 0)$. Clearly, (1) can be rewritten in the form:

$$f_1(x) = -kx + \min_{y \geq x} G_1(y), \quad \text{with} \quad G_1(y) = ky + r\alpha \int_{y/\alpha}^{\infty} \bar{F}(s)\, ds. \tag{2}$$

Now let $f_n(x)$ be the minimal expected costs during n periods and β the discount factor for future costs, $0 < \beta < 1$. Then using the dynamic programming (see, e.g., [3]), one easily obtains the following relation:

$$f_n(x) = -kx + \min_{y \geq x} G_n(y), \quad \text{with} \quad G_n(y) = G_1(y) + \beta E f_{n-1}(y + M - \alpha X). \tag{3}$$

2.2 Optimization Problem

Theorem 1. *There exists an increasing sequence of critical levels $\{y_n\}_{n\geq 1}$ such that*

$$f_n(x) = -kx + \begin{cases} G_n(y_n), & if \ x \leq y_n, \\ G_n(x), & if \ x > y_n. \end{cases} \tag{4}$$

Let $\bar{y}$ satisfy $H(\bar{y}) = 0$ where $H(y) = G_1'(y) - k\beta$, then $y_n \leq \bar{y}$.

Proof. Consider $G_1'(y) = 0$ where $G_1'(y) = k - r\bar{F}(y/\alpha)$. Since $r > k$, it follows immediately that $y_1 = \alpha F^{-1}(1 - k/r)$ exists and is the unique solution of the equation under consideration. Further results are obtained by induction.

Since $f_1(x)$ is given by (4),

$$f_1'(x) = -k + \begin{cases} 0, & if \ x \leq y_1, \\ G_1'(x), & if \ x > y_1, \end{cases}$$

$$= \begin{cases} -k, & if \ x \leq y_1, \\ -r\bar{F}(y/\alpha), & if \ x > y_1. \end{cases} \tag{5}$$

Hence, it is clear that $f_1'(x) < 0$ for all x. Moreover, on the one hand,

$$G_2'(y) = G_1'(y) + \beta \int_0^\infty f_1'(y + M - \alpha s)\varphi(s)\,ds \leq G_1'(y),$$

on the other hand, $G_2'(y)$ has the form

$$H(y) + \beta \int_0^{\frac{y+M-y_1}{\alpha}} G_1'(y + M - \alpha s)\,ds, \ \text{with } H(y) = G_1'(y) - \beta k.$$

That means $y_1 < y_2 < \bar{y}$. Furthermore, $f_2'(x) < 0$ for all x. Thus, the base of induction is established. Assuming that (4) is true for the number of periods less or equal to n we prove its validity for $n+1$ and deduce that $G_{n+1}'(x) < G_n'(x)$, so $y_n < y_{n+1}$.

Since

$$G_{n+1}'(y) = H(y) + \beta \int_0^{\frac{y+M-y_n}{\alpha}} G_n'(y + M - \alpha s)\,ds,$$

one easily gets $G_{n+1}'(y) > H(y)$. This entails the needed relation $y_{n+1} < \bar{y}$. □

Corollary 1. *There exists $\hat{y} = \lim_{n\to\infty} y_n$.*

Remark 1. It is interesting that $\hat{y} = \bar{y}$ only for $M = 0$, whereas $\hat{y} < \bar{y}$ for $M > 0$.

2.3 Model's Stability

Now we turn to sensitivity analysis and prove that the model under consideration is stable with respect to small perturbations of the underlying distribution. For this purpose we introduce two variants of the model (with claims X_i and Y_i). In the first one all functions has subscript X and in the second one Y.

For random variables X and Y defined on some probability space and possessing finite expectations, it is possible to define their distance on the base of Kantorovich metric, see, e.g., [30], in the following way

$$\kappa(X,Y) = \int_{-\infty}^{\infty} |F_X(t) - F_Y(t)|\, dt$$

where F_X and F_Y are the distribution functions of X and Y, respectively.

The distance between the cost functions is measured in terms of Kolmogorov uniform metric. Thus, we study

$$\Delta_n = \sup_x |f_{n,X}(x) - f_{n,Y}(x)|.$$

To this end we need the following

Lemma 1. *Let functions $g_i(y)$, $i = 1, 2$, be such that $|g_1(y) - g_2(y)| < \delta$ for some $\delta > 0$ and any y, then $\sup_x |\inf_{y \geq x} g_1(y) - \inf_{y \geq x} g_2(y)| < \delta$.*

Now we are able to estimate Δ_1.

Lemma 2. *Assume $\kappa(X,Y) = \rho$, then $\Delta_1 \leq \alpha r \rho$.*

Proof. According to Lemma 1 we need to estimate $|G_{1,X}(y) - G_{1,Y}(y)|$ for any y. The definition of these functions gives $G_{1,X}(y) - G_{1,Y}(y) = r[E(\alpha X - y)^+ - E(\alpha Y - y)^+] = r\alpha \int_{y/\alpha}^{\infty} (\bar{F}_X(t) - \bar{F}_Y(t))\, dt$. This leads immediately to the desired estimate. □

Next, we prove the main result demonstrating the model's stability.

Theorem 2. *If $\kappa(X,Y) = \rho$, then $\Delta_n \leq D_n \rho$ where $D_n = \alpha(\frac{r(1-\beta^n)}{1-\beta} + \frac{k(\beta-\beta^n)}{1-\beta})$.*

Proof. As in Lemma 2 we estimate $|G_{n,X}(y) - G_{n,Y}(y)|$ for any y. Due to definition (3) we have

$$|G_{n,X}(y) - G_{n,Y}(y)| \leq |G_{1,X}(y) - G_{1,Y}(y)|$$

$$+\beta \left| \int_0^{\infty} f_{n-1,X}(y + M - \alpha s)\varphi_X(s)\, ds - \int_0^{\infty} f_{n-1,Y}(y + M - \alpha s)\varphi_Y(s)\, ds \right|.$$

Obviously, the first term on the right-hand side of inequality is less than $\alpha r \rho$. To estimate the second term we rewrite it adding and subtracting an integral with different indices of f_{n-1} and φ. Then integrating by parts and using $\max_y |f'_{n-1,Y}(y)| \leq k$ for all n, we get

$$\Delta_n \leq \alpha(r + k\beta)\rho + \beta \Delta_{n-1}.$$

Solving this recurrent relation leads to the necessary form of D_n. □

Corollary 2. *$\Delta_n \leq \frac{r+k\beta}{1-\beta} \alpha \rho$ for any n.*

In other words, we established stability of the model with respect to small perturbations of claim distribution.

3 Model with Investment in a Non-risky Asset

3.1 Model Description

We consider the following generalization of the model introduced in [12] and further treated in [13]. For certainty, we proceed in terms of insurance company performance. During the ith period the company gets a fixed premium amount c and pays a random indemnity X_i, $i = 1, 2, \ldots$. It is supposed that $\{X_i\}_{i \geq 1}$ is a sequence of non-negative independent r.v.'s having Erlang$(2, \lambda_i)$ distributions. Let the initial capital $S_0 = x > 0$ be fixed. It is possible to place a quota δ of this amount in a bank for m periods, the interest rate being β per period. Thus, the surplus at the end of the first period has the form $S_1 = (1 - \delta)S_0 + c - X_1$. The same procedure is repeated each period. Putting $u_m = (1 + \beta)^m$ we get the following recurrent formula

$$S_n = \min[(1 - \delta)S_{n-1}, S_{n-1}] + c + u_m \delta S^+_{n-(m+1)} - X_n. \tag{6}$$

Here and further on it is assumed that $S_k = 0$ for $k < 0$. The expression (6) is useful if it is permitted to delay the company insolvency (Parisian ruin), for definition see [18]. If we use the classical notion of ruin, then the ruin time τ is defined as follows

$$\tau = \inf\{n > 0 : S_n \leq 0\}. \tag{7}$$

To calculate the ultimate ruin probability

$$P(\tau < \infty) = \sum_{n=1}^{\infty} P(S_1 > 0, \ldots, S_{n-1} > 0, S_n \leq 0)$$

one can use instead of (6) the relation

$$S_n = (1 - \delta)S_{n-1} + c - X_n + u_m \delta S_{n-(m+1)}, \tag{8}$$

treated in [12] for $m = 1$.

There is no need to find the explicit form of S_n (although it is possible to obtain it). For further investigation we need only the following result.

Lemma 3. *Put $S_0 = x$ and*

$$S_n = f_n - \sum_{i=1}^{n} g_{n,i} X_i, \quad n \geq 1, \tag{9}$$

then $f_0 = x$, whereas

$$f_n = (1 - \delta)f_{n-1} + c, \ n = \overline{1, m}, \quad f_n = (1 - \delta)f_{n-1} + u_m \delta f_{n-(m+1)} + c, \ n > m, \tag{10}$$

and

$$g_{n,i} = (1 - \delta)g_{n-1,i} + u_m \delta g_{n-(m+1),i}, \quad i = \overline{1, n - (m + 1)},$$
$$g_{n,i} = (1 - \delta)g_{n-1,i}, \quad i = \overline{n - m, n - 1}, \quad g_{n,n} = 1. \tag{11}$$

Proof. Obviously, combination of (8) and (9) provides the desired recursive relations (10) and (11). □

3.2   Algorithm for Ruin Probability Calculation

Our aim is investigation of ultimate and finite-time ruin probabilities. To this end, we reformulate Lemma 3 as follows. The company surplus S_n, $n \geq 1$, has the form $S_n = f_n - Y_n$ with f_n defined by (10) and vector $\mathbf{Y}_n = (Y_1, \ldots, Y_n)$ given by $\mathbf{Y}_n = \mathbf{G}_n \cdot \mathbf{X}_n$. Here vector $\mathbf{X}_n = (X_1, \ldots, X_n)$ and matrix $\mathbf{G}_n = (g_{k,i})_{k,i=1,\ldots,n}$, where $g_{k,i} = 0$ for $i > k$, the others being specified by (11).

Theorem 3. *The ultimate ruin probability is represented by*

$$\sum_{n=1}^{\infty} \int_0^{f_1} \cdots \int_0^{f_{n-1}} \int_{f_n}^{+\infty} \prod_{k=1}^{n} p_{X_k}(v_k(y_1, \ldots, y_n)) dy_1 \ldots dy_n,$$

where f_n, $n \geq 1$, and $g_{n,i}$, $n \geq 1$, $i = \overline{1,n}$, are given by (10) and (11), respectively. The function $v_k(y_1, \ldots, y_n)$, $k = \overline{1,n}$, is the kth component of vector $\mathbf{G}_n^{-1} \mathbf{y}_n$.

Proof. The ruin time is defined by (7). So, we calculate the probability of ruin at the nth step obtaining the distribution of the ruin time. To this end, we use Lemma 3. Hence, the probability under consideration $P(\tau = n)$ is given by $P(U_n)$ with

$$U_n = \left\{ g_{1,1} X_1 < f_1, \ldots, \sum_{i=1}^{n-1} g_{n-1,i} X_i < f_{n-1}, \sum_{i=1}^{n} g_{n,i} X_i \geq f_n \right\}.$$

Performing the change of variables $Y_k = \sum_{i=1}^{k} g_{k,i} X_i$, $1 \leq k \leq n$, we obtain a one-to-one correspondence between $\mathbf{Y}_n = (Y_i)_{i=1}^n$ and $\mathbf{X}_n = (X_i)_{i=1}^n$. Since $\det \mathbf{G}_n = 1$, we get the desired form of ruin probability.   □

Another interesting problem is calculation of finite-time ruin probability. In other words, we want to get the expression of probability

$$\phi_N = P(\tau \leq N)$$

for a fixed number N.

Corollary 3. *The following relation holds*

$$\phi_N = 1 - \int_0^{f_1} \cdots \int_0^{f_{N-1}} \int_0^{f_N} \prod_{k=1}^{N} p_{X_k}(v_k(y_1, \ldots, y_N)) dy_1 \ldots dy_N. \tag{12}$$

Proof. Clearly, using the expression which provides $P(\tau = n)$ we get ϕ_N as a sum of corresponding summands for $n \leq N$. On the other hand, it can be written as $1 - P(\tau > N) = 1 - P(S_1 > 0, \ldots, S_N > 0)$. Thus, proceeding along the same lines as in the proof of Theorem 3, we immediately obtain expression (12).   □

4   Model with Investment in Risky and Non-risky Assets Combined with Reinsurance

Problems of optimal reinsurance and investment in financial market attract increasing attention from academics and insurers during the last decade. There exist different approaches to solving these problems. Thus, the paper [32] studies excess-of-loss reinsurance and investment in risky and non-risky assets in order to maximize the expected S-shaped utility from the terminal wealth in the framework of classical Cramér-Lundberg model. The study of ruin probability bounds under proportional reinsurance and investment one can find in [27].

4.1   Model Description

We study a discrete-time analog of classical Cramér-Lundberg model with following modifications. It is assumed that the company's capital is controlled only at regular intervals. The space between two subsequent control times is called period. At the beginning of each period, the company enters into a reinsurance contract and invests part of the available capital in a risk-free asset, the remaining capital being invested in a risky one. Let Z_k be the claim amount during the k-th period. The premium amount p is accumulated by the company at the beginning of each period. Since the proportional reinsurance with quota α is applied, the company gains αp each period, paying only αZ_k during the k-th period. Let Y_k be the capital at the end of the k-th period and initial capital $Y_0 = y$. If amount x is invested in non-risky asset, the return at the end of period is equal $g(x)$, the return of investment $y - x$ in a risky asset is given by $h(y - x, \eta)$ where r.v. η characterizes the asset profitability.

The one-step company profit (loss) is the difference between the capital at the end and the beginning of the period under consideration. Thus, we can express the optimal one-step expected profit as follows

$$\Pi_1(y) = \max_{0 \leq x \leq y} [f_1(y, x)] + \alpha(p - EZ_1) - y, \quad f_1(y, x) = g(x) + Eh(y - x, \eta_1)$$

The capital at the end of the first period, as a function of the initial capital Y_0, is given by

$$Y_1 = Y_0 + \Pi_1(Y_0),$$

if an optimal investment is carried out.

We are going to consider the n-step model. The difference between the capital of company at the end of the n-th period and initial capital will be called the profit (loss) of insurance company during n periods. Using the methods of dynamic programming (see, e.g., [3]) we establish that the maximal expected n-step profit satisfies the functional equation

$$\Pi_{n+1}(y) = \max_{0 \leq x \leq y} [f_{n+1}(y, x)] + \alpha(p - EZ_1) - y,$$

$$f_{n+1}(y, x) = f_1(y, x) + E\Pi_n(g(x) + h(y - x, \eta_1) + \alpha(p - Z_1)). \quad (13)$$

It is possible to express the capital at the beginning of the n-th period as a function of initial capital

$$Y_n = Y_0 + \Pi_{n-1}(Y_0).$$

One-step survival probability is

$$P(Y_1 \geq 0 | Y_0 = y).$$

Similarly, n-steps survival probability is given by

$$P(\bigcap_{i=1}^{n} (Y_i \geq 0) | Y_0 = y).$$

4.2 Optimal Investment Strategy

First of all, it is necessary to define the optimal strategy of investment.

Definition 1. *Consider a model describing n subsequent steps of company functioning. Let $x_i(Y_{n-i})$ denote the amount invested in non-risky asset at the beginning of the $(n - i)$-th step, whereas Y_{n-i} is the capital at that time. Then the investment strategy is a sequence of the following n functions*

$$D_{x_n,\ldots,x_1}(Y_0, \ldots, Y_{n-1}) = (x_n(Y_0), \ldots, x_1(Y_{n-1})),$$

Definition 2. *The optimal strategy, that is, strategy maximizing the expected n-step profit, has the following form*

$$\overline{D}_{\hat{x}_n,\ldots,\hat{x}_1}(Y_0, \ldots, Y_{n-1}) = (\hat{x}_n(Y_0), \ldots, \hat{x}_1(Y_{n-1})),$$

where $\hat{x}_k(Y_{n-k})$ is the amount one has to invest in the non-risky asset at the $(n - k)$-th step to obtain the desired maximum.

Hence, for establishing the optimal strategy it is necessary to calculate $\hat{x}_i(Y_{n-i})$ for each i.

Further on we suppose that $\{Z_k\}_{k \geq 1}$ and $\{\eta_k\}_{k \geq 1}$ form two independent sequences of i.i.d. random variables. Introducing an additional notation

$$E\eta_k = \nu, \quad EZ_k = \mu,$$

we formulate the main result under additional assumption that the gain functions are linear.

Theorem 4. *Put*

$$1) h(y - x, \eta_k) = (y - x)\eta_k, \quad 2) g(x) = xI,$$

where I is the profitability of non-risky asset, η_k determines the profitability of risky investment during the k-th period and y is the initial capital. Then, for $I > \nu$,

$$\overline{D}_{\hat{x}_n,\ldots,\hat{x}_1}(Y_0, \ldots, Y_{n-1}) = (Y_0, \ldots, Y_{n-1}).$$

On the contrary, for $I < \nu$,

$$\overline{D}_{\hat{x}_n,\ldots,\hat{x}_1}(Y_0, \ldots, Y_{n-1}) = (0, \ldots, 0).$$

Proof. For simplicity, we put further $Y_0 = y$. Maximal expected one-step profit is given by

$$\Pi_1(y) = \max_{0 \le x \le y} [xI + (y-x)\nu] + \alpha(p-\mu) - y.$$

To find $\hat{x}_1(y)$ maximizing $f_1(y,x)$, we calculate its derivative in x

$$f_1(y,x)'_x = I - \nu.$$

Obviously, if $I > \nu$, then $f_1(y,x)'_x > 0$, and $f_1(y,x)$ increases in x, that means, $\hat{x}_1(y) = y$. On the contrary, if $I < \nu$, then $f_1(y,x)'_x < 0$, and $f_1(y,x)$ decreases in x, that is, $\hat{x}_1(y) = 0$. Hence, we get the explicit form of the maximal one-step profit

$$\Pi_1(y) = \begin{cases} \alpha(p-\mu) - y + yI, \ I > \nu, \\ \alpha(p-\mu) - y + y\nu, \ I < \nu. \end{cases}$$

Thus, $\overline{D}_{\hat{x}_1}(y) = y$ for $I > \nu$ and $\overline{D}_{\hat{x}_1}(y) = 0$ for $I < \nu$. Now consider two-step model

$$\Pi_2(y) = \max_{0 \le x \le y} [f_1(y,x) + E\Pi_1(xI + (y-x)\eta_1 + \alpha(p-Z_1))] +$$

$$+ \alpha(p-\mu) - y.$$

As in one-step case, we investigate $f_2(y,x)$ having the form

$$f_1(y,x) + E\Pi_1(xI + (y-x)\eta_1 + \alpha(p-Z_1)) =$$

$$= xI + (y-x)\nu + \int_0^\infty \int_0^\infty \Pi_1(xI + (y-x)s + \alpha(p-q))\Psi(s,q)dsdq,$$

where $\Psi(s,q)$ is the joint distribution density of random variables η_1 and Z_1. Put $b(y,x,s,q) = xI + (y-x)s + \alpha(p-q)$. Then, using the explicit form of $\Pi_1(y)$, we get

$$f_2(y,x) = xI + (y-x)\nu + \int_0^\infty \int_0^\infty (\alpha(p-\mu) - b + bI)\Psi(s,q)dsdq =$$

$$= xI + (y-x)\nu + \alpha(p-\mu) - Eb(y,x,\eta_1,Z_1) + I \int_0^\infty \int_0^\infty b\Psi(s,q)dsdq =$$

$$= I \int_0^\infty \int_0^\infty b\Psi(s,q)dsdq = IEb = I(f_1(y,x) + \alpha(p-\mu)). \qquad (14)$$

Here we have written $\Pi_1(y)$ for the case $I > \nu$, for the opposite inequality we obtain the following expression for $f_2(y,x)$

$$f_2(y,x) = \nu(f_1(y,x) + \alpha(p-\mu)),$$

that is, they differ by positive factor.

Calculating the derivative in second argument x, we can investigate behavior of $f_2(y,x)$

$$(f_2(y,x)'_x = I \int_0^\infty \int_0^\infty b'_x \Psi(s,q)dsdq =$$

$$= I \int_0^\infty \int_0^\infty (I - s)\Psi(s,q)dsdq = I(I - \nu).$$

Clearly the signs of $(f_2(y,x))'_x$ and $(f_1(y,x))'_x$ are the same. Therefore, if $I > \nu$, then $\hat{x}_2(y) = y$, whereas for $I < \nu$, we have $\hat{x}_2(y) = 0$. In the same way, it is possible to get

$$\Pi_2(y) = \begin{cases} f_2(y,y) + \alpha(p - \mu) - y, I > \nu, \\ f_2(y,0) + \alpha(p - \mu) - y, I < \nu. \end{cases}$$

So, $\overline{D}_{\hat{x}_1,\hat{x}_2}(Y_0, Y_1) = (Y_0, Y_1)$ for $I > \nu$ and $\overline{D}_{\hat{x}_1,\hat{x}_2}(Y_0, Y_1) = (0,0)$ for $I < \nu$. We use the mathematical induction in order to obtain the optimal strategy for the $(n + 1)$-step case. Induction base is given by (14) Now suppose that for n-step case we have

$$f_n(y,x) = I(f_{n-1}(y,x) + \alpha(p - \mu)),$$

if $I > \nu$.

In particular, that means

$$(f_n(y,x))'_x = I^{n-1}(f_1(y,x))'_x.$$

Therefore

$$\Pi_n(y) = \begin{cases} f_n(y,y) + \alpha(p - \mu) - y, I > \nu, \\ f_n(y,0) + \alpha(p - \mu) - y, I < \nu. \end{cases}$$

Now, for the $(n + 1)$-step model

$$\Pi_{n+1}(y) = \max_{0 \le x \le y} [f_1(y,x) + E\Pi_n(xI + (y - x)\eta_1 + \alpha(p - Z_1))] + \alpha(p - \mu) - y.$$

Our aim is to find $\hat{x}_{n+1}(y)$ maximizing $f_{n+1}(y,x)$. So we obtain the expression of $f_{n+1}(y,x)$ using the induction assumption

$$f_{n+1}(y,x) = f_1(y,x) + E\Pi_n(xI + (y - x)\eta_1 + \alpha(p - Z_1)) =$$

$$= f_1(y,x) + \int_0^\infty \int_0^\infty (f_n(b,b) + \alpha(p - \mu) - b)\Psi(s,q)dsdq =$$

$$= Ix + (y - x)\nu + \alpha(p - \mu) - (xI + (y - x)\nu + \alpha(p - \mu)) +$$

$$+ \int_0^\infty \int_0^\infty f_n(b,b)\Psi(s,q)dsdq = \int_0^\infty \int_0^\infty f_n(b,b)\Psi(s,q)dsdq = Ef_n(b,b),$$

where, as previously, $b = xI + (y - x)\eta_1 + \alpha(p - Z_1)$. According to induction assumption

$$f_n(y,x) = I(f_{n-1}(y,x) + \alpha(p - \mu)) =$$

$$= I(I(...(I(f_1(y,x) + \alpha(p - \mu)) + \alpha(p - \mu))...) + \alpha(p - \mu)).$$

Hence,

$$Ef_n(b,b) = EI(I(...(I(f_1(b,b) + \alpha(p-\mu)) + \alpha(p-\mu))...) + \alpha(p-\mu)) =$$

$$EI(I(...(I(Ib + \alpha(p-\mu)) + \alpha(p-\mu))...) + \alpha(p-\mu)) =$$
$$EI(I(...(I(I\underbrace{(xI + (y-x))\eta_1 + \alpha(p - Z_1))}_{b(y,x,\eta_1,Z_1)} +$$

$$\alpha(p-\mu)) + \alpha(p-\mu))..) + \alpha(p-\mu)) =$$
$$= I(I(...(I(I(xI + (y-x))\nu + \alpha(p-\mu)) + \alpha(p-\mu)) + \alpha(p-\mu))..) + \alpha(p-\mu)) =$$
$$= I(I(...(I(I(f_1(y,x) + \alpha(p-\mu)) + \alpha(p-\mu)) + \alpha(p-\mu))..) + \alpha(p-\mu)) =$$
$$= I(f_n(y,x) + \alpha(p-\mu)).$$

Thus, we established that $f_{n+1}(y,x) = I(f_n(y,x) + \alpha(p-\mu))$. It follows immediately that

$$(f_{n+1}(y,x))'_x = I^n(f_1(y,x))'_x = I^n(I-\nu).$$

Here we have written all functions f_i under assumption $I > \nu$, in the case of opposite inequality, ν^n will stand instead of I^n.

It follows from the expression of $(f_{n+1}(y,x))'_x$ that $\hat{x}_{n+1}(y) = y$ for $I > \nu$, whereas for $I < \nu$, we have $\hat{x}_{n+1}(y) = 0$. Finally, we get

$$\overline{D}_{\hat{x}_{n+1},...,\hat{x}_1}(Y_0,...,Y_n) = \begin{cases} (Y_0,...,Y_n), I > \nu, \\ (0,...,0), I < \nu. \end{cases}$$

$\square$

4.3   Survival Probability

For investigation of the survival probability we need to know the distributions of r.v.'s η_i and Z_i. We suppose that Z_i has exponential distribution with parameter λ. For the risky asset assume $\eta_i = \frac{S_i}{S_{i-1}}$, where S_i is the price of asset at time i, and $\{S_t\}_{t \in R}$ is a geometric Brownian motion with drift a and volatility σ.

Lemma 4. *The density of r.v. η_i has the following form*

$$f_\eta(s) = \begin{cases} \frac{1}{s\sigma} f_\xi(\frac{\ln s - (a - \sigma^2/2)}{\sigma}), & s > 0 \\ 0, & s \leq 0, \end{cases}$$

here f_ξ is the density of a standard normal random variable ξ.

Proof. Begin by finding he explicit form of S_t. It is well-known that S_t satisfies the following stochastic differential equation

$$dS_t = aS_t dt + \sigma S_t dW_t,$$

where W_t is standard Wiener process. Using Ito's formula we obtain

$$d \ln S_t = \frac{dS_t}{S_t} - \frac{1}{2S_t^2}(dS_t)^2 = \frac{dS_t}{S_t} - \frac{\sigma^2}{2}dt.$$

Substituting dS_t in the last expression, after some transformations we get

$$S_t = S_0 e^{\sigma W_t + (a - \sigma^2/2)t}.$$

Thus,

$$\eta_k = \frac{S_k}{S_{k-1}} = e^{\sigma \xi + (a - \sigma^2/2)}, \quad \xi = W_k - W_{k-1}$$

Next, we use the definition of distribution function

$$F_\eta(x) = P(\eta \le x) = P(e^{\sigma \xi + (a - \sigma^2/2)} \le x)$$

$$= P(\xi \le \frac{\ln x - (a - \sigma^2/2)}{\sigma})$$

$$= F_\xi(\frac{\ln x - (a - \sigma^2/2)}{\sigma}).$$

After differentiation of $F_\eta(x)$ in x, we obtain the density of r.v. η. □

Density of Z_i is

$$f_{Z_i}(x) = \begin{cases} \lambda e^{-\lambda x}, & x \ge 0, \\ 0, & x < 0. \end{cases}$$

Recall the expression of the capital at the end of the first step in terms of initial capital

$$Y_1 = \hat{x}_1(Y_0)I + (Y_0 - \hat{x}_1(Y_0))\eta_1 + \alpha(p - Z_1).$$

Obviously, the one-step survival probability can be expressed as follows

$$P(Y_1 \ge 0 | Y_0 = y) = P(\hat{x}_1(y)I + (y - \hat{x}_1(y))\eta_1 + \alpha(p - Z_1) \ge 0) =$$

$$= 1 - P(-\alpha Z_1 + (y - \hat{x}_1(y))\eta_1 + \hat{x}_1(y)I \le -\alpha p).$$

Denote the obtained result as $G(y, \hat{x}_1(y), \alpha)$.

Using the formula for distribution of the sum of independent random variables we can get the explicit expression for

$$G(y, v, \alpha) = 1 - \int_{vI}^{\infty} f_{(y-v)\eta_1 + vI}(s) f_{-\alpha Z_1}(-\alpha p - s) ds =$$

$$= 1 - \frac{\lambda}{\alpha \sqrt{2\pi}\sigma} \int_{vI}^{\infty} \frac{1}{s - vI} e^{-\frac{(\ln(\frac{s-vI}{y-v}) - (a - \sigma^2/2))^2}{2\sigma^2}} e^{\frac{\lambda}{\alpha}(-\alpha p - s)} ds.$$

It is useful to study the dependence of $G(y, v, \alpha)$ on v, however it is not needed in order to prove the following result.

Theorem 5. *If* 1)$\overline{D}_{\hat{x}_1}(Y_0) = (Y_0)$ *either* 2)$\overline{D}_{\hat{x}_1}(Y_0) = (0)$, *then* $G(y, \hat{x}_1(y), \alpha)$ *decreases in* α *for a fixed* y.

Proof. Under assumption 1) we have

$$G(y, \hat{x}_1(y), \alpha) = P(yI + \alpha(p - Z_1) \geq 0) = P(Z_1 \leq p + I(y/\alpha)).$$

Since $0 < \alpha \leq 1$, it is clear, that this function decreases in α.

In case of assumption 2) the function under consideration has the form

$$G(y, \hat{x}_1(y), \alpha) = P(y\eta_1 + \alpha p - \alpha Z_1 \geq 0) = P((y/\alpha)\eta_1 + p - Z_1 \geq 0).$$

Thus, obviously, it decreases in α. □

Remark 2. If y is not fixed, it is possible to state that $G(y, \hat{x}_1(y), \alpha)$ is an increasing function of y/α.

5   Conclusion and Further Research Directions

We have studied three discrete-time models. In the first one we used the cost approach and obtained the loan strategy minimizing expected n-step costs. The model's stability with respect to small distribution perturbations is established.

In the second model we used the reliability approach, namely, studied the ruin probability (ultimate and finite-time) under assumption that claims distributions have Erlang2 type instead of being exponential. The sensitivity analysis can be performed along the lines traced in [31], see also [11]. It is planned to carry out the investigation of dividends problems.

The third model is devoted to performance of insurance company with reinsurance and investment in risky and non-risky assets. Optimal investment policy maximizing the company profit is established under some additional assumptions. One-period survival probability is also considered. The next step is calculation of multi-period survival probability under optimal investment policy.

An interesting task is the study of similar problems for non-homogeneous models. It is also important to investigate the case of incomplete information as in [10].

References

1. Alfa, A.-S., Drekic, S.: Algorithmic analysis of the Sparre Andersen model in discrete time. ASTIN Bull. **37**, 293–317 (2007)
2. Asmussen, S., Albrecher, H.: Ruin Probabilities, 2nd edn. World Scientific, New Jersey (2010)
3. Bellman, R.: Dynamic Programming. Princeton University Press, Princeton (1957)
4. Bulinskaya, E.: New dividend strategies. In: Dimotikalis, Y., et al. (eds.) Applied Modeling Techniques and Data Analysis 2, chapter 3, pp. 39–52. ISTE Ltd., London (2021)
5. Bulinskaya, E.: Asymptotic analysis and optimization of some insurance models. Appl. Stochast. Mod. Bus. Ind. **34**(6), 762–773 (2018)
6. Bulinskaya, E.: New research directions in modern actuarial sciences. In: Panov, V. (ed.) MPSAS 2016. SPMS, vol. 208, pp. 349–408. Springer, Cham (2017). https://doi.org/10.1007/978-3-319-65313-6_15

7. Bulinskaya, E.-V.: Cost approach versus reliability. In: Proceedings of International Conference DCCN-2017, Technosphera, Moscow, pp. 382–389 (2017)
8. Bulinskaya, E.: Asymptotic analysis of insurance models with bank loans. In: Bozeman, J.-R., Girardin, V., Skiadas, Ch. (eds.) New Perspectives on Stochastic Modeling and Data Analysis, ISAST, Athens, Greece, pp. 255–270 (2014)
9. Bulinskaya, E.: On the cost approach in insurance. Rev. Appl. Ind. Math. **10**(2), 276–286 (2003). In Russian
10. Bulinskaya, E., Gusak, J.: Insurance models under incomplete information. In: Pilz, J., Rasch, D., Melas, V.B., Moder, K. (eds.) IWS 2015. SPMS, vol. 231, pp. 171–185. Springer, Cham (2018). https://doi.org/10.1007/978-3-319-76035-3_12
11. Bulinskaya, E., Gusak, J.: Optimal control and sensitivity analysis for two risk models. Commun. Stat. Simul. Comput. **45**(5), 1451–1466 (2016)
12. Bulinskaya, E., Kolesnik, A.: Reliability of a discrete-time system with investment. In: Vishnevskiy, V.M., Kozyrev, D.V. (eds.) DCCN 2018. CCIS, vol. 919, pp. 365–376. Springer, Cham (2018). https://doi.org/10.1007/978-3-319-99447-5_31
13. Bulinskaya, E., Shigida, B.: Discrete-time model of company capital dynamics with investment of a certain part of surplus in a non-risky asset for a fixed period. Methodol. Comput. Appl. Probabil. **23**(1), 103–121 (2021). https://doi.org/10.1007/s11009-020-09843-5
14. Bulinskaya, E., Yartseva, D.: Discrete-time models with dividends and reinsurance. In: Proceedings of SMTDA 2010, Chania, Greece, 8–11 June 2010, pp. 155–162 (2010)
15. Bulinskaya, E., Gusak, J., Muromskaya, A.: Discrete-time insurance model with capital injections and reinsurance. Methodol. Comput. Appl. Probabil. **17**, 899–914 (2015)
16. Castañer, A., Claramunt, M.-M., Lefèvre, C., Gathy, M., Mármol, M.: Ruin probabilities for a discrete-time risk model with non-homogeneous conditions. Scand. Actu. J. **2013**(2), 83–102 (2013)
17. Chan, W., Zhang, L.: Direct derivation of finite-time ruin probabilities in the discrete risk model with exponential or geometric claims. North Am. Actu. J. **10**(4), 269–279 (2006)
18. Czarna, I., Palmovsky, Z., Świątek, P.: Discrete time ruin probability with Parisian delay. arXiv:1403.7761v2 [math.PR] (2017)
19. Damarackas, J., Šiaulys. J.: Bi-seasonal discrete time risk model. Appl. Math. Comput. **247**, 930–940 (2014)
20. Diasparra, M., Romera, R.: Inequalities for the ruin probability in a controlled discrete-time risk process. Eur. J. Oper. Res. **204**(3), 496–504 (2010)
21. Dickson, D.-C.-M., Waters, H.-R.: Some optimal dividends problems. ASTIN Bull. **34**, 49–74 (2004)
22. Gajek, L.: On the deficit distribution when ruin occurs, discrete time model. Insur. Math. Econ. **36**, 13–24 (2005)
23. Gallager, R.-G.: Stochastic Processes: Theory for Application. Cambridge University Press, Cambridge (2013)
24. Gerber, H.-U.: Mathematical fun with compound binomial process. Astin Bull. **18**(2), 161–168 (1988)
25. Grandell, J.: Aspects of Risk Theory. Springer, New York (1991). https://doi.org/10.1007/978-1-4613-9058-9
26. Li, S., Lu, Y., Garrido, J.: A review of discrete-time risk models. Revista de la Real Academia de Ciencias Naturales. Serie A Matemàticas **103**, 321–337 (2009)

27. Luesamai, A.: Lower and upper bounds of the ultimate ruin probability in a discrete time risk model with proportional reinsurance and investment. J. Risk Manag. Insur. **25**(1), 1–10 (2021)
28. Mikosch, T.: Non-Life Insurance Mathematics. Springer, Heidelberg (2006). https://doi.org/10.1007/978-3-540-88233-6
29. Paulsen, J.: Ruin models with investement income. arXiv:0806.4125v1 (math. PR) (2008)
30. Rachev, S.T., Klebanov, L., Stoyanov, S.V., Fabozzi, F.: The Methods of Distances in the Theory of Probability and Statistics. Springer-Verlag, New York (2013). https://doi.org/10.1007/978-1-4614-4869-3
31. Saltelli, A., Tarantola, S., Campolongo, F.: Sensitivity analysis as an ingredient of modeling. Stat. Sci. **15**(4), 377–395 (2000)
32. Sun, Q.Y., Rong, X.M., Zhao, H.: Optimal excess-of-loss reinsurance and investment problem for insurers with loss aversion. Theor. Econ. Lett. **9**, 1129–1151 (2019). https://doi.org/10.4236/tel.2019.94073

The Markov Model
of the Needham-Schroeder Protocol

A. A. Shukmanova[1], A. S. Yermakov[2], T. T. Paltashev[3],
A. K. Mamyrova[4(✉)], and A. A. Ziro[4]

[1] Kazakh National Technical University, Almaty, Kazakhstan
[2] Caspian University, Almaty, Kazakhstan
[3] ITMO University, Saint Petersburg, Russia
[4] Turan University, Almaty, Kazakhstan
a.mamyrova@turan-edu.kz

Abstract. A user-friendly Markov model of procedures for the Needham-Schroeder information security algorithm is developed. To implement the development, auxiliary system models of three asynchronous execution algorithms are used, one of which considers systems of processing-synchronous input and processing-asynchro-nous output, while the other considers input-asynchronous processing and synchronous information output. A model of processing-asynchronous input and output is taken as the final model. Finally, the closed model of a three-phase queuing system is given.

Keywords: Markov · Smodel · Needham-Schroeder

1 Needham-Schroeder Protocol for Public Key Authentication on Sample

The Controller owns the public keys of all the clients it serves. In addition, each client has an authenticated copy of the Controller's public key [1]. Alice starts the protocol execution. First, she generates a random session key and encrypts it with a shared key known to her and Trent. Then she sends Trent an encrypted text with giving her and Bob's names. Upon receiving Alice's request to deliver the session key Trent finds in his database the two shared long-term keys mentioned in Alice's request. Then Trent decrypts the received text using Alice's key, encrypts it again with Bob's key, and sends the result to Bob. Finally, having received and decrypted the delivered session key, Bob acknowledges its receipt by sending Alice a message encrypted with this key. In this protocol, Alice is the **initiator** (the "Initiator") and Bob is the **responder** (the "Responder"). INITIAL CONDITIONS:

The protocol flowchart is shown in Fig. 1.

Kazakh National Technical University.

The Sender has the public key K_0. The Receiver has public key K_P, the Controller has public key K_k. OBJECTIVE: the Sender and the Receiver wish to generate a new shared secret key.

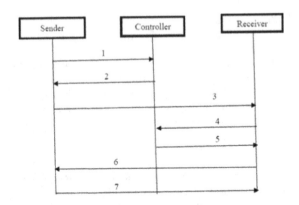

Fig. 1. Structure diagram of the Alice and Bob key exchange algorithm

1. The Sender to the Controller: *The Sender, the Receiver.*
2. The Controller to the Sender:$\{K_P, Rec\}_{K_K^{-1}}$
3. The Sender verifies the Controller's signature on the message $\{K_P, Rec\}_{K_K^{-1}}$
4. *The Receiver* decrypts the message, identifies the Sender, and sends the message $\{Rec, Send\}$ *Controller*
5. The Controller to the Receiver $\{K_0, Send\}_{K_K^{-1}}$:
6. The Receiver verifies the Controller's signature on the message $\{K_0, Send\}_{K_k^{-1}}$: generates a random number N_O and sends the message $\{N_0, N_P\}_{K_0}$ to the Sender.
7. The Sender decrypts the message and sends the information to the

Receiver:$\{N_P\}K_P$

Step 1. The protocol initiator is *the Sender.* He sends a message to *the Controller*, requesting the public key of *the Receiver.*

Step 2. *the Controller* sends the Sender the key K_p along with the Recipient's name in order to prevent an attack. The key and number are encrypted with *the Controller's* secret key K_K^{-1}. This message is a digital signature of *the Controller*, which must convince the Sender that the message he receives in Step 2 is sent by *the Controller*, i.e. the Sender verifies it with *the Controller's* public key.

Step 3. *the Sender.* generates a one-time random number N_0 and sends it to the Receiver together with his name (item 3), encrypting the message with the Receiver's public key.

Step 4. Upon receiving the message from *the Sender.*, the Receiver decrypts it and receives the number N_0. He then requests (item 4) and receives (item 5) an authentic copy of *the Sender's* public key. It then returns to the Sender the number N_0 and its own one-time random number N_P, encrypted with *the Sender's* public key (item 6).

Step 5. Upon receiving the message from the Receiver, the Sender makes sure that he is communicating with the Receiver, because only the Receiver can decrypt the message at step 3, which contains the number N_0. Furthermore, the Receiver could only decrypt this message after the Sender sent it back (a recent action). The Sender then returns the number N_0 to the Receiver by encrypting it with the Receiver's public key.

Step 6. Upon receiving this message from *the Sender.*, the Receiver also makes sure that he is communicating with *the Sender.*, since only he could decrypt the message at step 6, which contains the number N_P (another recent action).

Step 7. Thus, the successful execution of the protocol results in two random numbers N_0 and N_P, known only to *the Sender.* and the Recipient and, since these numbers are generated anew each time, the novelty property is guaranteed. In addition, each client must be convinced of the complete randomness of these numbers, since it generates one of these numbers itself.

The following is a discussion of mass service models at the level of processing operations in the processor and input/output through external devices. Interpretation of operations at the level of assembler commands.

2 Protocol "Alice → Bob" 2

The protocol flowchart is shown in Fig. 2.

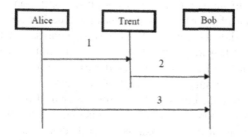

Fig. 2. Flowchart of the Alice and Bob key exchange algorithm

INITIAL CONDITIONS:
Alice and Trent share the key K_{AT} ; Bob and Trent share the key K_{BT}.
OBJECTIVE: Alice and Bob want to create a new secret key K.
Markov model of procedures for the Needham-Schroeder information security algorithm is developed.

1. Alice generates the random key K, creates the message $\{K\}_{K_{AT}}$ and sends the following information to Trent: $Alice, Bob, \{K\}_{K_{AT}}$.
2. Trent finds the keys K_{AT} and K_{BT} in the database, recovers the key K, creates the message $\{K\}_{K_{BT}}$ and sends the following information to Bob: $Alice, Bob, \{K\}_{K_{BT}}$
3. Bob decrypts the message $\{K\}_{K_{BT}}$ recovers the key K and sends the message to Alice $\{Hi, Alice, I'mBob!\}_K$.

Device communication provides three models:

1. Synchronous communication - synchronous input and synchronous output;
2. Synchronous-asynchronous communication - part of the actions is performed synchronously, the rest is performed in asynchronous mode;
3. Asynchronous interaction - all pairwise operations are performed in parallel.

From the standpoint of protection, the greatest depth of hiding secrets is provided in the first mode when users interact with the environment only at an auxiliary level. The use of technical means to improve performance reduces the level of protection.

3   Message Transferring Methods Between the Sender and the Receiver of Messages

As already mentioned, both the synchronous and asynchronous principle of message exchange between different network objects is possible. Various combinations of these methods are also possible.

The merely synchronous method of message transmission is characterized by the maximum depth of information protection. This is due to the direct connection between the data receiver and the data transmitter, when the maximum depth of protection of the transmitted information from unauthorized access can be achieved. This access method is referred to as the method of "synchronous input and syn- chronous output" (**SISO**) of messages. In this case, the transfer of data from hand to hand between the transmitter and receiver is provided.

The maximum parallelism of information exchange is ensured when using the asynchronous principle of data transmission and reception. This is the method of "asynchronous input and asynchronous output" (**AIAO**) of messages. The depth of parallelism depends on the complexity of the support material control and depends on the communication distance between the transmitter and the receiver of messages. Combinations are possible in receiving and transmitting messages. So, in the case of receiving and transmitting messages between the transmitter and the receiver through the intermediate processing point, a possible option is alternations of synchronous and asynchronous principles of data reception/processing. In the considered method of exchanges and message processing, a distinction is made between "synchronous input and asynchronous output" or "asynchronous input and synchronous output" in combination with intermediate processing of information.

These will be referred to as the method of "synchronous input/asynchronous output" (**SIAO**) and the method of "asynchronous input/synchronous output" (**AISO**) of messages with intermediate processing.

Another way to reduce the complexity of system analysis is the theorem on the sum of average stochastic values, which is replaced by an equivalent sum of the indicated values. This theorem provides an equivalent replacement of two or more phases with one common one with an intensity equal to the sum of the replaced phases.

3.1 Development of the General Case of Models with Places in AESD Phases

At the first stage, the communication scheme Alice - Bob - Trent in the AESD mode (Asynchronous Encryption - Synchronous Description) [2]. The model is built with byte-by-byte communication between the microprocessor and the device in the direct transmission mode with buffering and is shown in Fig. 3. The structure shows three phases of information exchange: input, processing and output of results with service rates $\mu_i, i = \overline{1,3}$. In the input queue IQ there is an infinite queue of requests with exponential distribution laws of service times. Service in the first and second phases is asynchronous, in the second and third phases that is synchronous. Allowable queue is limited by the value k.

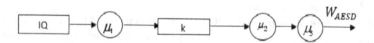

Fig. 3. Structure of the Markov model of asynchronous input and synchronous output of ASDE.

The generalized scheme of the Markov model of states and transitions for the ASDE mode is shown in Fig. 4. The generalized scheme shows states and transitions in accordance with the accepted conditions and denotations.

It is denoted here:

$P_{100} = X_0$; $P_{110} = Y_0$; $Y_1^0 = \rho_2 X_0$; $P_{101} = Y_1^0$; $P_{110}(m) = X_m$, $P_{101}(m) = Y_m$

for $m = (1, k-1)$; $P_{\beta 10} = X_\beta$; $P_{\beta 01} = Y_\beta$.

The following states are provided in the mode under consideration: The first phase can be in states $i = \{1, \beta\}$:

$i = 1$—Alice prepares an initial message for Bob;

$i = \beta$—locking the transmission of the prepared message by Alice due to the buffer being busy with serviced requests.

The second service phase can be in one of the following states $j = \{1, 0\}$:

$j = 1$—encryption by Trump and transmission of the original message;

346   A. A. Shukmanova et al.

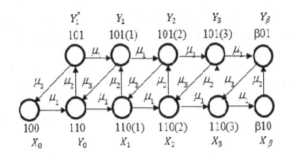

Fig. 4. Markov Graph of AESD-3 model

$j = 0$—Trump waits release from Bob.

The third phase can be in one of two states - $h = \{1, 0\}$:

$h = 1$—corresponds to reception of the encrypted message by Bob with its subsequent decryption;

$h = 0$—indicates no messages.

The following probabilities of states are denoted in the system:

P_{100}—state of preparation of the original message by Alice;

P_{110}—state of preparation of the next message by Alice and encryption of the original message by Trump;

P_{101}—preparation of the next message by Alice and transmission of the encrypted message by Trump;

$P_{110}(m)$—state of preparation of the next message by Alice, encryption of the current message by Trump as well as waiting in the buffer m previous messages for encryption and subsequent transmission;

$P_{101}(m)$—state of preparation of the next message by Alice, transmission of the encrypted message by Trump as well as waiting in the buffer m previous messages for encryption and subsequent transmission. Here $m = \overline{1, k}$.

And then these two states are repeated with a variation of the buffer content up to the value $m = k - 1$, which corresponds to the "buffer pre-locking" state.

These states are followed by two locking states of the first phase, which occur when all k places in the buffer are filled and Trump has completed the transmission of the last encrypted message.

$P_{\beta10}$—locking the preparation of the next message by Alice and encryption of the previous message by Trump;

$P_{\beta01}$—locking the preparation of the next message by Alice and transmission of the encrypted message by Trump to Bob.

Based on the foregoing, the system of balance equations is compiled for the ASDE mode under consideration.

$$\left.\begin{array}{r}
\mu_1 P_{100} = \mu_3 P_{101} \\
(\mu_1 + \mu_3)P_{101} = \mu_2 P_{110} \\
(\mu_1 + \mu_2)P_{110} = \mu_1 P_{100} + \mu_3 P_{101}(1) \\
(\mu_1 + \mu_2)P_{110}(1) = \mu_1 P_{110} + \mu_3 P_{101}(2) \\
(\mu_1 + \mu_3)P_{101}(1) = \mu_1 P_{101} + \mu_2 P_{110}(1) \\
(\mu_1 + \mu_2)P_{110}(2) = \mu_1 P_{110}(1) + \mu_3 P_{101}(3) \\
(\mu_1 + \mu_3)P_{101}(2) = \mu_1 P_{101}(1) + \mu_2 P_{110}(2) \\
(\mu_1 + \mu_2)P_{110}(3) = \mu_1 P_{110}(2) + \mu_3 P_{\beta 01} \\
(\mu_1 + \mu_3)P_{101}(3) = \mu_1 P_{101}(2) + \mu_2 P_{110}(3) \\
\mu_3 P_{\beta 01} = \mu_1 P_{101}(3) + \mu_2 P_{\beta 10} \\
\mu_2 P_{\beta 10} = \mu_1 P_{110}
\end{array}\right\} \tag{1}$$

It is denoted here:

$P_{100} = X_0$; $P_{110} = Y_0$; $Y_1^0 = \rho_2 X_0$; $P_{101} = Y_1^0$; $P_{110}(m) = X_m$; $P_{101}(m) = Y_m$ for $m = \overline{1, k-1}$; $P_{\beta 10} = X_\beta$; $P_{\beta 01} = Y_\beta$. The system of equations (1) in these denotations has the following form (2):

$$\left.\begin{array}{r}
\mu_1 X_0 = \mu_3 Y_1^0 \\
(\mu_1 + \mu_3)Y_1^0 = \mu_2 Y_0 \\
(\mu_1 + \mu_2)Y_0 = \mu_1 X_0 + \mu_3 Y_1 \\
(\mu_1 + \mu_2)X_1 = \mu_1 Y_0 + \mu_3 Y_2 \\
(\mu_1 + \mu_3)Y_1 = \mu_1 Y_1^0 + \mu_2 X_1 \\
(\mu_1 + \mu_2)X_2 = \mu_1 X_1 + \mu_3 Y_2 \\
(\mu_1 + \mu_3)Y_2 = \mu_1 Y_1 + \mu_2 X_2 \\
(\mu_1 + \mu_2)X_3 = \mu_1 X_2 + \mu_3 Y_\beta \\
(\mu_1 + \mu_3)Y_3 = \mu_1 Y_2 + \mu_2 X_3 \\
\mu_3 Y_\beta = \mu_1 Y_3 + \mu_2 X_\beta \\
\mu_2 X_\beta = \mu_1 X_3.
\end{array}\right\} \tag{2}$$

The solution is given below and shown in Table 1

Alternate solution for the case $\mu_1 = \mu_2 = \mu_3 = 2$, $k = 3$. It is denoted here: $P_{100} = X_0$; $P_{110} = Y_0$; $P_{101} = Y_1^0$; $P_{110}(m) = X_m$; $P_{101}(m) = Y_m form = \overline{1, k-1}$; $P_{\beta 10} = X_\beta$; $P_{\beta 01} = Y_\beta$.

System solution :

$Y_1^0 = X_0$;

$Y_0 = 2Y_1^0 = 2X_0$; $Y_1 = 2Y_0 - X_0 = 2 \times 2X_0 - X_0 = 3X_0$;

$Y_2 = 7Y_0 - 6X_0 = 7 \times 2X_0 - 6X_0 = (14 - 6)X_0 = 8X_0$;

$X_1 = 4Y_0 - 3X_0 = 4 \times 2X_0 - 3X_0 = 8X_0 - 3X_0 = 5X_0$;

$Y_3 = 20Y_0 - 19X_0 = 20 \times 2X_0 = 40X_0 - 19X_0 = 21X_0$;

$X_2 = 12Y_0 - 11X_0 = 12 \times 2X_0 - 11X_0 = 24X_0 - 11X_0 = 13X_0$;

$Y_\beta = 54Y_0 - 53X_0 = 54 \times 2X_0 - 53X_0 = 108X_0 - 53X_0 = 55X_0$;

$X_3 = 33Y_0 - 32X_0 = 33 \times 2X_0 - 32X_0 = 66X_0 - 32X_0 = 34X_0$;

$X_\beta = 34Y_0 - 34X_0 = 34 \times 2X_0 - 34X_0 = 68X_0 - 34X_0 = 34X_0$;

348   A. A. Shukmanova et al.

Normalization condition

$$P_{100} = X_0 = \frac{1}{1+1+2+3+8+5+21+13+55+34+34} = \frac{1}{20+55+102} = \frac{1}{177}.$$

The probabilities of the AESD-3 model states are defined in Table 1.

Table 1. AESD-3 model parameters

P_{100}	P_{110}	P_{101}	$P_{110}(1)$	$P_{101}(1)$	$P_{110}(2)$
0.0056	0.0113	0.0056	0.0168	0.0448	0.028

$P_{101}(2)$	$P_{110}(3)$	$P_{101}(3)$	P_{β_01}	P_{β_10}
0.1176	0.0734	0.1176	0.308	0.192

Based on the calculation performed using formula (3), the phase-to-phase balance is determined as well as the performance of the AESD mode is assessed.

$$W_s = \mu_1(1 - (P_{\beta_01} + P_{\beta_10})) = \mu_2(1 - (P_{100} + P_{101} + P_{101}(1) + P_{101}(2) + P_{101}(3) + P_{\beta_01})) = \mu_3(1 - (P_{100} + P_{110} + P_{110}(1) + P_{110}92) + P_{110}(3) + P_{\beta_10}))$$

$$W_s = \mu_1(1 - 0,5) = \mu_2(1 - 0,5) = \mu_3(1 - 0,5)$$

3.2 Development of the General Case of Models with Places in *SEAD* Phase)

The queuing model for the Synchronous Encryption - Asynchronous Description (SEAD) scheme is shown in Fig. 5. In the schemes, when service is allowed in one of the two phases, synchronous communications between phases are determined without buffer interaction. Asynchronous service in phases associated with the buffer of ultimate capacity k.

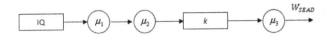

Fig. 5. Model SIAO.

As previously noted, the input buffer stores the infinite queue of requests, and input, processing and output operations are determined by service rates in phases. In the course of servicing requests, the system performs certain operations related to the forwarding of requests.

Service states for the SEAD mode are interpreted as follows.

The first phase can be in one of three states $i = \{1, 0, \beta\}$:

$i = 1$ – Trump receives the encrypted message from Bob and proceeds to decrypt it;

$i = 0$ – Trump locks receiving the next message
$i = \beta$ – locking at the end of receiving the next message from Bob when servicing the previous message in the third phase.

The second service phase can be in one of the following states $j = \{1, 0, \beta\}$:

$j = 1$ – Trump decrypts the message received from Bob;
$j = 0$ – locking the decryption of the message received from Bob;
$j = \beta$ – locking after the end of the decryption of the message received from Bob when servicing the previous message λ in the third phase.

Finally, the third phase can be in one of two states - $h = \{1, 0\}$. The state $h = 1$ corresponds to Alice's receiving a message from Trump on the channel agreed with him, and the state $h = 0$ indicates no messages. The graph of states and transitions of the SDAE mode is shown in Fig. 6.

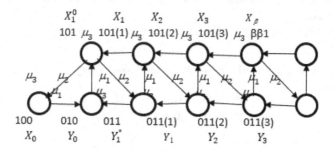

Fig. 6. Scheme of three-phase character-oriented query servicing according to the SDAE principle of overlapping processing with output

Assumed that $X_n = P_{101}(n), Y_n = P_{011}(n)$ for $n = (1, 3)$ $X_0 = P_{100}; Y_0 = P_{010}; X_1^0 = P_{101}; Y_1^0 = P_{011}; X_\beta = P_{\beta\beta1}$. These denotations are used in the structure as shown in Fig. 6 and further in the text. The system of balance equations for the presented scheme is shown below

$$\left.\begin{aligned}
c\mu_1 P_{100} &= \mu_3 P_{101}\\
(\mu_2 P_{100} &= \mu_1 P_{100} + \mu_3 P_{011}\\
(\mu_1 + \mu_3)P_{101} &= \mu_2 P_{010} + \mu_3 P_{011}\\
(\mu_2 + \mu_3)P_{011} &= \mu_1 P_{101} + \mu_3 P_{101}(1)\\
(\mu_2 + \mu_3)P_{011}(1) &= \mu_1 P_{101}(1) + \mu_3 P_{011}(1)\\
(\mu_1 + \mu_3)P_{101}(1) &= \mu_2 P_{011} + \mu_3 P_{101}(2)\\
(\mu_2 + \mu_3)P_{011}(2) &= \mu_1 P_{101}(2) + \mu_3 P_{011}(3)\\
(\mu_1 + \mu_3)P_{101}(2) &= \mu_2 P_{011}(1) + \mu_3 P_{101}(3)\\
(\mu_1 + \mu_3)P_{101}(3) &= \mu_2 P_{011}(2) + \mu_3 P_{\beta\beta1}\\
(\mu_2 + \mu_3)P_{011(3)} &= \mu_1 P_{101}\\
\mu_3 P_{\beta\beta1} &= \mu_2 P_{011}(3)
\end{aligned}\right\} \quad (3)$$

This system is presented in a more convenient form for the case $k = 3$:

$$
\left.
\begin{aligned}
\mu_1 X_0 &= \mu_3 X_1^0 \\
\mu_2 Y_0 &= \mu_1 X_0 + \mu_3 Y_1^0 \\
(\mu_1 + \mu_3) X_1^0 &= \mu_2 Y_0 + \mu_3 X_1 \\
(\mu_2 + \mu_3) Y_1^0 &= \mu_1 X_1^0 + \mu_3 Y_1 \\
(\mu_1 + \mu_3) X_1 &= \mu_2 Y_1^0 + \mu_2 X_1 \\
(\mu_2 + \mu_3) Y_1 &= \mu_1 X_1^0 + \mu_3 Y_2 \\
(\mu_2 + \mu_3) Y_2 &= \mu_1 X_2 + \mu_2 Y_3 \\
(\mu_1 + \mu_3) X_2 &= \mu_2 Y_1 + \mu_3 X_3 \\
(\mu_1 + \mu_3) X_3 &= \mu_2 Y_2 + \mu_3 X_\beta \\
(\mu_2 + \mu_3) Y_3 &= \mu_1 X_3 \\
\mu_3 X_\beta &= \mu_2 Y_3
\end{aligned}
\right\}
\tag{4}
$$

Based on the system considered, the solution of the system of equations for $k = 3$ is derived.

The solution is shown below in Table 2

Table 2. Normalization parameters

P_{100}	P_{010}	P_{101}	P_{011}	$P_{101}(1)$	$P_{011}(1)$
34	55	21	34	13	8

$P_{101}(2)$	$P_{011}(2)$	$P_{101}(3)$	$P_{011}(3)$	$P_{\beta\beta_1}$
5	3	2	1	1

Option $\mu_1 = \mu_2 = \mu_3 = 2, k = 3$ for SEAD model. The state probability coefficients for the SEAD model are shown in Table 2, 3.

Normalization condition

$P_{\beta\beta_1} = (34 + 55 + 21 + 34 + 13 + 8 + 5 + 3 + 2 + 1 + 1) = 1.$

The probability of locking the system is

$$
P_{\beta\beta_1} = \frac{1}{177}.
$$

That is, the probability of blocking is equal to the probability P_{100}, which is defined above in Sect. 3.1. The other probabilities for the case in question are given in Table 3.

Table 3. SDAE-3 model parameters

X_0	Y_0	X_1^0	Y_1^0	X_1	Y_1
P_{100}	P_{010}	P_{101}	P_{011}	$P_{101}(1)$	$P_{011}(1)$
34/177	55/177	21/177	34/177	13/177	8/177

X_2	Y_2	X_3	Y_3	X_β
$P_{101}(2)$	$P_{011}(2)$	$P_{101}(3)$	$P_{011}(3)$	$P_{\beta\beta_1}$
5/177	3/177	2/177	1/177	1/177

The normalization conditions in the case in question are given below: probability of input phase downtime

$$\theta_0^1 = P_{011} + P_{011} + P_{011}(1) + P_{011}(2) + P_{011}(3) + P_{\beta 01} = 89/177;$$

Probability of processing phase downtime

$$\theta_0^2 = P_{100} + P_{101} + P_{101}(1) + P_{101}(2) + P_{101}(3) + P_{\beta\beta 1} = 89/177;$$

probability of output phase downtime

$$\theta_0^2 = P_{100} + P_{010} = 89/177.$$

Hence the performance and the condition of balanced mode of operation of the protection system in SEAD mode with parameters $\mu_1 = \mu_2 = \mu_3 = 2, k = 3$ are equal

$$W_{SDAE} = 2(1-89/177) = 2(1-89/177) = 2(1-89/177) = 2 \times 0,4972 \approx 2 \times 0,5.$$

Hence, the equality of AESD and SEAD modes in terms of performance in the case of equality of parameters of I/O and processing request service intensities.

4   Three-Phase Model of the Presented Interaction of the Considered Objects

Alice is represented as the first phase generating requests. Requests come from Alice to Trent for subsequent transmission to Bob with a rate μ_1 and involve the transmission of messages to the Receiver Bob through Trent. Trent encrypts prepared messages with a rate of μ_2 and sends them to Bob with a rate of μ_3. The presence of buffers is assumed in the corresponding three phases. The model chart is presented in Fig. 7.

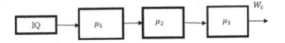

Fig. 7. Three-phase queuing model based of Alice and Bob key exchange

It should be noted that the presented model is very general in both schematic and algorithmic terms.Thus, the service of requests by modes and schemes for the forwarding of requests is not noted, there is no information about buffering, etc. The mentioned shortcomings will be eliminated when developing specific models.

As for the request service formats, two service modes are considered - byte-by-byte and word-byword - with finite request sizes. Between the phases the presence of buffers of a finite size is provided, if necessary.

Performance W_S is measured in the number of requests per unit of time. The phase service rates are determined by the speed of the corresponding service in the phases. The system performance is derived from the system normalization condition and is calculated by the following formula 5:

$$W_s = \mu_i \eta_i \tag{5}$$

where $i = \overline{1,3}$; μ_i is service rate by phases, η_i is phase load factor. It is advisable to determine these factors through the model concepts of "probability of downtime eliminating" due to technological reasons. These include, first of all, downtime due to waiting for requests from previous phases and blocking the start of service in subsequent phases.

The calculation of formula 6:

$$W_s = \mu_i (1 - (P_{\beta 10} + P_{Beta01})) = \mu_2 (1 - (P_{100} + P_{101} + P_{\beta 01})) =$$
$$= \mu_3 (1 - (P_{100} + P_{110} + P_{\beta 10}), \tag{6}$$

interpreting "transmission phase balance" determines the AESD key exchange load factors.

The first phase can be in states: $i = \{1, \beta\}$:

$i = 1$ – Alice prepares the initial message for Bob;

$i = \beta$ – locking the transmission of the prepared message by Alice.

The second service phase can be in one of the following states:$j = \{1, 0\}$:

$j = 1$ – encryption by Trump and transmission of the original message;

$j = 0$ – Trump waits release from Bob.

The third phase can be in one of two states:$h = \{1, 0\}$:

$h = 1$ – corresponds to reception of the encrypted message by Bob and its subsequent decryption;

$h = 0$ – indicates no messages.

For the SDAE mode, similar states are as follows.

The first phase can be in one of three states:$i = \{1, 0, \beta\}$:

$i = 1$ – Trump receives the encrypted message from Bob and proceeds to decrypt it;

$i = 0$ – locking reception of the next message by Trump

$i = beta$ – locking at the end of receiving the next message from Bob when servicing the previous message in the third phase.

The second service phase can be in one of the following states: $j = \{1, 0, \beta\}$:

$j = 1$ – Trump decrypts the message received from Bob;

$j = 0$ – locking the decryption of the message received from Bob;

$j = \beta$ – blocking after the end of the decryption of the message received from Bob when servicing the previous message λ in the third phase.

Finally, the third phase can be in one of two states: $h = \{1, 0\}$. The state $h = 1$ corresponds to reception of the Trump's message by Alice via the channel agreed with him, whereas the state $h = 0$ indicates no messages.

Finally, the third phase can be in one of two states: $h = \{1, 0\}$. The state $h = 1$ corresponds to reception of the Trump's message by Alice via the channel agreed with him, whereas the state $h = 0$ indicates no messages.

As a result of the analysis of the option of the three-phase QS, the considered option of the system appears to be equivalent to a single-phase QS, with an equivalent service rateW_{WAESD} (see Fig. 8), where the arrival rate is denoted by λ.

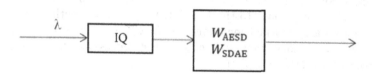

Fig. 8. Single-phase Alice and Bob key exchange model according to AESD/SDAE type

It seems that the single-phase QS model according to AESD/SDAE is similar, which is shown in the same Fig. 8.

5   Closed Stochastic Model of the Needham-Schroeder Algorithm

The network consists of M user systems S_0. The other two systems are the mediator (S_1 system) and the recipient (S_2 system) of messages (Fig. 9).

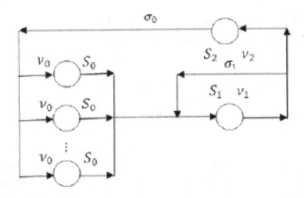

Fig. 9. Closed network of the Needham-Schroeder model

Three systems are noted in the structure: S_0 is the group of M_0 users (≥ 1); S_1 is the Alice and Bob messaging system; S_2 is the system for transmitting response messages. Calculation formulas for the specified structure are given in [2].

Thus, Fig. 9 shows the structure of a closed stochastic system, composed according to the Needham-Schroeder protocol, where ν_0 - input requests, which are formed by a group of input sources of quantity M_0.

S_1 - system of request servicing, in which Alice sends a request to Bob and this system belongs to the type **AESD**, S_2 - system of Bob's responses to Alice's requests and it also belongs to the type **SEAD**, σ_2 where with probability σ_1 Alice's requests are returned for additional servicing in case of completion of time interval - ν_1 for one stage of Alice's request processing, σ_0 - probability of Alice's request servicing completion.

Table 4. Load factors for AISO and SIAO models

μ_1	μ_2	μ_3	ρ_{AISO}	ρ_{SIAO}	ω_{max} $(AISO$	ω_{max} $(SIAO)$	ω_{SISO}
1	4	1	1,25	0,8	0,8	0,8	0,4444
1	2	3	0,43	0,22	1	0,67	0,5454
3	2	1	4,5	1,2	0,67	1	0,5454
2	2	2	2	0,5	1	1	0,6666

In Table 4, for some combinations of values μ_i, satisfying the condition

$$\sum_{i=1}^{3} \mu_i = 6 \tag{7}$$

the load factors ρ of equivalent two-phase models, the corresponding values of the maximum throughput $\omega_{max} = min\{\mu_1, \mu_2\}$, (AISO model) and $\omega_{max} =$

$min\{\mu_1, \mu_3\}$ (SIAO model) as well as the performance for the synchronous case, calculated by the formula (10) are given. In this formula, the number 6 is a close assemblage of the first 3 numbers of the Fibonacci series, where the enumeration goes as 1,2,3, then as 3,2,1, and in our case the first assembly has the form $\mu_i = 2$, that is $\mu_1 = \mu_2 = \mu_3 = 2$.

6 Conclusion

The presented material presents the results of a two-stage representation of the Needham-Schroeder algorithm as an implementation of Markov processes procedures. The representation is related to the consideration of three stages of the implementation of the Needham-Schroeder algorithm at different levels of combining input/output operations with its encryption. Buffering procedures for input or output information are defined. A method for exact performance calculation is presented and the conditions for balance of input/output and information encryption processes are determined. Calculation of system characteristics for a particular case is given.

References

1. Solution for the management, protection and control of pipeline leaks. https://spb-xxi.ru/index.php?option=com_content&view=article&id=39&Itemid=41.7. 5april2022
2. Vishnevsky V.M., Mukhtarov A.A., Pershin O.Yu. The problem of optimal placement of base stations of a broadband network to control a linear territory with a limitation on the amount of end-to-end delay. In: Vishnevsky, V.M., Samuilov, K.E. (eds.) Distributed Computer and Communication Networks: Control, Computation, Communications (DCCN-2020): Materials of XXIII International Scientific Conference, Moscow, 14–18 September 2020, pp. 148–155. IPU RAN (2020). https://2020.dccn.ru/downloads/DCCN-2020_Proceedings.pdf
3. Yermakov A.S.: Models for evaluating performance in synchronous/asynchronous execution of input/output processes and information processing. KazNU Bull. 1(60), 16–23 (2009). https://bm.kaznu.kz/index.php/kaznu/article/view/27
4. Modern Cryptography: Theory and Practice, p. 768. Williams (2005). https://www.ic.unicamp.br/~rdahab/cursos/mo421-mc889/Welcome_files/Stinson-Paterson_CryptographyTheoryAndPractice-CRC%20Press%20%282019%29.pdf
5. Rob, H.B.: Parallel Scientific Computation: A Structured Approach using BSP and MPI, p. 342. Oxford (2004)

Analysis of Functioning Photonic Switches in Next-Generation Networks Using Queueing Theory and Simulation Modeling

Elizaveta Barabanova[1]([✉]) [ID], Konstantin Vytovtov[1] [ID],
Vladimir Vishnevsky[1] [ID], and Iskander Khafizov[2] [ID]

[1] V.A. Trapeznikov Institute of Control Sciences of RAS,
Profsoyuznaya 65 Street, Moscow, Russia
elizavetaalexb@yandex.ru
[2] Moscow Institute of Physics and Technology, 9 Institutskiy per.,
Dolgoprudny, Moscow, Russia

Abstract. In this paper authors propose the approach for investigation of photonic switches performance metrics in modern all-optical networks. The approach includes photonic switches analytical and simulation modeling. The analytical model of the 4×4-photonic switch is based on the $M/M/1/4$ queuing model using so-called probability translation matrix method that allows to investigate functioning of the system in transient mode. The analytical expressions of such performance metrics as a loss probability, a buffer size and a transient time are obtained. A comparison of analytical and simulation modeling results was carried out. The analysis show that the steady-state loss probability obtained in results of analytical modeling is greater than the analogous parameter obtained in results of simulation modelling by 8%. The study of simulation models of 4×4 and 16×16-photonic switches show that the buffer size of these systems does not exceed one packet. Analysis of the 16×16-photonic switch buffer show that the buffer size of the first cascade is four times larger than the buffer of the second cascade. The results of the investigation show that the proposed type of photonic switches can be used in the high-performance all-optical network designed to transmit information from the ground module to tethered high-altitude unmanned module and vice versa.

Keywords: photonic switch · single-channel queuing system · Kolmogorov system of differential equations · simulation modeling · optical buffer

1 Introduction

An all-optical switch, or photonic switch, is a promising solution for use in next generation networks [1–7] and photonic supercomputer systems [8,9]. They have

---

The reported study was funded by Russian Science Foundation, project number 23-29-00795, https://rscf.ru/en/project/23-29-00795/.

such advantages as high-performance, energy efficiency, small size, etc. However, there are some features of photonic switch construction which inhibit the further development of such systems. The first problem of the existing all-optical switching systems is using of centralized electronic control principles. As photonic control circuits have not yet been developed the conversion of an optical signal inside an electronic processor reduces a performance of photonic switches [2]. Another disadvantage of existing high-capacity all-optical circuits is their high complexity, since a basic cell of an optical switch usually has only two inputs and two outputs [3]. Therefore, development of low complexity all-optical switches with decentralized control is an effective solution for perspective photonic networks.

The photonic switches based on 4×4 elements with decentralized control have been developed and described by the authors of this paper [4]. The methods of complexity and performance calculation and non-blocking estimation were proposed for this type of switches [4,10,11]. However, an important step before the practical implementation of all-optical switches is investigation of there functioning within a real optical network. Such a study can be carried out using analytical and simulation methods.

The analytical methods of optical switch functioning investigation are considered in [12–14]. The most of them are based on queuing theory and allow to calculate the main parameters of switches in stationary mode. Since the switching time of photonic switches can reach several picoseconds [11] that is usually much less than the transient time it is important to study not only a stationary mode of their operation, but also their transient behavior. The analytical method of the all-optical switches reliability indicators calculation in transient mode have been proposed by the authors of this paper [15]. But performance metrics of photonic switches in transient and stationary modes have not been calculated by using queuing theory and simulation modeling methods yet.

The aim of this work is development of analytical and simulation models of low complexity photonic switches with decentralized control for investigating the performance metrics of this type of switches in stationary and transient modes. The results of this approach can provide useful information about the limits of applicability of this type of switch in the next-generation networks and reduce the cost of its production.

The paper is organized as follows. In Sect. 2 the analytical model of 4×4 - photonic switch is given. Section 3 presents the simulation models of 4×4 and 16×16 photonic switches. Numerical simulation results are presented in Sect. 4.

2   The Analytical Model of 4×4 - Photonic Switch

Let us describe the process of packet switching in an all-optical network using queueing theory. Taking into account the peculiarities of functioning and construction of the proposed 4×4 - photonic switch [4], it has been used the $M/M/1/n$ model.

Taking into account that acceptable length fiber delay line using as optical buffer in photon switches can store approximately one or two packets it has been assumed that the 4×4 switching cell optical buffer capacity is equal to four packets. This buffer size corresponds to one packet per each input. So the 4×4-photonic switch can be described by the $M/M/1/4$ model. According to this model the system can be in one of six states with some probability p_i, were i corresponds to the number of one of the system states:

1. the buffer and the switching element are empty;
2. the buffer is empty, a packet is processed by the switching element;
3. one packet is processed by the switching element, and there is one packet in the buffer;
4. the switching element is occupied, and there are two packets in the buffer (arriving from different inputs);
5. the switching element is processing a packet, and there are three packets in the buffer;
6. the buffer is full, a new arriving packet will be lost (Fig. 1).

Fig. 1. The states graph of $M/M/1/4$ system

The $M/M/1/4$ system has been described by the Kolmogorov system of differential equations [16]:

$$
\begin{cases}
\dfrac{dp_i}{dt} = -\lambda p_i(t) + \mu p_{i+1}(t), i = 1 \\[2mm]
\dfrac{dp_i}{dt} = \lambda p_{i-1}(t) - (\lambda + \mu)p_i(t) + \mu p_{i+1}(t), i = \overline{2,5} \\[2mm]
\dfrac{dp_i}{dt} = \lambda p_{i-1}(t) - \mu p_i(t), i = 6
\end{cases}
\tag{1}
$$

were λ is the average number of packets arriving to the input of the switch per unit time (the arrival rate), μ is the average number of packets processed by the switch per unit time (the service rate).

According to the probability translation matrix method [17] the analytical solution of the system (1) can be found in the form

$$
p_i(t) = \sum_{j=1}^{5} p_{ij}(t) = \sum_{j=1}^{5} \xi_{ij} A_i exp(\gamma_j t)
\tag{2}
$$

where γ_i are the characteristic roots of the system, A_i are the constants given by the initial conditions, ξ_{ij} are the coefficients relating the system state probability in the initial moment of time with the probability of the system state at time t: $p_{ij}(t) = \xi_{ij}p_{0j}(t)$.

The vector of state probabilities $\mathbf{P}(t)$, can be written in terms of the probability translation matrix method [17]:

$$\mathbf{P}(t) = \mathbf{M}(t)\mathbf{P}(0) \tag{3}$$

where $\mathbf{M}(t)$ is the probability translation matrix [17] and $\mathbf{P}(0)$ is the vector of state probabilities in the initial moment of time.

In a stationary mode ($t \rightarrow \infty$) the steady-state probabilities can be calculated by the formula:

$$p_i(t \rightarrow \infty) = \frac{\lambda^i}{5\mu^i} \tag{4}$$

The important characteristic of the photonic switching system in transient mode is the transition time [17]. In [17] it is shown that the transition time $\tau_{tr} = (3 \div 5)\tau$, where $\tau = \max\{\tau_i, i = 1..6\} = \max\{\frac{1}{|\gamma_i|}, i = 1..6\}$), here γ_i are characteristic roots of the system.

3 The Simulation Models of 4×4 and 16×16 Photonic Switches

For simulation model creation Matlab & Simulink package was chosen. Simulink's SimEvents library was used as the main toolchain.

The simulation model of 4×4 photonic switch (Fig. 2) consists of two main logical blocks. The first block is the buffer (Buffer) and the second block is the switching device (Output Switch).

To simulate the delay lines, the FIFO Queue block was used. This module includes four blocks corresponded to four delay lines. Each of delay lines is able to store a predetermined number of packets arriving to corresponding input. The optical integrated device which controls the delay lines is modeled by the Path Combiner and the Single Server blocks (Fig. 3). The packet service time is used for the average packet waiting time calculation in the simulation model can be found as:

$$S_t = L \cdot 8 \cdot 0.8 \cdot 10^{-12} + m \cdot 3 \cdot 10^{-12} \tag{5}$$

where L is the packet size in bytes, m is the number of cascades ($m = 1$ for a 4×4-cell, $m = 2$ for a 16×16-cell). The first component of the expression (5) determines the transmission time of one packet for a laser pulse duration is equal to 0.8 ps. The second one allows to calculate the switching time of the $N \times N$ switching system (N is the number of system inputs), where $3 \cdot 10^{-12}$ s is the switching time of the 4×4-optical cell [11]. In according to the scheme of the photonic switch [4], the switching process is carried out without an external control based only on control optical signals arrived with the payload to the four inputs of the switch. The calculation of the average packet waiting time and the average buffer occupancy during simulation is organized inside the Buffer

360   E. Barabanova et al.

block. All this data is displayed in Matlab at the form of data structure. After passing through the Buffer block, the packet arrives to the Output switch block which forwards it to the desired output, depending on the specified "destination address" property. Output Switch produces a discrete signal, which value is equal to the number of packets passed through it (successfully switched). This value is also displayed in Matlab.

The simulation model of the 4×4 - photonic switch has been used to develop the simulation models of the 16×16 - system. The model of 16×16-switch (in Fig. 4) consists of two cascades of 4×4 - photonic switches according to the [4].

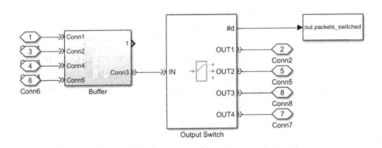

Fig. 2. Model of 4×4 photonic switch

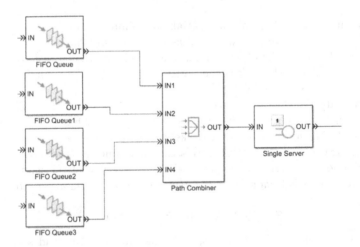

Fig. 3. The buffer model of the 4×4 photonic switch

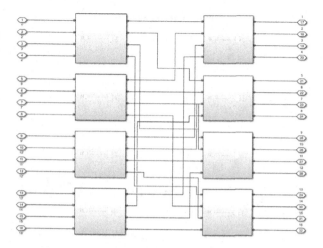

Fig. 4. Model of 16×16 photonic switch

4   Numerical Results

In this section the numerical results of analytical and simulation modelling are presented. The initial data for simulation modeling are:

- the packet size is 1500 bytes;
- the input channel throughput is 1.2 Tbps;
- the system load is 100%;
- the laser pulse duration is 0.8 ps

In accordance with the formula 5, since the 4×4-optical cell consists of one cascade, the service time, the service and arrival rates are: $S_t = 9.61 \cdot 10^{-9}$ s, $\mu = 1.04 \cdot 10^8$ packets/s, $\lambda = 1.07 \cdot 10^8$ packets/s correspondingly.

A comparison of the results of analytical (Fig. 5) and simulation modeling (Fig. 6) showed that the transient time in the cases of simulation and analytical modeling is $2\,\mu$s and $1.6\,\mu$s respectively. The steady-state loss probability obtained in results of analytical modeling is greater than the analogous parameter obtained in results of simulation modeling by 8%. The probabilities of $p_2(t) - p_5(t)$ states are not presented in Fig. 6, since they have a pronounced oscillatory character in the transient mode and it is hard to separate one from the other in the graph.

The dependencies of loss probability on packet size for 4×4 and 16×16-optical switching systems are presented in Fig. 7. As it can be seen from the Fig. 7 for a range of packet sizes $200 \div 3000$ bytes the loss probabilities is practically unchanged. It is less than 3% in the 4×4-optical system for the throughput of 1200 Gbps. The loss probabilities similar in their values are obtained in the case of using the 16×16-optical system but for the channel speed of 1150 Gbps (Table 1). But the loss probability of the 16×16-optical system abruptly increases

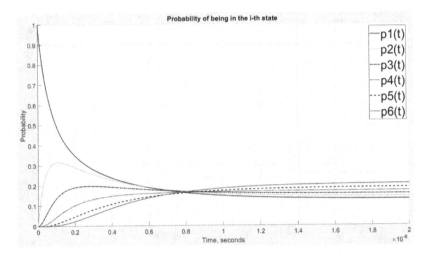

Fig. 5. Dependencies of the state probabilities on time - analytical results

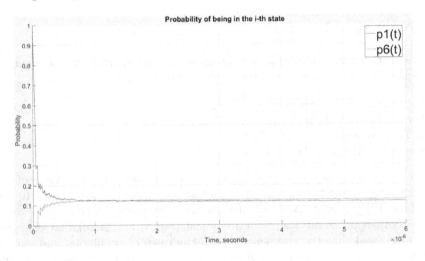

Fig. 6. Dependencies of the state probabilities on time - simulation results

up to 10% when the throughput of arrival channel is increased for 1200 Gbps (Table 2).

The results of 4×4 - optical switch loss probability investigation in depends of different network load are presented in Fig. 8. Here, the network load is the number of occupied inputs in percent. As you can see, when the network load is 100% and the channel speed is 1200 Gbps, the probability of loss does not exceed 3%. Simulation modeling showed that the analogous value of the loss probability can be obtain in the case of the network load is 90% and the channel speed is 1335 Gbps. The increasing of the channel speed to 1600 Gbps requires to decrease the network load to 75%. Figure 8 shows that the system functions without losses in following cases:

1. the channel speed is 1165 Gbps, network load is 100%;
2. the channel speed is 1290 Gbps, network load is 90%;
3. the channel speed is 1550 Gbps, network load is 75%;

The results of the photonic switches buffer occupancy investigation are presented in Fig. 9 and Fig. 10.

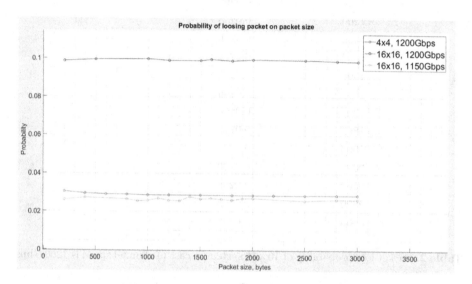

Fig. 7. Dependencies of loss probabilities on packet size

As you can see in Fig. 9 and Fig. 10 the 4×4 -photonic switches with a small size buffer (no more than one packet on one input) are perfectly acceptable for using in a high performance next-generation all-optical networks because the occupancy of their buffer in the results of simulation modelling does not exceed 50% in a stationary and transient modes. It also was found that the buffer of the first cascade of the 16×16 - photonic switch is in four times more occupied than the buffer of the second cascade (Fig. 10) and does not exceed one packet.

From the obtained results it can be seen that the system has a good performance metrics in a highly loaded all-optical network because the losses of data do not exceed to 3% and practically do not depend on the packet size. The result that the size of the optical buffer according to the simulation results does not exceed one packetis also a positive for perspective all-optical next-generation networks (Tables 3, 4 and 5).

Table 1. Packet loss probability of 16×16 system in case of channel speed is 1150 Gbps

Packet size, bytes	Packet loss probability
200	0.0261
400	0.0272
800	0.0265
900	0.0255
1000	0.0258
1100	0.0269
1200	0.0258
1300	0.0256
1400	0.0278
1500	0.0265
1600	0.0269
1700	0.0264
1800	0.0260
1900	0.0269
2000	0.0270
2500	0.0258
2800	0.0263
3000	0.0262

Table 2. Packet loss probability of 16×16 system in case of channel speed is 1200 Gbps

Packet size, bytes	Packet loss probability
200	0.0987
500	0.0995
1000	0.0998
1200	0.0990
1500	0.0990
1600	0.0996
1800	0.0989
2000	0.0995
2500	0.0993
2800	0.0988
3000	0.0986

Table 3. Dependence of packet loss probability on the channel speed, 100% load

Channel speed, Gbps	Packet loss probability
10	0.0000
1000	0.0000
1165	0.0000
1175	0.0079
1180	0.0121
1190	0.0204
1200	0.0286
1210	0.0366
1225	0.0484
1300	0.1033
1400	0.1673
1500	0.2229
1600	0.2714
1700	0.3143
2000	0.4171

Table 4. Dependence of packet loss probability on the channel speed, 90% load

Channel speed, Gbps	Packet loss probability
1000	0.0000
1295	0.0000
1300	0.0037
1325	0.0225
1350	0.0406
1375	0.0580
1400	0.0748
1500	0.1365
1600	0.1905
1700	0.2381

366   E. Barabanova et al.

Table 5. Dependence of packet loss probability on the channel speed, 75% load

Channel speed, Gbps	Packet loss probability
1200	0.0000
1550	0.0000
1575	0.0132
1600	0.0286
1625	0.0435
1650	0.0580
1700	0.0857

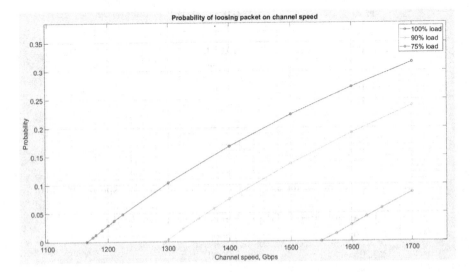

Fig. 8. The dependence of channel speed characteristics by different load

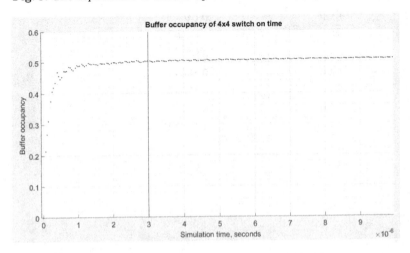

Fig. 9. The dependence of buffer occupancy of 4 × 4-switch on time

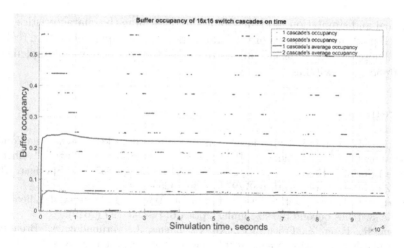

Fig. 10. The dependencies of buffer occupancy of 16×16-switch cascades on time

5 Conclusion

The analytical and simulation models of photonic switches based on 4×4 elements with decentralized control have been proposed. The analytical model of 4×4-photonic switch is considered as the $M/M/1/4$ queuing system model and investigated in stationary and transient modes. The expressions for the main performance metrics of photonic switch in stationary and transient modes have been obtained. The developed simulation models of 4×4 and 16×16-photonic switches allow to investigate their characteristics under various channel speeds and various packet sizes. The results of simulation modeling showed that 4×4 -photonic switch in case of 75% of network load functions without losses when the packet size is 1540bytes. The loss probabilities of 16×16 and 4×4-systems in the case of 100% of network load are approximately equal when the input channel speed is 1150 Gbps and 1200 Gbps correspondingly. A comparison of analytical and simulation modeling results showed that the transient modes time in simulation and analytical modeling is 2 μs and 0.16 μs respectively. The probability of loss in the case of analytical modeling is greater than in simulation modeling by 8%, which is a comparable result. The use of a small capacity optical buffer in simulation and analytical modeling showed that the proposed schemes of photonic switches with decentralized control can be used in promising all-optical networks. The results of an optical buffer occupancy investigation during simulation modeling showed that the photonic switches of small optical buffer capacity meet the requirements of perspective next-generation networks [19]. This is a very important result as the most of developed photonic switches can not be used in the real optical networks because of problems with high-capacity optical buffer physical implementation. Therefore, the proposed type of photonic switches can be used in an all-optical network designed to transmit information from the ground module to tethered high-altitude unmanned module and vice versa. The approach including analytical and simulation modeling allows to determine the characteristics of the proposed photonic switches more accurately and predict their behavior in real networks. Therefore, the approach allows to reduce the cost of physical implementation of photonic switches.

References

1. Argibay-Losada, P.J., Chiaroni, D., Qiao, C.: Optical packet switching and optical burst switching. In: Mukherjee, B., Tomkos, I., Tornatore, M., Winzer, P., Zhao, Y. (eds.) Springer Handbook of Optical Networks. SH, pp. 665–701. Springer, Cham (2020). https://doi.org/10.1007/978-3-030-16250-4_20
2. Seok, T.J., Kwon, K., Henriksson, J., Luo, J., Wu, M.C.: Wafer-scale silicon photonic switches beyond die size limit. Optica 6(4), 490–494 (2019)
3. Chen, R., Fang, Z., Fröch, J.E., Xu, P., Zheng, J., Majumdar, A.: Broadband nonvolatile electrically programmable silicon photonic switches. ACS Photon. 9(6), 2142–2150 (2022)
4. Barabanova, E.A., Vytovtov, K.A., Vishnevsky, V.M., Podlazov, V.S.: High-capacity strictly non-blocking optical switches based on new dual principle. J. Phys: Conf. Ser. **2091**, 012040 (2021)
5. Khlifi, Y.: Design, modeling and optimization using hybrid virtual topology for QoS provision over all-optical switched networks. Optik **258**, 168915 (2022)
6. Yang, M., Djordjevic, I.B., Tunc, C., Hariri, S., Akoglu, A.: An optical interconnect network design for dynamically composable data centers. In: Proceedings of the 2017 IEEE 19th International Conference on High Performance Computing and Communications; IEEE 15th International Conference on Smart City; IEEE 3rd International Conference on Data Science and Systems (HPCC/SmartCity/DSS), Bangkok, Thailand, 17595558 (2017)
7. De, S., Jyothish, M., Thunder, S.: Analytic modeling of optical recirculating loop to define the constraints of using it as optical buffer. In: 2017 International Conference on Computer, Electrical and Communication Engineering, ICCECE, Kolkata, India, 18248841 (2017)
8. Cheng, Q., Glick, M., Bergman, K.: Optical interconnection networks for high performance systems. Opt. Fiber Telecommun. **7**, 785–825 (2019)
9. Stepanenko, S.: Structure and Implementation Principles of a Photonic Computer. In: EPJ Web of Conferences, vol. 224, p. 04002 (2019)
10. Barabanova, E., Vytovtov, K., Podlazov, V.: Model and algorithm of next generation optical switching systems based on 8 × 8 elements. In: Vishnevskiy, V.M., Samouylov, K.E., Kozyrev, D.V. (eds.) DCCN 2019. LNCS, vol. 11965, pp. 58–70. Springer, Cham (2019). https://doi.org/10.1007/978-3-030-36614-8_5
11. Vytovtov, K.A., Barabanova, E.A., Vishnevsky, V.M.: Method for calculating performance of all-optical linear system of transmitting and processing information. In: XII International Conference "Digital Signal Processing and Its Application", 9213274 (2020) https://doi.org/10.1109/DSPA48919.2020
12. Jeong, H.-Y., Seo, S.-W.: Integrated model for performance analysis of all-optical multihop packet switches. Appl. Opt. **39**(26), 4770–4782 (2000)
13. Garofalakis, J., Stergiou, E.: Analytical model for performance evaluation of blocking Banyan switches supporting double priority traffic. In: 2008 International Conference on Communication Theory, Reliability, and Quality of Service, Bucharest, Romania, 10104675 (2008)

14. Gaballah, W.M.: Finite/infinite queueing models performance analysis in optical switching network nodes. Int. J. Sci. Res. Comput. Sci. Eng. Inf. Technol. 5(1), 163–170 (2019)
15. Barabanova, E.A., Vytovtov, K.A., Vishnevsky, V.M., Dvorkovich, A.V., Shurshev, V.F.: Investigating the reliability of all-optical switches in transient mode. J. Phys: Conf. Ser. 2091, 012039 (2021)
16. Dudin, A.N., Klimenok, V.I., Vishnevsky, V.M.: The Theory of Queuing Systems with Correlated Flows, 410p. Springer, Heidelberg (2020). https://doi.org/10.1007/978-3-030-32072-0
17. Vishnevsky, V.M., Vytovtov, K.A., Barabanova, E.A., Semenova, O.V.: Transient behavior of the MAP/M/1/N queuing system. Mathematics 9(20), 2559 (2021)
18. Chang, C.-S., Chen, Y.-T., Lee, D.-S.: Constructions of optical FIFO queues. IEEE Trans. Inf. Theory 52(6), 2838–2843 (2006)
19. Kozyrev, D., et al.: Mobility-centric analysis of communication offloading for heterogeneous Internet of Things Devices. Wirel. Commun. Mob. Comput. 2018, 3761075 (2018)

Two-Way Communication Retrial Queue with Markov Modulated Poisson Input and Multiple Types of Outgoing Calls

Anatoly Nazarov[1] , Tuan Phung-Duc[2] , Svetlana Paul[1] ,
and Olga Lizyura[1(✉)]

[1] Institute of Applied Mathematics and Computer Science, National Research Tomsk
State University, 36 Lenina ave., Tomsk 634050, Russia
oliztsu@mail.ru
[2] Faculty of Engineering Information and Systems, University of Tsukuba,
1-1-1 Tennodai, Tsukuba, Ibaraki 305-8573, Japan
tuan@sk.tsukuba.ac.jp

Abstract. In this paper, we consider single server two-way communication retrial queue with Markov modulated Poisson input. Such system
serves two types of calls: incoming and outgoing. We assume that incoming calls arrive according to the input process and retry in case the server
is busy upon arrival. If the server is occupied upon arrival, the incoming call joins the orbit and waits for some time before making the next
try to receive service. On the other hand, outgoing calls are initiated
by server and do not make redial. We present the analysis of the queue
using asymptotic diffusion method.

Keywords: retrial queue · two-way communication · incoming call ·
outgoing call · asymptotic-diffusion analysis · diffusion approximation

1 Introduction

Retrial queues with two-way communication were invented as a models of
blended call centers. In classical call centers, the operator has only one of two
options: making calls or answering calls. In blended call centers operator switches
between these two activities depending on the situation, which allows to lower
the idle time. In papers [7,8], the team of authors explore queues with two-way
communication from the profit maximization point of view.

The phenomenon of repeated calls also contributes to the modeling of call
centers. Papers [1,5] are devoted to studies of the influence of call delay on the
operation of telephone services. Authors consider delay announcement in call
centers and its impact on customers decision making.

Statistical studies of call centers are also of interest. In papers [12,15] authors
analyse the statistical data about call center operation and provide some advices
on performance improvement. Article [4] provides an analysis of arrival models

V. M. Vishnevskiy et al. (Eds.): DCCN 2022, CCIS 1748, pp. 370–381, 2023.
https://doi.org/10.1007/978-3-031-30648-8_29

suitable for modeling traffic in call centers. Paper [3] is devoted to the phenomena of lognormal service times in such systems.

Models with two-way communication were considered under the assumption that the system is closed, that is, when there is a limited number of external sources of incoming calls. The works devoted to simulation modeling of such systems are [6,13,14].

Two-way communication models were also considered under the assumption that the device calls requests not from third-party sources, but from the orbit [2,9].

In the current paper, we present asymptotic diffusion method for two-way communication retrial queue with Markov modulated Poisson input and multiple types of outgoing calls. Previously, we have provided the asymptotic diffusion analysis of retrial queue with two-way communication and renewal input [11] and multiserver version of the model with Poisson input [10]. Such method allows expanding the applicability area of the asymptotic approximation for distribution of the number of calls in the orbit.

The rest of our paper is structured as follows. In Sect. 2, we provide detailed description of mathematical model under investigation. Section 3 is devoted to asymptotic diffusion analysis of the system. Section 4 shortly describes how to use theoretic formulas. In Sect. 5, we provide a numerical example to demonstrate the accuracy of the asymptotic approximation. Finally, Sect. 6 is dedicated to the conclusion.

2   Mathematical Model

We consider a single server retrial queue with two-way communication and multiple types of outgoing calls, which is shown in Fig. 1. The input is Markov modulated Poisson process, which is defined by infinitesimal generator $\mathbf{Q}$ and diagonal matrix of conditional intensities $\mathbf{\Lambda}$. Service times of incoming calls are exponentially distributed with parameter μ_1.

If the server is busy upon arrival, an incoming call goes into orbit and makes some delay before redial. The duration of delay follows the exponential distribution with parameter σ.

If the server is idle, it initiates outgoing calls of type n with rate α_n, $n = 2, 3, ..., N$. We numerate types of outgoing calls from 2 to N to simplify the derivations. We assume that the server calls requests of all types at the same time, and the service is received by the request that arrived first. Service times of outgoing calls of type n are exponentially distributed with parameter μ_n, $n = 2, 3, ..., N$.

We denote the state of Markov chain underlying the input MMPP as $m(t)$, the number of calls in the orbit as $i(t)$ and the state of the server as $k(t)$. Process $k(t)$ has $N + 1$ states: 0 for the idle server, 1 for the service of incoming call, n for the service of outgoing call of type n, $n = 2, 3, ..., N$.

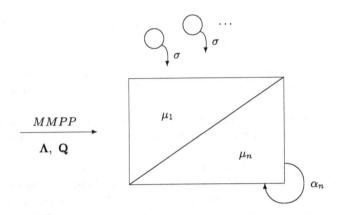

Fig. 1. Retrial queue with two-way communication and MMPP input

Process $\{i(t),\ k(t),\ m(t)\}$ is a three-dimensional Markov chain with probability distribution $P_k(i, m, t) = \mathbb{P}\{i(t) = i,\ k(t) = k,\ m(t) = m\}$. This distribution satisfies the following Kolmogorov system of equations

$$\frac{\partial P_0(i, m, t)}{\partial t} = -\left(\lambda_m + i\sigma + \sum_{n=2}^{N} \alpha_n\right) P_0(i, m, t) + \sum_{k=1}^{N} \mu_k P_k(i, m, t)$$

$$+ \sum_{\nu=1}^{M} P_0(i, \nu, t) q_{\nu m},$$

$$\frac{\partial P_1(i, m, t)}{\partial t} = -(\lambda_m + \mu_1) P_1(i, m, t) + \lambda_m P_1(i - 1, m, t) + \lambda_m P_0(i, m, t)$$

$$+(i + 1)\sigma P_0(i + 1, m, t) + \sum_{\nu=1}^{M} P_1(i, \nu, t) q_{\nu m},$$

$$\frac{\partial P_n(i, m, t)}{\partial t} = -(\lambda_m + \mu_n) P_n(i, m, t) + \lambda_m P_n(i - 1, m, t) + \alpha_n P_0(i, m, t)$$

$$+ \sum_{\nu=1}^{M} P_0(i, \nu, t) q_{\nu m},\ n = \overline{2, N}. \tag{1}$$

Solving system (1) under limit condition $\sigma \to 0$, we build the approximation of probability distribution of the number of calls in the orbit.

3   Asymptotic Diffusion Method

First, we transform system (1) to the system for characteristic functions using notations

$$H_k(u, m, t) = \sum_{i=0}^{\infty} e^{jui} P_k(i, m, t),\ k = \overline{0, N},$$

where j is the imaginary unit. Furthermore, we rewrite the system in matrix form, which yields

$$\frac{\partial \mathbf{H}_0(u,t)}{\partial t} = \mathbf{H}_0(u,t)\left(\mathbf{Q} - \mathbf{\Lambda} - \sum_{n=2}^{N} \alpha_n \mathbf{I}\right) + j\sigma \frac{\partial \mathbf{H}_0(u,t)}{\partial t} + \sum_{k=1}^{N} \mu_k \mathbf{H}_k(u,t),$$

$$\frac{\partial \mathbf{H}_1(u,t)}{\partial t} = \mathbf{H}_1(u,t)(\mathbf{Q} + (e^{ju} - 1)\mathbf{\Lambda} - \mu_1 \mathbf{I}) + \mathbf{H}_0(u,t)\mathbf{\Lambda} - j\sigma e^{-ju}\frac{\partial \mathbf{H}_0(u,t)}{\partial u},$$

$$\frac{\partial \mathbf{H}_n(u,t)}{\partial t} = \mathbf{H}_n(u,t)(\mathbf{Q} + (e^{ju} - 1)\mathbf{\Lambda} - \mu_n \mathbf{I}) + \alpha_n \mathbf{H}_0(u,t), \quad n = 2, ..., N, \quad (2)$$

where $\mathbf{H}_k(u,t) = \{H_k(u,1,t), H_k(u,2,t), ..., H_k(u,M,t)\}$, $k = 0, ..., N$, $\mathbf{I}$ is an identity matrix of corresponding dimension. We also derive an additional equation

$$\frac{\partial \mathbf{H}(u,t)}{\partial t}\mathbf{e} = (e^{ju} - 1)\left\{ j\sigma e^{-ju}\frac{\partial \mathbf{H}_0(u,t)}{\partial u}\mathbf{e} + \sum_{k=1}^{N} \mathbf{H}_k(u,t)\mathbf{\Lambda}\mathbf{e} \right\}, \qquad (3)$$

where $\mathbf{e}$ is the vector of ones.

3.1   First Step of Diffusion Analysis

In system (2) and Eq. (3), we introduce the following notations

$$\sigma = \varepsilon, \quad \tau = \varepsilon t, \quad u = \varepsilon w, \quad \mathbf{H}_k(u,t) = \mathbf{F}_k(w,\tau,\varepsilon), \quad k = \overline{0,N},$$

which yields

$$\varepsilon \frac{\partial \mathbf{F}_0(w,\tau,\varepsilon)}{\partial \tau} = \mathbf{F}_0(w,\tau,\varepsilon)\left(\mathbf{Q} - \mathbf{\Lambda} - \sum_{n=2}^{N} \alpha_n \mathbf{I}\right) + j\frac{\partial \mathbf{F}_0(w,\tau,\varepsilon)}{\partial w}$$

$$+ \sum_{k=1}^{N} \mu_k \mathbf{F}_k(w,\tau,\varepsilon),$$

$$\varepsilon \frac{\partial \mathbf{F}_1(w,\tau,\varepsilon)}{\partial \tau} = \mathbf{F}_1(w,\tau,\varepsilon)\left(\mathbf{Q} + (e^{jw\varepsilon} - 1)\mathbf{\Lambda} - \mu_1 \mathbf{I}\right) + \mathbf{F}_0(w,\tau,\varepsilon)\mathbf{\Lambda}$$

$$- je^{-jw\varepsilon}\frac{\partial \mathbf{F}_0(w,\tau,\varepsilon)}{\partial w},$$

$$\varepsilon \frac{\partial \mathbf{F}_n(w,\tau,\varepsilon)}{\partial \tau} = \mathbf{F}_n(w,\tau,\varepsilon)\left(\mathbf{Q} + (e^{jw\varepsilon} - 1)\mathbf{\Lambda} - \mu_n \mathbf{I}\right)$$

$$+ \alpha_n \mathbf{F}_0(w,\tau,\varepsilon), \quad n = 2, ..., N. \qquad (4)$$

$$\varepsilon \frac{\partial \mathbf{F}(w,\tau,\varepsilon)}{\partial \tau}\mathbf{e}$$

$$= (e^{jw\varepsilon} - 1)\left\{ je^{-jw\varepsilon}\frac{\partial \mathbf{F}_0(w,\tau,\varepsilon)}{\partial w}\mathbf{e} + (\mathbf{F}(w,\tau,\varepsilon) - \mathbf{F}_0(w,\tau,\varepsilon))\mathbf{\Lambda}\mathbf{e} \right\}. \qquad (5)$$

Solving (4) under the limit condition $\varepsilon \to 0$, we present Theorem 1.

Theorem 1.

$$\lim_{\sigma \to 0} \mathbb{E} e^{jw\sigma i \left(\frac{\tau}{\sigma} \right)} = e^{jwx(\tau)}, \tag{6}$$

where $x(\tau)$ is the solution of

$$x'(\tau) = -x(\tau) \mathbf{r}_0 \mathbf{e} + (\mathbf{r} - \mathbf{r}_0) \mathbf{\Lambda} \mathbf{e}. \tag{7}$$

Vector $\mathbf{r}$ is the probability distribution of underlying Markov chain $m(t)$. Vectors $\mathbf{r}_k = \mathbf{r}_k(x)$ satisfy

$$\mathbf{r}_0 \left(\mathbf{Q} - \mathbf{\Lambda} - \sum_{n=2}^{N} \alpha_n \mathbf{I} - x\mathbf{I} \right) + \sum_{k=1}^{N} \mu_k \mathbf{r}_k = 0,$$

$$\mathbf{r}_1 (\mathbf{Q} - \mu_1 \mathbf{I}) + \mathbf{r}_0 (\mathbf{\Lambda} + x\mathbf{I}) = 0,$$

$$\mathbf{r}_n (\mathbf{Q} - \mu_n \mathbf{I}) + \alpha_n \mathbf{r}_0 = 0, \ n = 2, ..., N,$$

$$\sum_{k=0}^{N} \mathbf{r}_k = \mathbf{r}. \tag{8}$$

Proof. Taking the limit by $\varepsilon \to 0$ in (4) and (5), we obtain

$$\mathbf{F}_0(w, \tau) \left(\mathbf{Q} - \mathbf{\Lambda} - \sum_{n=2}^{N} \alpha_n \mathbf{I} \right) + j \frac{\partial \mathbf{F}_0(w, \tau)}{\partial w} + \sum_{k=1}^{N} \mu_k \mathbf{F}_k(w, \tau) = 0,$$

$$\mathbf{F}_1(w, \tau)(\mathbf{Q} - \mu_1 \mathbf{I}) + \mathbf{F}_0(w, \tau)\mathbf{\Lambda} - j \frac{\partial \mathbf{F}_0(w, \tau)}{\partial w} = 0,$$

$$\mathbf{F}_n(w, \tau)(\mathbf{Q} - \mu_n \mathbf{I}) + \alpha_n \mathbf{F}_0(w, \tau) = 0, \ n = 2, ..., N, \tag{9}$$

$$\frac{\partial \mathbf{F}(w, \tau)}{\partial \tau} \mathbf{e} = jw \left\{ j \frac{\partial \mathbf{F}_0(w, \tau)}{\partial w} \mathbf{e} + (\mathbf{F}(w, \tau) - \mathbf{F}_0(w, \tau))\mathbf{\Lambda} \mathbf{e} \right\}, \tag{10}$$

where

$$\lim_{\varepsilon \to 0} \mathbf{F}_k(w, \tau, \varepsilon) = \mathbf{F}_k(w, \tau), \ k = 0, ..., N.$$

We seek the solution of the obtained system in the following form:

$$\mathbf{F}_k(w, \tau) = e^{jwx(\tau)} \mathbf{r}_k,$$

which yields (7) (taking into account normalization condition) and (8). The theorem is proved.

The right part of (7) is the drift coefficient of the future diffusion approximation, which we denote as

$$a(x) = -x\mathbf{r}_0(x)\mathbf{e} + (\mathbf{r} - \mathbf{r}_0(x))\mathbf{\Lambda} \mathbf{e}. \tag{11}$$

3.2   Second Step of Diffusion Analysis

Returning to system (2) and Eq. (3), we make the following substitution

$$\mathbf{H}_k(u,t) = e^{j\frac{u}{\sigma}x(\sigma t)}\mathbf{H}_k^{(2)}(u,t),$$

which give us the system for characteristic functions $\mathbf{H}_k^{(2)}(u,t)$ representing the centered number of calls in the orbit. After that, we introduce the notations

$$\sigma = \varepsilon^2, \ \tau = \varepsilon^2 t, \ u = \varepsilon w, \ \mathbf{H}_k^{(2)}(u,t) = \mathbf{F}_k^{(2)}(w,\tau,\varepsilon), \ k = 0, ..., N,$$

and obtain

$$\varepsilon^2 \frac{\partial \mathbf{F}_0^{(2)}(w,\tau,\varepsilon)}{\partial \tau} + j\varepsilon w a(x)\mathbf{F}_0^{(2)}(w,\tau,\varepsilon)$$

$$= \mathbf{F}_0^{(2)}(w,\tau,\varepsilon)\left(\mathbf{Q} - \mathbf{\Lambda} - \left(x(\tau) + \sum_{n=2}^{N}\alpha_n\right)\mathbf{I}\right) + j\varepsilon\frac{\partial \mathbf{F}_0^{(2)}(w,\tau,\varepsilon)}{\partial w}$$

$$+ \sum_{k=1}^{N}\mu_k\mathbf{F}_k^{(2)}(w,\tau,\varepsilon),$$

$$\varepsilon^2 \frac{\partial \mathbf{F}_1^{(2)}(w,\tau,\varepsilon)}{\partial \tau} + j\varepsilon w a(x)\mathbf{F}_1^{(2)}(w,\tau,\varepsilon)$$

$$= \mathbf{F}_1^{(2)}(w,\tau,\varepsilon)(\mathbf{Q} + (e^{ju} - 1)\mathbf{\Lambda} - \mu_1\mathbf{I})$$

$$+ \mathbf{F}_0^{(2)}(w,\tau,\varepsilon)(\mathbf{\Lambda} + e^{-jw\varepsilon}x(\tau)\mathbf{I}) - j\varepsilon\varepsilon e^{-jw\varepsilon}\frac{\partial \mathbf{F}_0^{(2)}(w,\tau,\varepsilon)}{\partial w},$$

$$\varepsilon^2 \frac{\partial \mathbf{F}_n^{(2)}(w,\tau,\varepsilon)}{\partial \tau} + j\varepsilon w a(x)\mathbf{F}_n^{(2)}(w,\tau,\varepsilon) =$$

$$\mathbf{F}_n^{(2)}(w,\tau,\varepsilon)(\mathbf{Q} + (e^{jw\varepsilon} - 1)\mathbf{\Lambda} - \mu_n\mathbf{I}) + \alpha_n\mathbf{F}_0^{(2)}(w,\tau,\varepsilon), \qquad (12)$$

$$\varepsilon^2 \frac{\partial \mathbf{F}^{(2)}(w,\tau,\varepsilon)}{\partial \tau}\mathbf{e} + j\varepsilon w a(x)\mathbf{F}^{(2)}(w,\tau,\varepsilon)\mathbf{e}$$

$$= (e^{jw\varepsilon} - 1)\left\{ j\varepsilon e^{-jw\varepsilon}\frac{\partial \mathbf{F}_0^{(2)}(w,\tau,\varepsilon)}{\partial w}\mathbf{e} \right.$$

$$\left. -x(\tau)e^{-jw\varepsilon}\mathbf{F}_0^{(2)}(w,\tau,\varepsilon)\mathbf{e} + \left(\mathbf{F}^{(2)}(w,\tau,\varepsilon) - \mathbf{F}_0^{(2)}(w,\tau,\varepsilon)\right)\mathbf{\Lambda}\mathbf{e} \right\}. \qquad (13)$$

Theorem 2. *Asymptotic characteristic function* $\mathbf{F}^{(2)}(w,\tau) = \lim_{\varepsilon \to 0}\mathbf{F}^{(2)}(w,\tau,\varepsilon)$ *is given by*

$$\mathbf{F}^{(2)}(w,\tau) = \Phi(w,\tau)\mathbf{r}, \qquad (14)$$

where $\mathbf{r}$ *is the stationary probability distribution of the states of underlying Markov chain* $m(t)$ *and* $\Phi(w,\tau)$ *is the solution of*

$$\frac{\partial \Phi(w,\tau)}{\partial \tau} = a'(x)w\frac{\partial \Phi(w,\tau)}{\partial w} + \frac{(jw)^2}{2}b(x)\Phi(w,\tau). \qquad (15)$$

$a(x)$ *is given by* (11) *and* $b(x)$ *is given by*

$$b(x) = a(x) + 2\{-x(\tau)\mathbf{g}_0\mathbf{e} + (\mathbf{g} - \mathbf{g}_0)\Lambda\mathbf{e} + x(\tau)\mathbf{r}_0\mathbf{e}\}, \quad (16)$$

and where vector $\mathbf{g} = \mathbf{g}(x)$ *is defined by*

$$\mathbf{g}_0\left(\mathbf{Q} - \Lambda - \left(x + \sum_{n=2}^{N}\alpha_n\right)\mathbf{I}\right) + \sum_{k=1}^{N}\mu_k\mathbf{g}_k = a(x)\mathbf{r}_0,$$

$$\mathbf{g}_0(\Lambda + x\mathbf{I}) + \mathbf{g}_1(\mathbf{Q} - \mu_1\mathbf{I}) = a(x)\mathbf{r}_1 - \mathbf{r}_1\Lambda + x\mathbf{r}_0,$$

$$\alpha_n\mathbf{g}_0 + \mathbf{g}_n(\mathbf{Q} - \mu_n\mathbf{I}) = a(x)\mathbf{r}_n - \mathbf{r}_n\Lambda, \quad n = \overline{2, N},$$

$$\sum_{k=0}^{N}\mathbf{g}_k\mathbf{e} = 0. \quad (17)$$

Proof. In (12), we expand the exponents using Taylor series and group the terms of order no higher than ε. Then we present the solution in the form of

$$\mathbf{F}_k^{(2)}(w, \tau, \varepsilon) = \Phi(w, \tau)\{\mathbf{r}_k + jw\varepsilon\mathbf{f}_k\} + O(\varepsilon^2). \quad (18)$$

We obtain

$$jw\varepsilon a(x)\Phi(w, \tau)\mathbf{r}_0 = jw\varepsilon\Phi(w, \tau)\mathbf{f}_0\left(\mathbf{Q} - \Lambda - \left(x + \sum_{n=2}^{N}\alpha_n\right)\mathbf{I}\right)$$

$$+ j\varepsilon\frac{\partial\Phi(w, \tau)}{\partial w}\mathbf{r}_0 + jw\varepsilon\Phi(w, \tau)\sum_{k=1}^{N}\mu_k\mathbf{f}_k + O(\varepsilon^2),$$

$$jw\varepsilon a(x)\Phi(w, \tau)\mathbf{r}_1 = jw\varepsilon\Phi(w, \tau)\mathbf{f}_1(\mathbf{Q} - \mu_1\mathbf{I}) + jw\varepsilon\Phi(w, \tau)\mathbf{r}_1\Lambda$$

$$- jw\varepsilon x\Phi(w, \tau)\mathbf{r}_0 + jw\varepsilon\Phi(w, \tau)\mathbf{f}_0(\Lambda + x\mathbf{I}) - j\varepsilon\frac{\partial\Phi(w, \tau)}{\partial w}\mathbf{r}_0 + O(\varepsilon^2),$$

$$jw\varepsilon a(x)\Phi(w, \tau)\mathbf{r}_n = jw\varepsilon\Phi(w, \tau)\mathbf{f}_n(\mathbf{Q} - \mu_n\mathbf{I}) + jw\varepsilon\Phi(w, \tau)\mathbf{r}_n\Lambda$$

$$+ jw\varepsilon\alpha_n\Phi(w, \tau)\mathbf{f}_0 + O(\varepsilon^2), \quad n = 2, ..., N.$$

Dividing equations by $jw\varepsilon\Phi(w, \tau)$ and taking the limit by $\varepsilon \to 0$ yields

$$\mathbf{f}_0\left(\mathbf{Q} - \Lambda - \left(x + \sum_{n=2}^{N}\alpha_n\right)\mathbf{I}\right) + \sum_{k=1}^{N}\mu_k\mathbf{f}_k = a(x)\mathbf{r}_0 - \frac{\partial\Phi(w, \tau)/\partial w}{w\Phi(w, \tau)}\mathbf{r}_0,$$

$$\mathbf{f}_1(\mathbf{Q} - \mu_1\mathbf{I}) + \mathbf{f}_0(\Lambda + x\mathbf{I}) = a(x)\mathbf{r}_1 - \mathbf{r}_1\Lambda + x\mathbf{r}_0 + \frac{\partial\Phi(w, \tau)/\partial w}{w\Phi(w, \tau)}\mathbf{r}_0,$$

$$\mathbf{f}_n(\mathbf{Q} - \mu_n\mathbf{I}) + \alpha_n\mathbf{f}_0 = a(x)\mathbf{r}_n - \mathbf{r}_n\Lambda, \quad n = 2, ..., N.$$

Presenting vectors $\mathbf{f}_k$ in the form of

$$\mathbf{f}_k = C\mathbf{r}_k + \mathbf{g}_k - \varphi_k\frac{\partial\Phi(w, \tau)/\partial w}{w\Phi(w, \tau)}, \quad (19)$$

we obtain two systems of equations for $\mathbf{g}_k = \mathbf{g}_k(x)$ and $\varphi_k = \varphi_k(x)$

$$\mathbf{g}_0\left(\mathbf{Q} - \mathbf{\Lambda} - \left(x + \sum_{n=2}^{N} \alpha_n\right)\mathbf{I}\right) + \sum_{k=1}^{N} \mu_k \mathbf{g}_k = a(x)\mathbf{r}_0,$$

$$\mathbf{g}_0(\mathbf{\Lambda} + x\mathbf{I}) + \mathbf{g}_1(\mathbf{Q} - \mu_1\mathbf{I}) = a(x)\mathbf{r}_1 - \mathbf{r}_1\mathbf{\Lambda} + x\mathbf{r}_0,$$

$$\alpha_n\mathbf{g}_0 + \mathbf{g}_n(\mathbf{Q} - \mu_n\mathbf{I}) = a(x)\mathbf{r}_n - \mathbf{r}_n\mathbf{\Lambda}, \ n = 2, ..., N, \tag{20}$$

$$\varphi_0\left(\mathbf{Q} - \mathbf{\Lambda} - \left(x + \sum_{n=2}^{N} \alpha_n\right)\mathbf{I}\right) + \sum_{k=1}^{N} \mu_k \varphi_k = \mathbf{r}_0,$$

$$\varphi_0(\mathbf{\Lambda} + x\mathbf{I}) + \varphi_1(\mathbf{Q} - \mu_1\mathbf{I}) = -\mathbf{r}_0,$$

$$\alpha_n\varphi_0 + \varphi_n(\mathbf{Q} - \mu_n\mathbf{I}) = 0, \ n = 2, ..., N. \tag{21}$$

System (20) together with additional condition $\sum_{k=0}^{N} \mathbf{g}_k\mathbf{e} = 0$ coincide with (17). Considering system (21), we note that the system is derivative of system (8), then we conclude that

$$\varphi_k(x) = \mathbf{r}_k'(x).$$

After that, we consider (3), expand the exponents using Taylor series and group the terms of order no higher than ε^2. Then we make decompositions (18) and (19) and in the limit by $\varepsilon \to 0$ to obtain

$$\frac{\partial \Phi(w, \tau)}{\partial \tau} = (jw)^2 \Phi(w, \tau)\left\{ -x(\tau)\mathbf{g}_0\mathbf{e} + (\mathbf{g} - \mathbf{g}_0)\mathbf{\Lambda e}\right.$$

$$-\frac{\partial \Phi(w, \tau)/\partial w}{w\Phi(w, \tau)}\left(-x(\tau)\varphi_0\mathbf{e}(\varphi - \varphi_0)\mathbf{\Lambda e} - \mathbf{r}_0\mathbf{e}\right)\Big\}$$

$$+\frac{(jw)^2}{2}\Big\{ a(x)\Phi(w, \tau) + 2x(\tau)\Phi(w, \tau)\mathbf{r}_0\mathbf{e}\Big\}.$$

We note that

$$-x(\tau)\varphi_0\mathbf{e} + (\varphi - \varphi_0)\mathbf{\Lambda e} - \mathbf{r}_0\mathbf{e} = a'(x).$$

Denoting

$$b(x) = a(x) + 2\left\{-x(\tau)\mathbf{g}_0\mathbf{e} + (\mathbf{g} - \mathbf{g}_0)\mathbf{\Lambda e} + x(\tau)\mathbf{r}_0\mathbf{e}\right\},$$

which coincides with (16), we obtain

$$\frac{\partial \Phi(w, \tau)}{\partial \tau} = a'(x)w\frac{\partial \Phi(w, \tau)}{\partial w} + \frac{(jw)^2}{2}b(x)\Phi(w, \tau),$$

which is the same as (15). The theorem is proved.

378 A. Nazarov et al.

Function $b(x)$ is the diffusion coefficient of the scaled number of calls in the orbit.

Making the inverse Fourier transform in (15), we obtain Fokker-Planck equation for probability density $p(y,\tau)$ of the scaled number of calls in the orbit $y(\tau)$ as follows:

$$\frac{\partial p(y,\tau)}{\partial \tau} = -\frac{\partial}{\partial y}\{a'(x)yp(y,\tau)\} + \frac{1}{2}\frac{\partial^2}{\partial y^2}\{b(x)p(y,\tau)\}. \tag{22}$$

The process $y(\tau)$ is the solution of stochastic differential equation

$$dy(\tau) = a'(x)yd\tau + \sqrt{b(x)}dw(\tau), \tag{23}$$

where $w(\tau)$ is Wiener process. We introduce

$$z(\tau) = x(\tau) + \varepsilon y(\tau), \tag{24}$$

where $\varepsilon = \sqrt{\sigma}$. We note that $z(\tau) = \lim_{\sigma\to 0}\sigma i(\tau/\sigma)$. Taking into account that $dx(\tau) = a(x)d\tau$, we obtain

$$dz(\tau) = d(x(\tau)+\varepsilon y(\tau)) = (a(x)+\varepsilon ya'(x))d\tau + \varepsilon\sqrt{b(x)}dw(\tau). \tag{25}$$

Using decompositions

$$a(z) = a(x+\varepsilon y) = a(x) + \varepsilon ya'(x) + o(\varepsilon^2),$$
$$\varepsilon\sqrt{b(z)} = \varepsilon\sqrt{b(x+\varepsilon y)} = \varepsilon\sqrt{b(x)+o(\varepsilon)} = \varepsilon\sqrt{b(x)} + o(\varepsilon^2),$$

we transform (25) up to $o(\varepsilon^2)$

$$dz(\tau) = a(z)d\tau + \sqrt{\sigma b(z)}dw(\tau). \tag{26}$$

We denote the stationary probability density of the process $z(\tau)$ as $S(z)$.

Theorem 3. *Stationary probability density $S(z)$ of diffusion process $z(\tau)$ is given as follows:*

$$S(z) = \frac{C}{b(z)}\exp\left\{\frac{2}{\sigma}\int_0^z \frac{a(x)}{b(x)}dx\right\}, \tag{27}$$

where C is normalizing constant.

Proof. Since the process $z(\tau)$ is the solution of (26), it is a diffusion process having drift coefficient $a(z)$ and diffusion coefficient $b(z)$. Thus, the stationary probability density is the solution of Fokker-Planck equation

$$-\frac{\partial}{\partial z}\{a(z)S(z)\} + \frac{1}{2}\frac{\partial^2}{\partial z^2}\{\sigma b(z)S(z)\} = 0.$$

The solution of the equation above has the following form:

$$S(z) = \frac{C}{b(z)} \exp\left\{ \frac{2}{\sigma} \int_0^z \frac{a(x)}{b(x)} dx \right\},$$

which coincides with (27). The theorem is proved.

To build the diffusion approximation of the number of calls in the orbit $i(t)$ we use formula

$$P(i) = \frac{S(\sigma i)}{\sum_{n=0}^{\infty} S(\sigma n)}. \tag{28}$$

We note that using formula (28), we do not need obtaining constant C.

4 Calculation Algorithm

1. Define parameters σ, μ_1, μ_n, α_n, $n = 2, ..., N$ of the system and matrices $\mathbf{Q}$ and $\mathbf{\Lambda}$ describing the MMPP.
2. Calculate vector $\mathbf{r}$ as the solution of linear system $\mathbf{rQ} = 0$, $\mathbf{re} = 1$.
3. Calculate vectors $\mathbf{r}_k(x)$ as a vectors of functions, which is the solution of system (8) for each x.
4. Define drift coefficient $a(x)$ as a function of x using formula (11).
5. Obtain vectors $\mathbf{g}_k$ as the solution of linear system (17) for each x.
6. Calculate diffusion coefficient $b(x)$ as a function of x using formula (16).
7. Build discrete approximation of probability distribution $P(i)$ of the number of calls in the orbit using formula (28).

5 Numerical Example

For the numerical example, we set $N = 4$, $\mu_1 = 1$, $\mu_2 = 2$, $\mu_3 = 3$, $\mu_4 = 4$, $\alpha_2 = 1$, $\alpha_3 = 2$, $\alpha_4 = 3$,

$$\mathbf{Q} = \begin{bmatrix} -1 & 0.2 & 0.8 \\ 0.5 & -1.5 & 1 \\ 0.8 & 1.2 & -2 \end{bmatrix}.$$

Matrix $\mathbf{\Lambda}$ is defined using formula

$$\mathbf{\Lambda} = \frac{\rho \mu_1 \mathbf{\Lambda}_1}{\mathbf{r}\mathbf{\Lambda}_1\mathbf{e}},$$

where

$$\mathbf{\Lambda}_1 = \text{diag}\{1, 2, 3\}.$$

We see how the load parameter ρ affects the accuracy of approximation (28)

$$\rho = \frac{\mathbf{r}\mathbf{\Lambda}\mathbf{e}}{\mu_1}.$$

In Table 1, we present Kolmogorov distance between our approximation and simulation results. Formula for Kolmogorov distance Δ is given by

$$\Delta = \max_{0 \leqslant n < \infty} \left| \sum_{i=0}^{n} (P(i) - Sim(i)) \right|, \tag{29}$$

where $Sim(i)$ is the distribution of the number of calls in the orbit obtained via simulation.

Table 1. Kolmogorov distance.

Δ	$\sigma = 10$	$\sigma = 5$	$\sigma = 3$	$\sigma = 1$	$\sigma = 0.5$
$\rho = 0.2$	0.064	0.068	0.063	**0.008**	0.031
$\rho = 0.5$	0.06	**0.029**	**0.022**	**0.019**	**0.016**
$\rho = 0.7$	**0.025**	**0.022**	**0.021**	**0.018**	**0.013**
$\rho = 0.9$	**0.016**	**0.016**	**0.013**	**0.011**	**0.006**

As we can see, the accuracy of approximation increases while $\sigma \to 0$ and also when load parameter ρ increases. We note that approximation (28) is good enough for values of $\sigma < 5$ far from the condition $\sigma \to 0$. For low levels of ρ, we have some gap at zero point between approximation and simulated results, which gives unexpected values of Kolmogorov distance. As we assume, it is due to the fact that mean number of calls in the orbit is low.

6   Conclusion

In this paper, we have derived the diffusion approximation for probability distribution of the number of customers in the orbit. The accuracy of the obtained approximation is good for medium and heavy load of the system and increases while the retrial rate decreases.

References

1. Akşin, Z., Ata, B., Emadi, S.M., Su, C.L.: Impact of delay announcements in call centers: an empirical approach. Oper. Res. **65**(1), 242–265 (2017)
2. Dragieva, V., Phung-Duc, T.: Two-way communication m/m/1 retrial queue with server-orbit interaction. In: Proceedings of the 11th International Conference on Queueing Theory and Network Applications, pp. 1–7 (2016)
3. Gualandi, S., Toscani, G.: Call center service times are lognormal: a Fokker-Planck description. Math. Models Methods Appl. Sci. **28**(08), 1513–1527 (2018)
4. Ibrahim, R., Ye, H., L'lashhcEcuyer, P., Shen, H.: Modeling and forecasting call center arrivals: a literature survey and a case study. Int. J. Forecast. **32**(3), 865–874 (2016)

5. Jouini, O., Akşin, O.Z., Karaesmen, F., Aguir, M.S., Dallery, Y.: Call center delay announcement using a newsvendor-like performance criterion. Prod. Oper. Manag. **24**(4), 587–604 (2015)
6. Kuki, A., Sztrik, J., Tóth, Á., Bérczes, T.: A contribution to modeling two-way communication with retrial queueing systems. In: Dudin, A., Nazarov, A., Moiseev, A. (eds.) ITMM/WRQ -2018. CCIS, vol. 912, pp. 236–247. Springer, Cham (2018). https://doi.org/10.1007/978-3-319-97595-5_19
7. Legros, B., Jouini, O., Koole, G.: Blended call center with idling times during the call service. IISE Trans. **50**(4), 279–297 (2018)
8. Legros, B., Jouini, O., Koole, G.: Should we wait before outsourcing? analysis of a revenue-generating blended contact center. Manuf. Serv. Oper. Manage. **23**(5), 1118–1138 (2021)
9. Morozov, E., Phung-Duc, T.: Regenerative analysis of two-way communication orbit-queue with general service time. In: Takahashi, Y., Phung-Duc, T., Wittevrongel, S., Yue, W. (eds.) QTNA 2018. LNCS, vol. 10932, pp. 22–32. Springer, Cham (2018). https://doi.org/10.1007/978-3-319-93736-6_2
10. Nazarov, A., Phung-Duc, T., Paul, S., Lizyura, O.: Diffusion approximation for multiserver retrial queue with two-way communication. In: Vishnevskiy, V.M., Samouylov, K.E., Kozyrev, D.V. (eds.) DCCN 2020. LNCS, vol. 12563, pp. 567–578. Springer, Cham (2020). https://doi.org/10.1007/978-3-030-66471-8_43
11. Nazarov, A., Phung-Duc, T., Paul, S., Lizyura, O.: Diffusion limit for single-server retrial queues with renewal input and outgoing calls. Mathematics **10**(6), 948 (2022)
12. Pehlivan, S.A., Pehlivan, C., Martinez, C., Cellier, N., Fontanili, F., Lamine, E.: Performance evaluation and statistical data analysis of a call center for the deaf community. In: 2021 IEEE/ACS 18th International Conference on Computer Systems and Applications (AICCSA), pp. 1–6. IEEE (2021)
13. Sztrik, J., Tóth, Á., Pintér, Á., Bács, Z.: Simulation of finite-source retrial queues with two-way communications to the orbit. In: Dudin, A., Nazarov, A., Moiseev, A. (eds.) ITMM 2019. CCIS, vol. 1109, pp. 270–284. Springer, Cham (2019). https://doi.org/10.1007/978-3-030-33388-1_22
14. Toth, A., Sztrik, J., Kuki, A., Berczes, T., Effosinin, D.: Reliability analysis of finite-source retrial queues with outgoing calls using simulation. In: 2019 International Conference on Information and Digital Technologies (IDT), pp. 504–511. IEEE (2019)
15. Wongvigran, S., Premchaiswadi, W.: Analysis of call-center operational data using role hierarchy miner. In: 2015 13th international conference on ICT and knowledge engineering (ICT & knowledge engineering 2015), pp. 142–146. IEEE (2015)

Author Index

Printed in the United States
by Baker & Taylor Publisher Services

Printed in the United States
by Baker & Taylor Publisher Services